Computer-Mediated Communication

A Theoretical and Practical Introduction to Online Human Communication

Caleb T. Carr

Illinois State University

ROWMAN & LITTLEFIELD
Lanham • Boulder • New York • London

Acquisitions Editor: Natalie Mandziuk
Acquisitions Assistant: Sylvia Landis
Sales and Marketing Inquiries: textbooks@rowman.com

Published by Rowman & Littlefield
An imprint of The Rowman & Littlefield Publishing Group, Inc.
4501 Forbes Boulevard, Suite 200, Lanham, Maryland 20706
www.rowman.com

6 Tinworth Street, London SE11 5AL, United Kingdom

British Library Cataloguing in Publication Information Available

Library of Congress Cataloging-in-Publication Data
Names: Carr, Caleb T., author.
Title: Computer-mediated communication : a theoretical and practical introduction to online human
    communication / Caleb T. Carr, Illinois State University.
Description: Lanham : Rowman & Littlefield, [2021] | Includes bibliographical references and index.
Identifiers: LCCN 2020055402 (print) | LCCN 2020055403 (ebook) | ISBN 9781538131701 (cloth) |
    ISBN 9781538131718 (paperback) | ISBN 9781538131725 (epub)
Subjects: LCSH: Telematics. | Social media. | Communication. | Social interaction—Computer
    simulation. | Computer literacy.
Classification: LCC TK5105.6 .C37 2021 (print) | LCC TK5105.6 (ebook) | DDC 302.23/1—dc23
LC record available at https://lccn.loc.gov/2020055402
LC ebook record available at https://lccn.loc.gov/2020055403

This book is dedicated to my wife and our daughter. They're the best and most long-lasting contributions to society I'll ever be a part of, and I couldn't be prouder.

This book is also a direct result of the inspiration, scholarship, guidance, and friendship that Joe Walther has gifted me over the years. Any good ideas you see herein are due to him.

Contents

Somehow, Palpatine returned. No, no. That's a stupid line and doesn't drive a narrative. ... So how *do* you begin a textbook?

This text started long, long ago in a country far, far away, late one afternoon on the steps of St. Peter's Basilica. I had been working in the field of computer-mediated communication (CMC) for a short time, but increasingly I found it frustrating there was no good, accessible, and current consolidated text that would convey all of these cool things I'd been experiencing and studying in a way undergraduates could quickly understand and apply. And so, on the tablet computer I'd carried while living in Europe for the summer, I began typing bits and fragments of chapters and outlines, saving them on a 64 GB thumb drive I could also connect to my mobile phone. While typing under the Roman sun, I realized a textbook about computers would be too quickly outdated: my tablet computer was a far cry from the first portable computer (to call it a "laptop" would be to stretch the meaning of the term) I first used in about 1992. Yet that tablet has since been technologically surpassed by my current tablet and smartphone. Anything I wrote about computers would be obsolete by the time it went to press, so I had to take a different approach. This book is a partial response to that concern, as I've worked to make its contents contemporary yet flexible.

This book's contents are contemporary as I've tried to reflect the online communication landscape as it looked as this book was being finished late in 2020 as much as I can. But I also realize the social media tools and online trends of today will be outdated and replaced a few years (at best) after publication. And so I've sought to make this text flexible, not focusing *too* much on specific technologies or practices, so that if you replace "Zoom" or "Skype" with some other audiovisual conferencing tool that's since emerged, the ideas discussed will still work on that new platform. Ultimately, that's my goal as a scholar—and here, an author—of CMC: to think about the fundamental building blocks of human communication in mediated environments and then apply those deep-rooted concepts to specific devices and situations. I've tried to do that here, and will leave the challenge to my fellow learners to apply these ideas to the generalized and specific communicative behaviors that pervade your world as you read this text.

The Organization of This Book

This book is broadly organized into four sections, whose working names for me have been "Introduction," "Theories," "Applications," and "Contexts." Though these sections certainly build off of each other, their approach to and emphasis on CMC differ slightly based on which facet of CMC they cover.

Introduction

The first two chapters serve as a foundation to the specific tools and technologies that enable CMC. These chapters set up the subdiscipline as well as focus on some of the physical properties and characteristics of digital-mediated channels. For many of us, computers likely pervade our lives. We have smartphones, tablets, and some desktop or laptop, and then we also have computers in our refrigerators, robotic pets, and attire. But how much do we know about these technologies and how they actually let us communicate? In some ways, these questions aren't terribly germane to how we'll approach CMC. The theories and applications of CMC are rarely dependent on specific programming languages, physical configurations of devices, or other technical aspects. And yet a brief history of mediated communication and communal understanding of what and where CMC is can help us understand the boundaries and contexts of the theories and applications we'll discuss. Some of these may seem already quaint and antiquated (Who still uses email!?!), but the understanding of human communication and behavior formed through understanding these older tools and forms of mediated interaction still resonate and hold true in today's more modern tools. A good theory of how people interact through email continues to help us understand how people interact through modern messengers.

Theories

Chapters 3, 4, and 5 articulate some of the key theories that inform CMC. The good news for us is that, even with nearly forty years of research, the CMC discipline still has few theories uniquely its own. We'll discuss most of the extant major theories of CMC, broadly understood as theories of the self, theories of dyadic interaction, and theories of large-scale interaction. To help make sense of the theories, examples are given and applications are made, but this second section of this text is primarily focused on exploring and delving deeply into the mechanisms of specific theories. Theories are particularly critical to understand in CMC as they serve as the building blocks of understanding how communication occurs, regardless of channel. Once you understand theories of online relationship development, they can help you understand how relationships develop via both older CMC channels (e.g., email, text chat) and channels that are still emerging (e.g., virtual reality, the latest social medium). Having a good working knowledge and ability to apply these theories is critical to being a good CMC scholar, as well as understanding how underlying human communication processes work and applying them to any particular channel or device.

Contexts

The third section of the text—chapters 6 through 9—focuses on specific contexts for interaction via CMC. One of the amazing things about communication technology work is that it spans and connects the communication field. All facets of communication—from interpersonal to organizational to intercultural—can be viewed and experienced through the lens of digital-mediated communication. Our third section is an attempt to reflect the breadth of CMC work within and across the discipline, using CMC as a focal point. Working backward from how the theories were introduced, we begin by understanding CMC in large-scale interactions of organizations and groups, then move to smaller scales of interactions of dyadic and self-communication. These chapters are an opportunity to take the theories we covered in the second section and apply them in specific communicative contexts, making the abstract theories more concrete by applying them to particular situations, interactions, and relationships.

Application

Chapters 10 through 13, the text's final act, focus on applications of the CMC theories and practices we've discussed. Beginning with an exploration of the proverbial gorilla in the equally-proverbial room—social media—we will look at how these still-emerging tools are shaping the nature of our interactions from small to large scales. The discussion of these tools was specifically pushed back a bit in the book's structure to let us understand CMC more broadly, and to try to avoid the trend I see in my own students of wanting to situate all our discussions of CMC within whatever big two social media they're using at the moment, to the exclusion of the many other channels that exist and that are just as important to consider. From there, we discuss two big applications of CMC: persuasion and politics. Great subjects in their own rights, these two topics are explored broadly in an attempt to understand how persuasion works from interpersonal to focused advertising efforts; we then look at how CMC tools have infiltrated the political landscape, affecting voters, elections, and political tactics. Finally, Chapter 13 explores some of the CMC trends on and just over the horizon, considering the landscape of CMC beyond the now. By looking at augmented and virtual, utopian and dystopian potentials for the application of CMC tools, and how CMC may reshape the relationships and interactions that we have typically considered as necessary to be offline, we'll conclude by trying to take a look into the future of CMC. It can be hard to guess what CMC will look like when you're in the fifth year of your career, which may still be six to nine years off, and therefore what knowledge and skills you'll need to be successful using CMC. But by taking a theoretical approach, we can build a foundation of understanding and application that will continue to serve you in the future.

How to Use This Book

This heading is misleading, as I know you know how to use a book: open it, read the words, learn the material. But whether you're an instructor or a student, what you want to get from this book likely differs. Let me therefore briefly offer some insights from the author regarding how I tried to write the book and some of the intentions I had and choices I opted for to help make this book more useful and valuable to you.

For Instructors

There's no "right" way to incorporate this text into your class, and I know many of you (like myself) will cobble together textbook readings with supplemental ones, or Frankenstein the way you use the text, reading chapters out of order and skipping some altogether. I encourage you to do the same here, using the book as best fits your approach to CMC. I've tried to structure this book as it maps to my own curriculum, but that doesn't mean you should do the same. Broadly, I'll suggest two ways to integrate this book into your coursework.

First, you can use it in order. As written, it seeks to move back and forth among the topics, trying to relate and synthesize them in a way that would be most accessible to students if read in order. This is why the Theories and Context sections are not parallel. Rather than just starting from micro-level processes and moving to macro-level processes again, the Context section begins with organizational CMC, picking up right where the Theories section left off in its discussion of organizational and group CMC theories. It then moves backward (relative to the order of the Theories section), ending with the intrapersonal CMC context. Ultimately, I wrote the book with the ebbs and flows that CMC scholarship and scholars often take, as many of these processes and ideas are naturally synergetic and interdisciplinary.

The other way this textbook may be helpful when integrated into classes is by pairing chapters. Rather than waiting until your class has gotten through all of the theories to go back and talk about how they are applied more directly, read pairs of chapters sequentially. Reading Chapter 3 and then Chapter 8, then Chapter 4 followed by 7, then 5 and 6, and concluding with Chapter 9 gives you and your students a chance to get the theory first and then immediately apply it while the theory is fresh in your mind. This uncoupling of chapters from sections in favor of tethering chapters by context may make some of these ideas less chaotic and make it easier to focus specific weeks or class sessions on ideas.

The other thing you may notice is that I've deliberately tried to keep examples broad and somewhat conceptual. An example with Instagram could readily be reread with "TikTok" substituted and likely remain accurate. But I've tried to not go in depth too much on specific platforms or technologies. That technology books are innately already somewhat outdated by the time they're published, and they don't reflect what current students are using and doing online. Let me also encourage you to early on in each academic year or semester do a quick, even informal poll of your students to get a general sense of what tools they're currently using and how. Then use this text as a jumping-off point for those discussions. Questions like "Sure, this study from 2007 that focused on Facebook may not reflect what Facebook or its users are doing now, but how would the same concepts apply in [New hotness channel]?" can help students extend and make inferences about the more stable processes that haven't changed just because the technology evolved. I'd like to think of this book as a foundation and conversation starter, and you are much better situated than this book to talk about how these ideas apply right now. To that end, I hope you use it to start conversations, spark ideas, initiate scholarly debates, and generally inspire the great insights and engagements your own students can offer.

For Students

Having been—and still thinking of myself as—a student, I know how intimidating textbooks can be. There's a lot of words, many of them big and some in bold font, and it's often clinical and dry. Particularly the nature of this textbook may be worrisome, as what's some old and broken dude going to tell me about being online that I—a member of the digital generation!—don't already know?! I've tried to proactively and innately address these problems in this work, in the hopes of making it an easier tool for you to use.

First, for those who know me, although I take my work very seriously, I rarely take myself seriously. Consequently, and somewhat against the recommendation of a few reviewers, you'll see bits of my personality sprinkled throughout the text to try to make it humorous, engaging, and somewhat more interesting to read. You'll see my love of *Star Wars* and my loathing of Justin Bieber's music, and (if you know where to look) a heavy peppering of my family and friends throughout to make it all more personal. These bad (sometimes dad) jokes are as much for you as they are for me, and if looking to see if that reference is the hidden jab you think it is helps you retain and recall that concept better, then I'll consider it a win.

Second, I find a lot of students have been weaned on digital channels since their childhood, and much of their life is conducted on a smartphone. So what can I, through this book, teach you that you don't already know? Though many of us now use computers ubiquitously in our lives, one of the things I've tried to do here is focus less on the channels and more on the human element of what's going on as you communicate through those channels. You, as students, will always be ahead of me with regard to adoption and use of digital channels—I'm always playing catch-up. But just as your interpersonal communication course can help you better understand your daily relational interactions, I hope this text can help you better understand your daily online interactions. That's why the applications are so important: you already know how you use CMC tools, but this book will hopefully help you understand why you use those tools as you do and to what effect.

You'll see a couple of things in this text that I hope you deliberately use, as tempting as it may be to skip over them. First, each chapter begins with learning objectives. Counter to what you may assume from other books, I wrote these more for you than your instructor. They likely already know what goals and outcomes they want from each class session or content section. The learning objectives are a way to both preview and review the big, sweeping ideas in each chapter. Presenting them before you read the chapter is intended to give you a good outline of the essential ideas and takeaways from each chapter, giving you a good set of highlights and a road map for your reading and studying. But I'll also encourage you to go back after reading each chapter and look at those learning objectives again. "Now that you've read the chapter, can you answer to speak to each of those points?" The learning objectives are one way to spot-check how well the information in the chapter actually landed, and potentially highlight parts you may need to take another read through to make sure you really get the ideas.

A related tool for you are the review questions that appear at the end of each chapter. These questions are not rhetorical. I already know their answers—or at least one way to answer them. Once you've finished the chapter, read through the questions with classmates or a study buddy. Don't just answer them; but talk through them. Debate each other's responses. Raise counterpoints. Support your claims! Review questions have been deliberately written to be broad and somewhat ambiguous, with the hope they can spur some good and deep reflection of the course content beyond key words. Studies show that simply repeating information isn't as good for learning as actually interacting with ideas, so use the review questions herein as tools to interact with others about these ideas and how they may manifest more personally and directly in your own life.

Finally, each chapter has a corresponding *Research in Brief* breakout section. I remember as a student often being mystified by the black box of research. Some scientist (I assume wearing a lab coat) at Big School U did some study and consequently we know this little datum. Science can be like this, as we hear these claims in texts and trust the authors and reviewers that the claims are accurate and faithful, but without spending much time wondering how we arrived at those conclusions. The *Research in Brief* sections are my attempt to pull back the curtain on communication science, taking time to profile one particular study that contributed to that particular area of knowledge. By unpacking what the researchers did, with whom (i.e., participants), and how they drew and supported their conclusions, I hope you get a chance to better understand how the science was actually made.

Final Thoughts

That's some of the mindset and ideology behind this textbook. Thanks to those of you who read it, as I know most just skip right by prologues and forwards, mostly because there are no test questions on these sections. As your reward, I'll include the most cliché phrase to ever appear in a book: *Mitochondria are the powerhouse of the cell*! Seriously, this book was a passion project, helping fill a need in most colleges and universities, a textbook to introduce a form of communication in which we almost all engage but has not been well codified into an accessible text. Ultimately, if there is one thing I wanted this textbook to do, it was to discuss CMC in terms of how we interact online rather than what online tools we use. If there are two things I want this book to do, the second is to do the first in a way that is both accessible to you as a current scholar yet applicable to you in a few years as an engaged member of society. (The third thing is to tie more *Buffy the Vampire Slayer* into my scholarly body of work.) Ultimately, this textbook has been written with the love I have for this exciting and timely field, and I hope it will engage your own interests about the innovations technology affords to us to communicate with each other.

Acknowledgments

Scholarship is always a collaborative effort, and everything I've ever done, including this book, is a testament to the many brilliant, strong, collaborative individuals with whom I've had the pleasure to associate. Rebecca Hayes has been my stalwart rock as a friend, scholar, and partner; I couldn't have done this without her, and I'm deeply grateful for and appreciative of all she's shared with me. Every piece I've ever written has been done with an audience of four in mind, even if they didn't know it. Thanks to Joe Walther, Stephanie Tong, Dave DeAndrea, and Brandon Van Der Heide for inspiring and motivating me since our days in the CAS conference room, and for continuing to instill in me the love of this field. They challenge me to be a better scholar. I remain indebted to the faculty of Michigan State University's College of Communication Arts and Sciences, and the knowledge, rigor, and collegiality they fostered. Nicole Ellison, Vernon Miller, Cliff Lampe, Steve Lacy, Jih-Hsuan (TA-M-M-Y) Lin, Jessica Vitak, Nick Bowman, Paul Zube, Eric Dickens, Andy Smock, Jake Liang, Bob LaRose, Charles Atkin, Sandy Smith, Mandy Holmstrom, John Sherry, and many others continue to inspire me. Nicole Krämer and her amazing colleagues from the University of Duisburg-Essen continue to be my German krewe and colleagues and I appreciate the *freundschaft* of Astrid Rosenthal-von der Pütten, Tina Ganster, Stephan Winter, German Neubaum, Laura Hoffmann, and Leonie Rösner. If the MSU research team is my family, the Communication and Technology Division of the International Communication Association is my home, and I couldn't ask for better colleagues. Katy Pearce, Jessie Fox, Grace Ahn, Bree McEwan, Marjolijn Antheunis, Alex Schouten, Sonja Utz, Norah Dunbar, Amy Gonzales, Andrew High, Andrew Gambino, and all the other members of CAT who continue to foster a spirit of scholar collegiality make it a wonderful community, and I always look forward to our May meet-ups. Deep gratitude to my amazing coauthors for continuing to energize (and tolerate) me: Cameron Piercy, Brianna Lane, Adam Mason, Yvette Wohn, Erin Sumner, Jaime Banks, Chad Stefaniak, Alex Hinck, Ashley Foreman, Patrick O'Sullivan, and all of you who've helped mitigate my run-on sentences. I also deeply appreciate my second home, Coffee Hound, for the copious amounts of caffeine it has provided me over the years, with the hope that one day soon my jitters will stop. Finally, thanks to Elizabeth Swayze for seeing the potential of my work and for starting this journey with me. These individuals, as well as several whom I'm sure I've unintentionally forgotten to bring up and credit here, have been my inspiration and allowed me to stand on their shoulders. I wouldn't be anything without them, and I'm deeply indebted to them for their contributions to the field as well as to me. What follows is really the culmination of the hard work of hundreds of global scholars, and I'm simply humbled to do the best I can to share their great scholarship and science with you, the reader.

PART I

Introduction and Infrastructure

CHAPTER 1

What Is Computer-Mediated Communication?

LEARNING OBJECTIVES

After reading this chapter, you should be able to …

- Define *computer-mediated communication* (CMC) and distinguish it as a subdiscipline within the communication field.
- Consider how CMC is helping connect different areas of communication.
- Differentiate how digital technologies are actually changing human

communication processes as well as some processes that are fundamental and not changed by CMC.

- Identify and apply technodeterministic and social deterministic perspectives to explain how new media are adopted and used.

> "I think there is a world market for maybe five computers."
> —Thomas Watson, IBM president (1943)

> "Almost all of the many predictions now being made about 1996 hinge on the Internet's continuing and exponential growth. But I predict the Internet will soon go spectacularly supernova and in 1996 catastrophically collapse."
> —Robert Metcalfe, founder of 3Com and inventor of Ethernet

When Dick Tracy's wrist radio watch debuted in a 1946 comic strip, it became one of the most iconic and seemingly futuristic forms of mediated communication, untethering the detective from corded telephones to allow seemingly magic interaction while he was out of the office on a case. Today, having *only* a telephone on your wrist seems quaint and underpowered. Now, individuals have access to telephone calls, emails and text messages, video, and haptic interfaces through their Pebble, Gear, or Apple smartwatches, which are in turn linked to ever smarter mobile phones that would make Tracy green with envy. As the two quotations that open this chapter highlight, many would not have expected the spread and seeming ubiquity of

FIGURE 1.1 Dick Tracy's prescient wristwatch communicator was just a phone then, but it closely resembles the full-fledged computers that are our current smartwatches.
Cartoonist: Scrivan, Maria.

modern computing, or its exponential rise and integration into our lives. In a time we increasingly talk of the "Internet of Things," it seems the opportunities to interact with others online are unlimited, both in devices to use and in modes of interaction. Yet how do these mediated interactions affect the way we communicate, our interactions with others, and how we perceive others and ourselves?

Computer-mediated communication, or *CMC*, can be hard to define given the rapid changes in technologies, but it may generally be considered as the process of the exchange of meaning among two or more humans through digital channels (Carr, 2020a). The study of CMC within the communication discipline is relatively new: CMC has emerged in its own right only within the past four decades, compared to the century of formalized scholarship into speech and human communication—and add on a few extra millennia if we go back to Aristotle and the other Greek orators. For its relative novelty, CMC has quickly become one of the dominant areas of research for scholars. As we try to understand how individuals change and are changed by the ability to interact online, the nature of self-presentation and online identity management, the cues we use to encode messages and how cues are decoded differently online, and how we perceive and communicate with others and ourselves can make it challenging to study CMC. The rapid pace at which technologies and social uses and norms around these technologies develop further challenges us to keep up, as how we use CMC often changes faster than we can study and understand it. To effectively consider how humans are adapting to virtual interactions, we therefore need to consider CMC from a theoretical approach, understanding how elements and properties of computer mediation affect communicative processes.

Approaching CMC as a subdiscipline through the more stable theories that underlie how we use our various communication technologies, this text explores CMC from a functional approach. By "functional approach," I mean that we will seek to understand how mediation influences our interactions in a general sense, all without being too entrenched in or tied to specific technologies. Though we will use some current (at least at the time of writing) technologies to exemplify and illustrate concepts, we won't focus on specific technologies for two reasons. First, there are just too many technologies now to give due attention to each. Second, and much more importantly, to ground our conversation in a specific technology limits what we can do with that information. Imagine learning how to communicate specifically using MySpace or Club Penguin: how many of the skills for those specific services are applicable to communication via Facebook or TikTok (or even newer and trendier communication technologies)? To avoid the pitfall of learning about CMC only to have your knowledge and understanding become outdated by the time you graduate, we will use theory to understand what communicative, psychological, and social processes are driving CMC in a way that will keep your knowledge and ability to apply that knowledge relevant long after you have entered the job market.

With that said, one of the most fundamental and currently frustrating issues in the area of CMC is to define a "computer," upon which *computer*-mediated communication seems to hinge. As Mr. Watson and Mr. Metcalfe would not begin to recognize the computing environment of 2021, we too must be careful not to constrain our conceptualization of computers or let it grow too broad to the point where everything is CMC. Consequently, before getting into the theories, processes, and effects of CMC, we begin by exploring what constitutes a "computer."

CMC: Beyond Desktop Computers

What is computer-mediated communication? Were you to be asked this question in the late 1980s or early 1990s, the answer may have been much easier and self-evident than it is now. In an age when personal computing was limited to universities and some workplaces (Rice, 1980), CMC was the study of what happened when an individual communicated via a terminal or computer at a fixed location or desktop. You typed in a message, and someone else (often somewhere else and at some other time) read that message after it was sent through the Internet. Early CMC scholars asked questions about the ways that communicating through chat or email—in which interactants could not hear each other's voices or see nonverbal cues and were limited to text exchange—were different from **face-to-face** (FtF) exchanges. In many ways, early CMC was merely the study of pen pals that typed (rather than handwrote) their messages. With few people having access to computers, let alone computers connected to the Internet, a safe and standard set of assumptions about CMC came into play any time someone communicated via messages typed into a computer's keyboard. However, as computers became more accessible and devices increasingly connected to the Internet, this definition became more problematic.

Now, we find ourselves asking additional questions to probe and better answer the initial question, "What is CMC?" We must consider terms fundamental to its conceptualization, such as "computer." Gone are the days we access the Internet only via a desktop. Now, we use desktop computers, mobile laptops, tablets, smartphones, smartwatches, and many other emergent (often wearable) technologies to access the Internet. Other questions must be asked, like what constitutes "computer-mediated." As devices converge, more channels are now available to us through a single tool. Your mobile phone can likely send an email or text just as readily as a voice call, so which of those three messages are CMC? If you use voice over internet

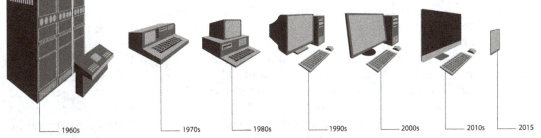

FIGURE 1.2 "Computers" for a long time were well defined. Even as they evolved and advanced, they remained a configuration of a monitor, a keyboard and mouse, and a desktop processor that sat atop a table or desk. Though these still exist, is this what you think of—either first or at all—when you hear the word "computer"?
Valeriy_Katrevich.

protocols (VOIP) to place a call from your computer to someone else's telephone, is that CMC because you used your computer, or merely voice communication like a telephone call? Even more complex are questions about mediation, as computers offer greater bandwidth for communication—that is, they allow us to communicate through more cues than just our verbal message. In the era of Zoom and Google Hangouts, is a videoconference closer to CMC or FtF communication? It quickly becomes apparent that as communication media and devices continue their steady, relentless march forward these technological advances will increasingly raise fundamental questions about CMC as a subdiscipline within communication. The good news for us is that all of this upheaval and questioning means CMC is an exciting, volatile, and engaging area in which to research and learn. Good for us! We'll get back to defining CMC in a bit, but first let's take a look at what CMC *can* and *can't* do effectively by looking at how it is situated within the communication discipline more broadly.

Situating CMC within the Communication Discipline

CMC can be considered as a subdiscipline situated within the broader field of communication. Like interpersonal communication, group communication, organizational communication, and other areas within communication science, CMC focuses on a specific subset of questions, variables, and theories of human communication. But unique among the communication subdisciplines, CMC readily spans other subdisciplines and fosters new ways to ask new questions about the fundamental nature of human interaction. In addition to augmenting what is going on in other parts of the field, CMC is increasingly distinguishing itself as its own area and with its own unique sets of concerns that we, as CMC scholars, are best prepared to address. One way to begin to understand what CMC *is* is to consider our place within the communication discipline as a whole, and the distinct role of CMC both as a means of augmenting other subdisciplines and as a subdiscipline in its own right.

CMC as Convergence

Certainly, there are elements of all of communication's subdisciplines in others. For example, when studying two friends talking—an interpersonal interaction—the cultures of the two friends play a role in the communicative exchange, even if the intercultural element of the interaction is not the focus of the examination. However, CMC offers a particularly unique opportunity to converge with other parts of the communication discipline and to help understand, isolate, and probe facets of human interaction.

Foundational CMC research in the early 1990s actually started with studies of nonverbal and verbal communication (Walther, 2011). As people started becoming more comfortable using computers for messages, text-based tools like email and chats presented an opportunity for researchers to study how to isolate, manipulate, and explore specific parts of a communicative exchange that wouldn't be possible even over the telephone. How do we study whether someone's message is persuasive regardless of how they look, act, or sound when delivering it? We simply send it via an email message, which innately strips out all the extra information and cues about the message sender, leaving only the message itself. Early CMC research often took this approach, using mediated channels to isolate particular variables of interest to manipulate just them and to understand their subsequent effects.

More recently, as computers and their various means of mediating interactions have pervaded our personal and professional lives, we see the theories, processes, and effects of computer mediation integrated into other interactions and types of communication, transcending subdisciplines. As job applicants turn to services like LinkedIn and GlassDoor to learn about organizations to which they are considering applying, CMC becomes integrated into organizational communication and its processes of assimilation and information seeking. When dating partners meet and initially communicate via dating sites like Match or Tindr, moving their communication offline after they feel comfortable meeting face-to-face, CMC is integrated into interpersonal communication and its exploration of relational formation and romantic relationships. As social media like Facebook and Twitter allow us to keep in touch not just with close friends but also with those whom we met in study abroad programs, CMC allows us to transcend geography and culture, offering new opportunities for understanding intercultural communication. Questions about the function of an online forum for fans of a particular sports team, or about the differences in fandom processes and effects between fans in a Discord server for Red Wing fans and fans together in person at a Red Wing hockey game, are both CMC questions, yet they are also innately questions of intergroup and intragroup communication.

Thus CMC can be and often is used as a lens to understand other parts of the field, particularly as their communicative processes are maintained or changed by the nature of being online. As we'll see in later chapters, CMC can help explain and unpack phenomena within relationships, organizations, groups, public relations, education, and politics. In this way, CMC is a great bridge within our field, serving a distinct role of spanning and connecting several often different subdisciplines and fostering understanding, collaboration, and deeper exploration. If you attend an academic conference on communication and sit on a panel focused on CMC, you will likely hear discussion that connects several areas within the communication field, bound together by the nature of the mediated processes that facilitate the communication. In this way, CMC has helped converge previously separate subdisciplines, allowing multiple contexts and relationships to collapse online (Marwick & Boyd, 2011) and giving us an opportunity to more holistically study and understand the fundamental processes of human communication.

CMC as Divergence

And yet CMC is also unique, sometimes divergent from other subdisciplines. Computer-mediated communication is much more than just taking questions from other parts of the discipline and putting them online to ask anew. Indeed, approaching CMC as simply moving offline processes online is rarely a fruitful or interesting way to advance our knowledge. There is little reason to think that if you insult me online, that threat is relationally different than the same offline. Particularly as more communication occurs online and both channels (e.g., email, Facebook, Snapchat) and devices (e.g., tablets, smartphones, smartwatches) become more ubiquitous and integrated into the social fabric of our lives, there are numerous facets of CMC that are distinct in their own right. Far beyond the nature of merely "being online," the study of CMC diverges from other subdisciplines, setting itself apart in several ways.

First, CMC has set itself apart from other parts of the communication discipline by exploring how humans adapt to a new means of communication and take advantage of new channels and cues. Just as media (including print, radio, and television) studies before it, CMC research explores how people adopt and adapt to new and emerging channels for communication. As people increasingly interact via computer systems, it changes the way that we interact by giving us new channels, new affordances, and new opportunities for interaction. For example, unlike when we interact offline, social network sites like Facebook and LinkedIn make our social networks immediately visible and readily navigable (Ellison & Boyd, 2013). Thus we don't have to remember that old classmate from elementary school we have not talked to in a while: the system displays that connection and occasionally reminds us of that connection. Likewise, if you are interested in a job at a particular company but are unsure who to ask about working conditions or possible job openings, tools like LinkedIn can help you identify people you know who know someone who works for that organization (Piercy & Lee, 2019), which is good because people don't walk around offline wearing a list of people they know.

Second, CMC is unique as a subfield due to some of the communicative possibilities that, while possible offline, are readily evident and usable online. Some processes appear in many places, but are much more common and natural online, making CMC an effective means to explore them. For example, **masspersonal communication** was put forward to describe (a) interpersonal communication over a mass medium, (b) mass communication via an interpersonal medium, or (c) the convergence and combination of the two (O'Sullivan & Carr, 2018). This definition could include a marriage proposal at a sporting event broadcast on the Jumbotron: the message is certainly interpersonal, as one person is (hopefully) asking only one other person to marry her/him, but the interpersonal interaction is publicly viewable as it is broadcast on the stadium's television. Although these masspersonal processes can occur offline (Walther, Carr, et al., 2010), the technological affordances of many CMC tools make masspersonal communication much more common. Now, when you post "Happy birthday" to a friend's Facebook timeline, the message is partially interpersonal, as it a message from you to them on their birthday. And yet, at the same time, the message is *also* a mass message in that it is public and accessible by anyone visiting your friend's Facebook page. Even more complex is that your friend could use the same social media tool, Facebook, to send a private message back to you thanking you for your birthday thoughts. Similarly, a user can now broadcast a mass message via a YouTube video and receive interpersonal feedback tailored to the poster and video via comments below the post. Tensions such as the public-or-private, mass-or-interpersonal, or broad-or-personalized natures of messages online make CMC a unique subfield as it facilitates processes that, while possible in other media, are not common, and thus are of interest to CMC scholars as they investigate these processes in a much more natural setting.

What Is CMC?

What, then, is CMC? Though many conceptualizations for CMC exist, there has surprisingly not been a concise definition offered to help us clearly distinguish CMC from mass media or other mediated forms of communication. For the purposes of this book, **computer-mediated communication** is considered to be the transmission of meaning between two or more humans via digital technologies and emphasizing the effects of mediation on human communication processes over specific technological processes (Carr, 2020a). How we understand "computers" continues to change, as computers are no longer just bulky desktop systems and can now include smartphones, wearable technologies, and Internet-based programs. However, acknowledging the role of "computers" as some form of digital intermediary helps distinguish CMC from television or radio, which transmit meaning via radio waves. Additionally, this definition helps distinguish email from a written letter. Though both may have similar effects and processes (as we will discuss throughout this text), a written letter may be mediated communication, but it does not meet our criteria for consideration as a computer-mediated communication as messages are not transmitted by digital technology.

Though limited in the early days of computing, the channels available for CMC continue to expand rapidly. What used to be limited to email, teleconferencing, and chat rooms has given way to complex, multimodal CMC tools that allow users to share text, photos, and audiovisuals in collaborative ways. Now millions of users can come together to contribute to Wikipedia, and do so more accurately than the authors and editors of textbooks and encyclopedias (Kittur & Kraut, 2008). We still email and chat online, but we can now do so with computer programs rather than having a human on the other end of the conversation. And our daily interactions occur online, with billions of emails, text messages, and social media messages generated every day worldwide. So although the history of computing can be traced back to the 1940s (see next chapter), the topic of computer-mediated communication is an obvious and increasingly important subject to discuss.

Approaching CMC

Though it may seem like a cornucopia of CMC channels is already available, that number will likely continue to increase even after you graduate. It is because of this exponential growth in computer-mediated channels that we will take a very functional and theoretical approach to CMC, including its processes and effects. This approach means that we will use theory to understand the communicative processes that underlie the practical uses and outcomes of CMC, explaining them first at the conceptual level and then connecting them to specific examples and experiences. But why approach and discuss CMC in this vague, conceptual way? Why not just talk about YouTube and Reddit directly and right away? The answer is that by the time this book is published or the next cohort of students uses it, the technologies discussed may already be passé. Though you may be a big Snapchat and Instagram user now, in the next five years students reading this book will think of Snapchat as a dated program, old and busted and used by elderly people. Moreover, if we talk in specifics now—like how to use TikTok for social purposes—if you get into your first or second job and find out the company doesn't use TikTok, then what can you do? Working conceptually helps us be flexible and broad and apply ideas to specific technologies and devices.

There is a common temptation in the field to study specific technologies or online tools: we want to look at Snapchat and WhatsApp because they are the newest hotness. But there is a challenge to focusing on a specific tool in CMC research. Part of communication in a specific channel like Snapchat or WhatsApp is a function of the tool itself: its design and user interface, the purpose of the tool, the types of people who use the tool, and the affordances within that tool (e.g., message systems, upvoting tools, self-presentation opportunities). The more entrenched you get in the site's functions to describe how users communicate, the less applicable your findings and ideas are in other sites. A study conducted in the mid-1990s to describe how users communicated via Bebo or Friendster may not necessarily be useful to explain how people communicate today via Facebook or Whisper. However, the same initial study, done from a theoretical perspective, would help explain the psychological, sociological, and communicative processes that serve as

FIGURE 1.3 Working theoretically means that as the specific channels you (and others) use change, your understanding of how those mediated channels affect human communication processes does not. You still know the processes and effects in Facebook, Instagram, and GroupMe.

the foundation for interactions on Bebo or Friendster, and it is likely those theoretical processes—fundamental attitudes and behaviors—have not changed in the twenty years since and that they are just as applicable today to Facebook and Whisper, and will remain so another twenty years from now. Thus, taking a theoretical approach lets us understand how humans communicate via computer-mediated tools in a way that is timeless and that will be just as utile to you in the thirtieth year of your career as it is in the third month. By understanding the communicative processes going on within an interaction rather than focusing on a specific website or online tool, our knowledge and understanding of CMC can be much richer and more robust, and thus valuable to understanding how humans communicate.

Technology and Humans: Who Controls Whom?

An initial question when studying CMC may be how we actually use computers for mediated communication. How do we choose which computers to use for which interactions and in what ways? One way to broadly consider how individuals use mediated communication is to consider the mutual influence of technologies and their users on how those technologies are used. Two dominant approaches can be taken to explore how individuals adopt, co-opt, and are influenced by the devices and channels they use, and scholars and practitioners have utilized both to explain and explore CMC and its processes. Below, we discuss technodeterminism and social determinism as means of understanding the mutual influence exerted between humans and the communicative tools we use.

Technodeterminism A technodeterministic approach to CMC takes the position that the mere presence of a technology changes society. Technological determinism, first suggested by Karl Marx, suggests technology shapes a society's development and culture, so that society adapts to technology. In other words, because the technology is there, society inevitably is changed because of it. A technodeterminist, for example, may take the position that because we now have widespread diffusion of mobile phones, we have less of a need to communicate in person and face-to-face as we can utilize the technology to interact. Many organizations currently seem to take a technodeterministic perspective as they seek to jump onto the bandwagon of that latest social medium, developing a presence in the channel because it is there rather than because either (a) they have a strategic use for the medium or (b) that channel lets the organization do something other social media or channels may not. For example, consider a staff member at an assisted-living home for elderly residents who realizes that TikTok is becoming popular, and adopts the new channel first, and only after joining tries to figure out what to do with it and whom to communicate with via TikTok. Technodeterminism suggests the medium comes first and we adapt our communication to it.

Social Determinism Alternately, a **socially deterministic** approach to CMC takes the position that society guides how a technology is adopted and used. Social determinism suggests a society collectively constructs a technology and its uses, so that society adapts the technology to its own purposes and goals. In other words, because people use a technology, they collectively shape how that technology is used. A social determinist, for example, may note that we have appropriated mobile phones to allow us more ways to keep in contact with diverse people we may not have been as likely to reach out to over just landline phones or through written letters. Particularly those focused on user-design and end-user experience (often denoted as "UX" in industry parlance) take a social deterministic perspective as they seek to explore user behavior to understand how users are actually using a particular device, platform, or service, regardless of whether users are utilizing the technology as its designers intended.

Which Is It? At times, either, both, or neither of these two approaches may be useful to understanding how individuals utilize CMC. For example, exploring the meaning of a Facebook "Like," Hayes, Carr, and Wohn (2016) found that some Facebook users utilized the Like button simply because it was there. As the only one-click means of communication, users clicked Like to explicitly communicate to another user they liked content the user posted to Facebook, using the button because it was there as a channel to send a message. However, in the same study, Hayes et al. found that users also sometimes used the Like button more selectively within certain groups or subsets of their friends because of social norms that had developed around the button: some users Liked something a friend posted because that's what their network had decided was the means of indicating a post had been seen or valued, thereby socially constructing the meaning of a Like. In these cases, users had perceived the Like button as a way to communicate more meaning than simple appreciation of a post, constructing meaning amongst themselves.

FIGURE 1.4 We often hear laments that smartphones disconnect us from our surroundings. Yet similar concerns were expressed about personal music devices, the telephone, books, and even writing in general. Perhaps the concern is one of technodeterminism rather than how we actually use our media in comparison to how we used earlier channels.

Because different perspectives can help make sense of different situations, it is probably not fair to say that either technodeterminsm or social determinism is correct. Rather, both are useful means of thinking about the relationship between CMC and the people who use computer technologies to interact. Sometimes people use the technology just because it's there, sometimes people use the technology because that's how they've been shown to use it or how their peers use it, and sometimes (and likely often) it is a combination of the two. Take a moment and think about the CMC tools you have used today and how you have used them. What tools did you use and why did you use them the way you did? Likely you will find an interesting mix of technodeterminism and social determinism in your own use.

Technology and Others: Beyond Our Focus

If all this is CMC, what is *not* CMC? Just as "What is CMC" has become an increasingly complex question, so too is what may *not* be considered CMC, and thus is beyond the focus of our exploration. The easiest part of this answer is FtF communication, which has likely been and will continue to be the focus in most of your other communication coursework. We will explore FtF communication here, but only insomuch as it serves as a comparison point to CMC. Another form of communication often confused with (but typically different from) CMC is mass-mediated communication: television, radio, newsprint, and other forms of one-to-many communication. Although these forms of communication are often mediated, they lack several of the fundamental elements of CMC, such as interactivity and identifiability. Related, several types of communication that occur via the Internet and computers, such as online newspapers, podcasts, and Skype phone calls do not have the distinct hallmarks of CMC. One way to identify if your interaction qualifies as CMC is to ask, "Can I look at the same phenomenon in another channel?" If so, you may be asking an interesting communication question, but not necessarily a CMC question. For example, what is the difference between a phone call and an audio-only call using Skype? If, for your interest, the research could just as easily be done via phone as via Skype, then you may have an interesting question about mediated communication, but not about CMC.

An additional area of increasing interest is how individuals communicate *with* computers and vice versa. This question is of both theoretical and practical interest as algorithms and computers are increasingly able to both send and receive messages. As robots, virtual agents, and adaptive programs make human-computer interaction easier, a field of research is emerging that explores how we communicate with machines. This is a very interesting and timely area of work, as there are times when communicating with computers may be more natural or less stressful. For example, digital and virtual assistants (including Alexa and Google Home smart speakers, the Siri and Google Assistant on smartphones, and Cortana on the Windows operating system) can help us schedule appointments, play music, and execute tasks for us if we ask them (Guzman, 2019). Another potential for interacting with computers as communicators may be

Research in Brief

Is this real life? Is this just fantasy?

What's the difference between *online* and *offline*? We often hear that offline is "real life," but does that mean that what occurs online is less real? In an early essay exploring and documenting the nature of online communities, Dibbell (1993) brought attention to the nature and consequences that online experiences could have for our offline selves.

LambdaMoo was a text-based community accessed via dial-up modems, in which users entered and moved about "rooms," or individual virtual spaces constructed by users. All spaces and users were described textually, as LambdaMoo did not use graphics, and so much of the interaction in LambdaMoo relied on users' storytelling and descriptions. The user "exu," for example, was self-described as "a Haitian trickster spirit of indeterminate gender, brown-skinned and wearing an expensive pear grey suit, top hat, and dark glasses." One March night, a user—Mr. Bungle, a self-described obese jester—entered LambdaMoo's common room and proceeded to describe exu performing degrading sexual acts on Mr. Bungle. exu was not typing in descriptions of these activities, rather Mr. Bungle was typing them on exu's behalf.

Though dramatic, was anyone hurt in this episode? Not physically, as again this all took place in an online space while users sat safely and privately in their homes. But to the LambdaMoo community, this transgression was considered a crime. exu felt violated by having Mr. Bungle forcibly control exu's online presence, and the LambdaMoo members took steps both to ban Mr. Bungle's presence and to create community standard guidelines to prevent future episodes. To LambdaMoo users, this incident was not some elaborate storytelling fantasy. It was very real.

So where is the boundary between offline and online? We often discuss and treat our interactions online as if they are siloed away, occurring only there. And yet profane or crass tweets have lost people jobs, and doxing (providing an online user's offline information to others for harassment) and cyberbullying constitute online communication with direct and evident offline consequences. Though the strange tale of a Haitian trickster and a voodoo clown may seem isolated and antiquated in a time of social media and audiovisual interaction, the questions and tensions it offers regarding the online-offline boundary are just as critical now.

Reference

Dibbell, J. (1993, December 23). *A rape in cyberspace or how an evil clown, a Haitian trickster spirit, two wizards, and a cast of dozens turned a database into a society.* The Village Voice. http://www.juliandibbell.com/texts/bungle_vv.html

when it is less stressful or stigmatizing to communicate with the machine than with a person, such as using a program for therapy or counseling, which cannot—by virtue of its programming—judge or stigmatize you (Olafsson et al., 2018). However, the field of human-computer interaction (HCI) exists a bit beyond the focus of our exploration, as it does not deal with interaction between two (or more) *humans*. We will explore some small parts of HCI as it ties into our notions of CMC, but we will not focus on it intensely given other venues and outlets can better focus on this area. For now, we will keep our focus on human-human interaction that occurs over computer channels, which has its own rich and growing body of work.

Summary

Computer-mediated communication is a still-developing area that has affected many facets of communication. Because of the radical speed at which new technologies emerge, it can be one of the more difficult areas of communication to study because you will always feel just a bit behind the newest developments. But the rapid development of the CMC environment also makes it exciting—there's always something new to explore and novel connections to make. In the following chapters, we will first explore some of the theories of CMC that can serve as the foundation to our understanding of how CMC affects communicative processes. Next, we will look at some of the contexts of communication (e.g., interpersonal, organizational, educational) to explore how mediated communication is influencing the diverse facets of the field. And across it all, we will seek to make contemporary applications of the theories and ideas we discuss to help connect this to your life. It is my hope that, as we probe some of the applications and implications of CMC, you have the opportunity to reconsider many things that may just be staples of your daily life—email, social network sites, texting, and more—and think about the processes that made you select a particular channel, craft a specific message, or understand another person, all through computer-mediated communication.

Key Terms

face-to-face (FtF) communication, 5

computer-mediated communication, 8

masspersonal communication, 8

technodeterminism, 10

social determinism, 10

Review Questions

1. What is "computer-mediated communication" and what is it not? When everything from a garage door to a refrigerator has a microchip, what makes computer-mediated communication distinct from offline communication?

2. How can CMC help articulate the intersection between communication subfields (e.g., organizational, intergroup, health, cultural)?

3. How do CMC tools enable masspersonal communication more readily than offline channels?

4. Does technology guide how society develops or does society decide how to use and integrate a new technology? Based on your response, explain the current role of Facebook in society.

CHAPTER 2

The Technologies and Users of Computer-Mediated Communication

LEARNING OBJECTIVES

After reading this chapter, you should be able to ...

- Define the Internet and discuss its historical roots.
- Distinguish between various evolutions of the World Wide Web.
- Critically discuss Internet "access" and address how communication technologies (including the Internet) can vary between countries and within your own country.
- Consider the changing demographics and means of using various CMC tools, and the influences of technology adoption.
- Explain how digital media are used either as intended or in novel ways to communicate.

"The Internet is not a big truck. It's a series of tubes." Spoken by former senator Ted Stevens in 2006 when discussing the regulation of Internet service providers (ISPs), the companies and organizations that make access to the Internet available to consumers, this statement has since become a cliché. But this statement also provides an interesting perspective on the hidden infrastructure of a system many of us rely on for our daily communication. We often look at our laptops, tablets, and smartphones as if their connection to other devices is magical, so that these signals and data just appear in our pockets or on our desks as we go about our day. But the physical tools and services that carry information from one device to another—whether it is from the phone of our friend next to us or a data center in Estonia—are connected by a complex global network of computers and other systems that help convey information worldwide. Though not tubes per se, a large part of the physical object that is the Internet is hundreds of thousands of miles of insulated tubes filled with fiber-optic cables (run under oceans, fields, and mountains as well as under cities' roads and buildings), supplemented by satellite and microwave signals and huge buildings filled with routers and cable wires (Blum, 2013). We so often vaguely allude to the Internet as a nebulous and intangible idea, we rarely think about the physical Internet that makes CMC possible.

What Is the Internet?

At its simplest, the **Internet** is the worldwide public network of networked computers used to transmit information. Though this definition may oversimplify the complexity and scope of the Internet, it will do for our purposes here. This network of networked computers is employed by every program, application, and extension you utilize to communicate online, from emailing to file transfers to websites to video sharing (Cerf, 2013). All of these services use various technical protocols to digitize, transmit, receive, and reconstitute information, which allows Internet users to quickly send and receive information anywhere in the world. Consequently, the Internet is essential for CMC because it serves as the system that connects our computers and allows us to use them to send and receive information.

For a long time, "the Internet" referred to the connection among large servers and desktop computers. As technology has advanced, the Internet created by these connected technologies has expanded too. The emergence of laptops, tablets, and smartphones extended the Internet even further, and now more than 5 billion computers are connected by more than 300 million servers to comprise the Internet (WhoIsHostingThis, 2019). Within the past decade, the shrinking of microprocessors (in both physical size and economic cost) has allowed us to equip even more devices with sensors and Internet connections, further expanding the Internet in both size and reach. This interlinking of everyday objects—including lamps, garage door openers, refrigerators, doorbells, printers, and mirrors—to the Internet in a way that makes them devices for communication, interaction, and task facilitation has been called the **Internet of Things** (see Kim, 2016). In 2020, an additional 15 billion devices—most of them physical objects rather than computers per se—were connected to the Internet, adding a host of everyday items to the network of networks that is the Internet. But how did this exponentially growing network of systems begin?

History of the Internet

The foundation of the Internet is closely tied to the foundation of computers during World War II. The first computer is often considered to be Charles Babbage's mechanical difference engine, which he designed (but never built) in 1849 (Computer History Museum, 2020). However, Babbage's machine was merely a mechanical abacus used to automate basic arithmetic. It wasn't until the 1940s that electronic computers as we now know them began to take form. World War II was filled with codes, as both sides involved in the conflict encoded secret messages to coordinate supplies and troop movements. For example, the Allied forces used the Navajo language, while the Axis forces took pride in their mechanical Enigma code engine. To crack these increasingly complex codes, cryptographers developed the earliest of computers—enormous systems of transistors and wires. In 1946, the first electronic computer, ENIAC (Electronic Numerical Integrator and Computer), was developed: a room-sized machine taking up 1,800 square feet and programmed using punch cards. Following the end of World War II, computers were engineered to be more compact and repurposed for more than analyzing coded messages, increasingly tasked with computing the complex math behind hydrogen bombs.

As the Cold War began in the 1950s, a new fear arose: the Intercontinental Ballistic Missile (ICBM). As ICBMs were able to attack global targets, the United States' Department of Defense (DOD) began to fear that a preemptive strike (primarily by the USSR) on communication centers located in select US locations could cripple the government and military's ability to coordinate and respond to the threat. At the time, all long-distance telephone communication was routed through a few major

  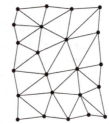

Centralized   Decentralized   Distributed

FIGURE 2.1 Various network structures. Note how the distributed network structure (used by the Internet) provides more paths by which any two points can be connected than centralized and decentralized structures.
hakule

hubs, including Chicago and Denver. This telecommunication infrastructure is referred to as **decentralized**, as a few key locations, or *nodes*, served to link all other local and distant points (see Figure 2.1). In these decentralized networks, a call from the East Coast to the West Coast was always routed through Chicago. The problem with a decentralized network, however, is that should a central node (for example, Chicago) be bombed, the loss of that central point in that telecommunication network could mean that calls could not be placed from Washington, DC, to Los Angeles to relay intelligence and coordinate civil defense. Concerned about the loss of key elements in this decentralized network infrastructure, members of the DOD's research branch, DARPA, followed an initiative to develop a distributed network of computers to ensure nationwide communication could not be disrupted, even if a communication center were disabled. A **distributed** network is one in which information can be routed through many different network connections, so that if one connection fails, the information can be routed through an alternate node. By developing this distributed network system, the DoD hoped to ensure that communication could be maintained even should specific communication hubs be lost. The first application of these distributed networks was through the ARPANET project, which networked several DoD computers tracking radar signals across the United States as an early-warning missile defense system. But beyond helping set up a global radar network, the ARPANET laid the technological foundation for the Internet.

In the early 1960s, the DoD funded universities and research firms to find a way to create and maintain these distributed networks. In October 1969, the first two nodes of the ARPANET went live, and packets of information were transferred between computers in Los Angeles and Menlo Park, California. Quickly, additional ARPANET nodes emerged in Santa Barbara, Salt Lake City, Boston, and Pittsburgh, growing the ARPANET and its ability to transmit information nationwide. Within five years, the structure of the ARPANET had grown to include dozens of institutions (mostly universities) and crisscross the United States (see Figure 2.2). In 1982, protocols were introduced to standardize how information was relayed across the ARPANET, giving us the now-standard TCP/IP (Transmission Control Protocol and Internet Protocol) that run the Internet, and the Internet as we now know it was truly born. A transatlantic cable was laid in 1988, connecting New Jersey and Sweden and expanding the fledgling Internet globally. Federal funding, particularly from the National Science Foundation, continued to advance the infrastructure (i.e., the series of cables, wires, and computers that create the physical Internet) as well as research into the transmission of digital information, including the Computer Science Network, the National Science Foundation Network, and the national supercomputing centers (the first at the University of Illinois Urbana–Champaign). The ARPANET was decommissioned in 1990, as commercial ISPs had emerged to bear the cost of laying and maintaining the infrastructure originally developed by universities and the military, but the protocols, procedures, and infrastructure the ARPANET had fostered had laid the foundation for the Internet.

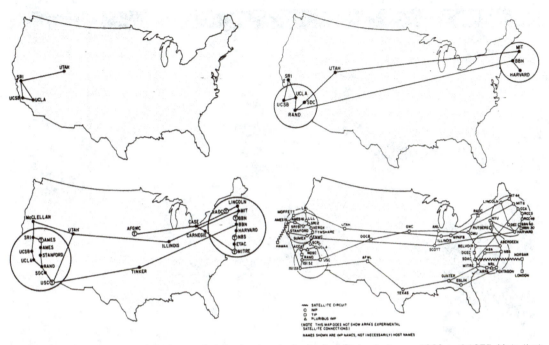

FIGURE 2.2 Development of the Internet's infrastructure in the United States between 1969 and 1975. Note that many of the early Internet hubs—the key connection points between systems—were housed at universities.

The Internet Now

A 1999 interview with Vice President (and former US senator) Al Gore on CNN's *Late Edition* provided one of history's more misconstrued quotations regarding the Internet. Asked to contrast himself from his competitor in his upcoming bid to be the Democratic presidential nominee, Mr. Gore replied, "During my service in the United States Congress, I took the initiative in creating the Internet." Often misquoted or misattributed as taking credit for personally inventing the Internet, this statement actually refers to Mr. Gore's role in guiding federal regulation and law regarding telecommunication in the United States. Authoring the High Performance Computing and Communication Act of 1991 (also known as the "Gore Bill"), then-Senator Gore advocated for public commitment to the infrastructure required for computer networks, intercomparable computing, and databases to make information more accessible and meaningful, particularly in the public sector. The Gore Bill fostered the laying of fiber-optic cable to enable high-speed, high-bandwidth data transmission, and also helped fund the development of Mosaic—the first commercially available and successful Web browser. By helping develop the technology to create and access the Internet and to make it possible for lay users (i.e., those of us not fluent in Cobalt, C++, or other programming languages) to access information via the Web, Mr. Gore did indeed play a critical role in committing the United States to the development and rise of the Internet as an independent and accessible communicative tool.

The spread of the Internet globally maps to its diffusion in the United States. Originally, nodes in large cities of developed nations (e.g., Stockholm, London, Tokyo) emerged, but diffused both throughout their region and to connect with each other. Although the Internet did not spring up overnight, it did emerge quickly (at least relative to prior communication technologies) to span the globe. Now, through satellites, microwave transmitters, and fiber-optic cables, the Internet spans

FIGURE 2.3 The current Internet infrastructure, which is a little more complex and expansive than its 1970s predecessor.

all seven continents (see Figure 2.3), and researchers in Antarctica can access information on servers in Reykjavik that were originally uploaded from New Delhi. However, it remains important to note that although the Internet's infrastructure spans the globe, not all are yet able to access it. By 2019, only about 51% of the Earth's population could use the Internet (*The Economist*, 2018), leaving some 4 billion people without access to the Internet. This inability of so many to access the Internet is often the result of what is often called the **last-mile problem**. Though the national or regional Internet infrastructure may exist, many individuals—particularly those in countries with low socioeconomic status—cannot access a computer or a satellite or cellular Internet connection. As the cost associated with getting from the main lines of the Internet to a community are often put upon the community, many developing countries and remote locations still cannot afford to access the Internet.

The World Wide Web

The Internet and the World Wide Web

If you take one thing from this section, let it be the knowledge that the Internet and the **World Wide Web** (WWW) are not the same thing. There is a common misperception that these two terms are interchangeable, reflecting the same set of processes and technologies. Whereas the Internet is a series of connected computers, the WWW, built on the backbone of the Internet, is one of many applications that uses the Internet to function. The WWW utilizes the structure of the Internet to send information, just like email and file transfer protocols (FTP). Like the Internet, the WWW has an interesting history that likely predates much of what we now think of as the WWW.

History of the World Wide Web

In July 1945, Vannevar Bush published his article "As We May Think" in *The Atlantic* magazine. Nearly four decades before the dawn of what we now think of as "the Web," Bush predicted a machine used to facilitate collective memory that would allow individuals to access knowledge mechanically by automatically creating connections between ideas. By linking ideas like a tightly interwoven card catalog that could replace the encyclopedia, Bush's device was intended to make knowledge more accessible to help solve world problems. Bush's idea can be seen today almost verbatim in Wikipedia. But in a more

FIGURE 2.4 Sir Tim Berners-Lee, originator of the World Wide Web.
Credit: Paul Clarke (CC BY-SA 4.0)

general sense, the way that text and ideas could be connected served as the foundation of the WWW when it was developed fifty years after Bush's article.

The WWW was developed as a way to link and access information. In 1989, at the European Organization for Nuclear Research (CERN) near Geneva, Switzerland, Tim Berners-Lee (Figure 2.4) developed the hypertext transfer protocol (HTTP), the protocol that standardizes and runs the WWW. The hypertext transfer protocol allowed Berners-Lee to develop **hypertext**—logical links between elements (e.g., text, images) of a web page. This hypertext brought Bush's vision to reality, as users could follow links to traverse not only within individual web pages but also across interconnected pages. Early websites used hypertext to connect ideas and interrelate content, with the hope of making information and navigation easier for users. The form and nature of the Web has radically evolved since its development, moving from static, primarily text-based pages to dynamic areas for interaction. Briefly, the evolution of the Web can be broken down into three phases: Web 1.0, Web 2.0, and Web 3.0.

Web 1.0: The Static Web The Web began as a relatively simple means of disseminating information. Early websites contained information generated by a single author: either an individual, a group, or an organization. These websites were considered static because the pages and their contents were relatively stable and fixed. What was on the page one day would likely be there the next, and could remain until the author went in and edited the **hypertext markup language** (HTML)—the code that runs the Web. Web pages in the first iteration of the Web were predominantly text-based; but over time, audio files, drawings and photographs, and video files were increasingly integrated to make for dynamic sources of information.

In many ways, Web 1.0 was just another mass medium helping individuals and organizations broadcast a single-authored message to an undifferentiated, passive audience, though admittedly an audience much more global than that of other channels. Just like television, newspapers, radio, and books, the Web let individuals reach out to wide audiences, often crossing geographic and temporal boundaries even more readily than older mass media channels had. A website can be accessed instantly at any hour of the day regardless of what country, culture, or time zone the user is in. However, websites—even static ones—were in some ways new and different from older information channels.

Web 1.0 was different from traditional mass media in three significant ways. First, the Web minimized the role of **gatekeepers**—those controlling information flows—by allowing anyone to post a web page. The birth of the Web was heralded as a democratizing force (Napoli, 1998), allowing anyone with access to a computer and a minimum of HTML knowledge to create and put a page online, rapidly expanding the number of views and perspectives others could access.

Second, the static Web became a source of rapid information on demand. Unlike books (which take time to edit, publish, and update with new information) or television and radio (which can be consumed only when broadcast), the Internet allowed organizations, individuals, and others to communicate at will and immediately with a large audience. Even though the static web page may need to be recoded and reposted, it was immensely faster and cheaper to do so than traditional media, and information could therefore be updated quickly. With web pages, news outlets were not limited to their next broadcast period or printing cycle to transmit news. Instead, CNN television, the *Telegraph* newspaper, and *Time* magazine could simply update their web pages and the latest breaking news was immediately disseminated and available online whenever consumers sought it.

Third, the early Web was different from traditional mass media because it allowed *individuals* to have a presence online. Personal web pages let individual people, not just organizations and corporations, have a place online. Early personal websites (at least until the early 2000s) were not strategic tools for professional advancement or browsed simply to pass time, in stark contrast to our more habitual use of social media now (Wohn et al., 2012). Papacharrisi (2002) found that individuals instead created their personal web pages to provide information to others about their interests or hobbies as well as themselves, and concurrently as a means of interacting with friends and family. Static websites remain today, even if overshadowed by social media, particularly for organizations as a means of providing narratives about themselves and their identities and making information available to those interested (Powell, Horvath, & Brandtner, 2016) as well as driving users to other online spaces that may be more dynamic and engaging (Nitschke, Donges, & Shade, 2016). In all, the static Web afforded many new opportunities for communication by reducing gatekeepers, speeding up the process of information dissemination, and facilitating self-presentation, all instantly and to a worldwide user base.

Web 2.0: The Social Web The rise of Web 2.0 is the rise of collaboration. Web 2.0 is denoted by collaboratively generated content on websites. Rather than a single author or entity dictating a website's contents, users of Web 2.0 came together to generate the content of a page, both by directly authoring content and by commenting on or notating content to modify how others perceive the site's primary content. Many of the modern Web tools we think of and use exemplify the Social Web. Users create the content of a Wikipedia entry, generating new articles and correcting errors or filling in omissions of current articles. Facebook and LinkedIn as services generate little information themselves—rather, they rely on you the user joining to generate content, creating and populating your page with information about you, and then interacting with others. YouTube allows users to post videos online but also allows viewers to comment on those videos, affecting what others think or how they view the video or its author (Walther, DeAndrea et al., 2010). Many of the websites and services we now frequent—including social media—are reflective of the Social Web taking advantage of the interactivity they afford.

It may surprise you to learn that, for many of the tools on the Social Web, only a minority of users actually generate their content. In discussing the **"long tail"** of media, Anderson (2004) noted that while the majority of users consume a large amount of media with little diversity, the minority of users is still very powerful, in large part due to the sheer volume of users at play online (Figure 2.5). For example, Top 40 artists like Taylor Swift may be appealing to 90% of radio listeners—hence,

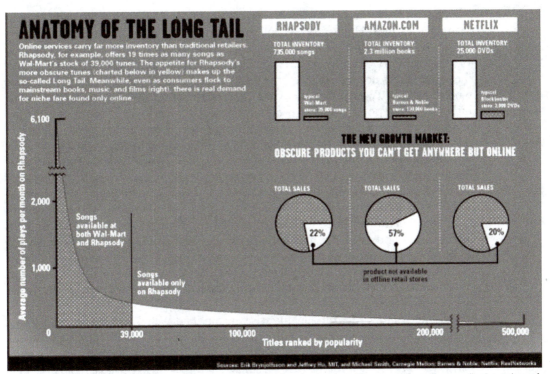

FIGURE 2.5 Anderson's "long tail" proffers that the majority of value in a site comes from the smallest number of users.

a Top 40 artist—whose music consequently makes up a lot of radio airplay. Listeners still like many other artists, but those third- and fourth-favorite artists are distributed across a much larger number of musicians and artists who are less well known. I may like Roger Clyne and the Peacemakers, you may like Comethazine, and our mutual friend may like Pink Martini. Although we all consume Taylor Swift's music, our different music tastes cumulatively account for the consumption of *much* more music than just T-Swift's latest album, even though *Lover* is the most frequently played.

The same that holds true for consumption also holds true for production: a small proportion of users generates the majority of the content. Though you have likely used Wikipedia in the past week to learn about something new, have you created or edited a Wikipedia entry? Typically, only 10% of a site's users are actual contributors, with 1% accounting for "heavy contributors" (Neilsen, 2006). Though many people may use Web 2.0 tools, often only a small fraction of individuals actually contributes to the information and value these Web 2.0 tools provide, again reflecting Anderson's long tail. Consequently, a popular question for many social media and Web managers is how to increase user engagement, converting the inactive 90% into the active 10%.

Web 3.0: The Semantic Web Web 3.0 offers an interesting solution (or perhaps a workaround) to increasing interaction and participation: finding something other than human users to do the interaction. Web 3.0 is marked by the role of computers, systems, and programs in making connections and generating new information online. Web 3.0 is often referred to as the Semantic Web, as it utilizes natural language processing (hence "semantics") to share and cross-apply information across systems, platforms, and databases.

Perhaps the most obvious examples of Web 3.0 at work are Netflix and Amazon recommendations. How does Netflix know with such accuracy what other movies and shows you may like? One of Netflix's greatest assets is its database of other viewers' habits (Gomez-Uribe & Hunt, 2015). As millions of viewers watch Netflix, the system takes note of what each viewer watches and it compiles profiles: viewers who often view *Firefly* often also view *Community*. By making associations among billions of data points, Netflix eventually starts to identify very complex patterns among its users' behaviors and uses them to make recommendations. You like historical documentaries and science fiction? Try *The Man in the High Castle*, a vision of a dystopian timeline in which the Axis forces won World War II. You may also like the animated show *Avatar: The Legend of Korra*, which follows the eponymous avatar—master of all four elements—in an Asian-inspired steampunk anime fantasy world. By analyzing huge amounts of data about users' demographics and patterns (including what programs they viewed and when), Netflix can identify not only its users' genres of interest but also potentially whether they likely have a young child at home (as they often watch more adult-content programming after 9 p.m., after putting the child to bed). These predictive analytics give Netflix an advantage—and thus a value to you—by sometimes guessing even better than you what other types of programs may interest you.

Online retailer Amazon has likewise exemplified Web 3.0 processes and ideas, integrating itself across various platforms through **cookies**—small applications that remain on your computer to track your behaviors and identifications to ease your use of various sites. As you browse on Amazon, it tracks what you've looked up and considered buying (including time spent lingering on a page). Taking note of these behaviors, interests, and preferences, Amazon then interfaces with other Web tools to make those suggestions on other sites, perhaps reminding you or attempting to otherwise elicit a sale of a product you had considered. This is why when you log on to Facebook, one column appears to be reminding you of an Amazon item you had considered a month ago, just to see if you are now interested in purchasing that item after all. That interface between Amazon and Facebook that provides complementary service is reflective of the computational processes underlying Web 3.0. The Semantic Web is using the technology and growing amount of information at its disposal to create the connections, information, and value that users provided in the Web 1.0 and 2.0 environments. If it all seems a little like *The Terminator*'s Skynet, it kind of is.

The Digital Divide

It's easy enough—especially for students—to think everyone everywhere has ubiquitous and universal access to the Internet, including the Web and the other applications that run atop the Internet. After all, many of us carry around Internet access via multiple devices at any given time: our laptops, tablets, smartphones, wearable watches and health monitors, and more. Though it may seem so, such a perspective is more a function of where you are and who you're around rather than a global reality. Worldwide, the percentage of the population even able to access the Internet just recently grew to more than half (*The Economist*, 2018). In other words, only in the past few years were more than half of the people on our planet able to go online (Figure 2.6). Even as we think the Internet is everywhere and CMC can connect us all, a large gap remains between individuals who can and cannot go online.

Broadly, the **digital divide** refers to the division in societies between those who have access to digital technologies and those who do not (Selwyn, 2004; Selwyn et al., 2003). As such, the digital divide can include access to non-Internet digital

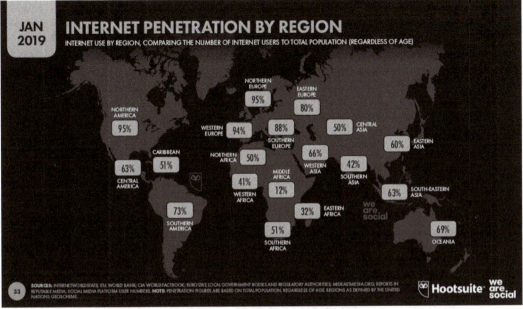

FIGURE 2.6 Where you live can have large implications for your ability to easily access the Internet. With just over 50% of the worldwide population now online, billions are still on the "have not" side of the digital divide, particularly those living in Africa, Central Asia, and Central America. What social, geographic, technological, and cultural factors do you think have led to these access patterns?
Credit: DataReportal. Hootsuite & We Are Social. "Digital 2019 Global Digital Overview" https://wearesocial-net. s3.amazonaws.com/wp-content/uploads/2019/01/Screen-Shot-2019-01-30-at-12.02.22.png

technologies, including screen readers, eBooks, and even audio players like iPods and iPhones (Luque et al., 2013). However, much of the focus on the digital divide has been on exploring the differences in experiences, opportunities, and effects between those who can and those who cannot access the Internet and its communication capabilities. As computers and the ability to go online to interact with others have become a central part of much of the human experience and the ability to operate in our current society, not having access to computers and the Internet can lead to inequalities in educational, occupational, and social opportunities.

Factors Leading to Differences in Access

The digital divide is often a result of topographical, technical, or political challenges, as well as economic and social factors, that can help or stymie the development of information communication technologies either together or in isolation. In the following subsections, we'll take a moment to explore each of these three factors individually. Though we address them separately, it's important to remember these factors often work together to even further separate those who can get online and those who cannot.

Geography The physical layout of where one lives can play an immediate role in the digital divide, as countries and regions with mountainous terrain or very distributed populations tend to take longer to build the physical infrastructure needed for Internet access. Though it's relatively easy to lay fiber-optic cable through a field or other open area, drilling through mountains or maintaining cables and wires in harsh physical environments is expensive and time-consuming, often leading to a lack of Internet access in remote or low-population-density populations (Buys et al., 2009; Zhang & Wolff, 2004). Alternately, the geography of urban environments can also cause problems with creating or upgrading infrastructure, as digging into highly trafficked or developed areas can be costly and difficult to do without substantive disruptions to existing infrastructure. For example, a major Internet upgrade in Paris may require digging up streets and around electrical and water lines, even halting vehicle and pedestrian traffic during construction, and it can often be cost prohibitive (Blum, 2013). Consequently, for several reasons—from topography to existing development—an area's geographic layout can affect the penetration of Internet access to residents.

Technical and Geopolitical Challenges In addition to physically wiring—including establishing wireless networks—an area, technical and political challenges can sometimes prevent the development of Internet accessibility. One of the ways this challenge manifests is when networks are not upgraded and modernized because system is already established and the cost of upgrading is perceived as prohibitive. One of the places in which we see this challenge is the adoption of new wireless transmission standards in the United States. Relative to other countries, the adoption of new standards with faster data-transfer capabilities (e.g., 3G, 4G, and 5G) has been slow in the United States, in large part because earlier networks had already been established. Companies, municipalities, and businesses may be cautious to pay to upgrade systems to the latest technologies when the current systems suffice. In such cases, the need for improved services and technological advances must be demonstrated and called for to justify the cost of technical upgrades.

Another challenge to the ability to go online is the effect of geopolitical forces on the availability of information. Perhaps the most notable examples of this effect are countries where access to the totality of information on the Internet is limited, filtering out content based on topic, bias, or values. For example, in a process of Internet censorship often dubbed "the Great Firewall," the government of China blocks certain information, audiovisual content, and even some emails based on content—particularly content the country's leadership deems undesirable—from Internet users in its country. Concerned about the influence of outside ideas on its citizens, the government of North Korea drastically limits its citizens' access to the global Internet with a few heavily monitored exceptions (Freedom House, 2013). In addition to an area's own local governmental limits on Internet access, international politics can challenge digital development and expansion. For example, the United States has prohibited the installation of cellular communication infrastructure made by Chinese telecommunication company Huawei, citing concerns of national security and privacy, which has led to the slow adoption of and accessibility to 5G mobile communication in the United States (Newman, 2019). As more systems and system components go international, geopolitics and technical standards may hamper access to new digital technologies.

Socioeconomics Finally, and perhaps most obviously, socioeconomics exerts a substantive influence on the penetration and accessibility of the Internet. Developing countries typically have lower adoption rates than developed nations (Van Dijk, 2020), and even within a given country, more affluent areas are more likely to have Internet infrastructure in place than are low-income areas (Warschauer, 2002). Wealthier countries and communities have more money to spend on the costly endeavor of building Internet infrastructure—laying fiber-optic cable and erecting transmitters—than poorer areas or those facing other economic challenges like war or political unrest. Consequently, some of the digital divide is a "rich get richer" phenomenon by which the countries, communities, and segments of the population that can afford to spend money or taxes on infrastructure and devices are more likely to have the Internet and the benefits that come with Internet use. Those living in areas with lower socioeconomic status are less likely to have home computers, and those communities find their access problems compounded by lower spending on public services like libraries that may serve as a means to access the Internet (Agosto, 2005). Even in countries where Internet access is relatively ubiquitous, richer schools, libraries, and community centers are more likely to make Internet-enabled computers available for public use than are those in poorer areas. Though Internet access has recently been considered a "basic human right" by the United Nations (Sandle, 2016), the reality is that some nations, regions, and communities are still working to afford the underlying infrastructure, the cost of having an Internet-enabled device, and maintaining a connection to use and take advantage of CMC tools.

Implications of Differences in Access

Lack of Internet access obviously limits the potential for individuals to do things those with Internet access may be able to do. Think of the things you have done on the Internet this week alone: paid a bill, bought a ticket, read a news article, learned about something new on Wikipedia, watched a movie, heard a song, or completed a job application. When no Internet access is available, individuals find their opportunities limited, putting them at a disadvantage.

Individuals without access to the Internet, and subsequently CMC, can feel the negative impact of being offline in many facets of their lives relative to their online peers, including economic and political participation, educational opportunities, and social inclusion (McLaren & Zappalà, 2002). Among other effects, not having Internet access can prevent individuals from:

- seeking jobs through online postings or job boards;
- using online commercial and governmental services, including taking advantage of discounts for buying online and the reduced time and costs associated with traveling to purchase or use services in person;
- using online tools to search for information, including for educational purposes, to better themselves, or even provide practical information such as how to change a tire;
- engaging in political activity and activism, such as learning about candidates or issues or engaging with others about political issues;
- joining and meaningfully taking part in information generation and dissemination, and participating in informational capitalism; and
- communicating with others, particularly those not nearby either geographically or socially.

When you cannot go online, you are prevented from accessing and engaging with large parts of contemporary society. Even where offline analog tools exist, use of these alternate tools has its own costs. For example, you can still check the movie times printed in a local paper rather than the theater's website when looking for something to watch tonight; but you still either need to pay for a subscription and await the paper's arrival or take time and money to go out and get a copy of the day's issue, and even then the local theater still needs to take out a daily ad to run movie times in the equally local paper, or else the newspaper idea doesn't work. In another example, many governments allow you to pay a discounted rate by going online to renew license plates, pay taxes, get passports, and other services; but those without Internet access now must pay a premium in addition to the time and cost of going to pay in person. The difference in advantages online access provides—the chasm of opportunity that separates those with and without Internet access—is at the heart of the digital divide.

Reconceptualizing the Digital Divide

As the world becomes more wired, the digital divide is being reformulated to consider not just "access" to the Internet, but "meaningful access" (Warschauer, 2002), which takes into account communities and individuals that may have access to the Internet but are not able to access it with the discretion and freedom with which many of us do. For example, many urban community members may not be able to afford a personal computer or Internet connection at home, but they do have access at their local library, church, or community center. However, when they use that public computer, they do so under several constraints: they may have limited time (imposed by the venue) to use the Internet-connected computer, they may not be able to freely browse the Internet as one can in the privacy and comfort of home, and they may feel stigmatized or uncomfortable even going into these spaces (Figure 2.7). There's a big difference between (1) using a personal computer in your own home, on your own time, and at your own leisure; and (2) having to access the Internet by going to a public location where your Google search "Is my rash an infection?" will be seen by others. Thus, a growing concern in *meaningful* Internet access is personal and private access to the Internet, particularly at home.

Even those fortunate enough to have home access to the Internet may not always have high-speed connections, which prevents a different form of *meaningful* Internet access. In 2017, about one in four homes in the United States with an Internet-connected home computer still did not connect to the Internet via a high-speed broadband connection (Vick, 2017). Instead, these millions of users connected via a dial-up modem or DSL connection. These connections provide Internet access, but at slower data transfer rates than broadband access. This means that downloading videos, music, or large files is implausible. As online content becomes increasingly data-intensive (e.g., videos, large data files), individuals without broadband Internet access have limited opportunities for online education, media consumption, and interactive technologies (File & Ryan, 2014). Thus, even for those 51% of the population that *can* access the Internet, not all can do so with the luxury of accessing it on their smartphone while in class to look up an unfamiliar term. Those without high-speed Internet access may represent a different component of the reconceptualized digital divide, able to access the Internet, but unable to meaningfully use and engage with the data-heavy multimedia content required of most online tools.

FIGURE 2.7 Just because you can access an Internet-enabled computer in a library or office does not mean you can use it the same way you would use your own. Would you be willing to pay bills or open your bank account on a public computer? Get the results of your medical test? Look at pornography? Some things we just don't want to do in public.
Credit: chelmsfordpubliclibrary (CC BY-NC-SA 2.0)

Though we'll revisit the divide again in Chapter 13, we have devoted some time to discussing this divide and its challenges now because the following sections and chapters operate under the assumption that individuals are online and can communicate through digital tools. But it is important to remember that this is an assumption, and one that disproportionately overlooks the experiences of minorities, remote communities, and developing nations. Moving forward, we discuss how digital tools are used to communicate with others, but the recognition that Internet use is not a universal opportunity should not be forgotten.

Getting Online to Communicate

Individuals able to go online have many opportunities in contemporary society, from emailing a friend to taking online classes to creating and distributing media content. But an important consideration is *how* individuals get online. In the 1980s, using the Internet—and therefore using CMC—was limited to a desktop computer, usually in a business office or university. Home adoption began to increase individuals' ability to get online in the 1990s, particularly with the introduction of affordable home computers and ISPs like Compuserve and America Online. Internet service providers allowed individuals to connect their computers to the Internet, and in doing so access and use CMC tools. Computers appeared in the home much faster than prior communication media like radio and television (see Figure 2.8), but ultimately they have plateaued. Since 2007, the percentage of homes in the United States with a desktop or laptop computer has stayed steady at about 75%

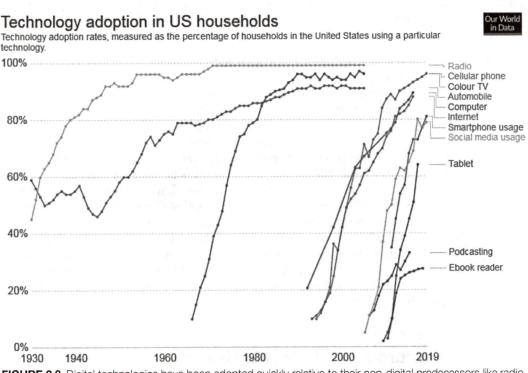

FIGURE 2.8 Digital technologies have been adopted quickly relative to their non-digital predecessors like radio and television. Most recently, even as desktop and laptop ownership and use have leveled out, smartphones and tablets continue to be used more to access the Internet and for CMC.

Credit: Technology Diffusion (Comin and Hobijn [2004] and others). Creative Commons, https://ourworldindata.org/exports/technology-adoption-by-households-in-the-united-states-bf751363f32b1ed7fa503f73b455511d_v6_850x600.svg

Technology adoption in US households, 1970 to 2019

Technology adoption rates, measured as the percentage of households in the United States using a particular technology.

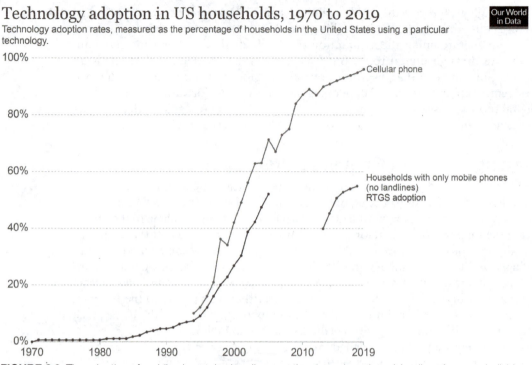

FIGURE 2.9 The adoption of mobile phones is also disproportionate and can be misleading. As many individuals in some countries have multiple devices (think a work and a personal cell phone), some national adoption rates exceed 100%, even as parts of the population still do not have a mobile device.

Credit: Technology Diffusion (Comin and Hobijn [2004] and others). Creative Commons, https://ourworldindata.org/exports/technology-adoption-by-households-in-the-united-states-bf751363f32b1ed7fa503f73b455511d_v6_850x600.svg

(Pew Research Center, 2019). At the same time, ownership of tablet computers (e.g., iPads) and using them to access the Internet have quickly risen from just 3% of homes in 2010 to about 52% of homes now (Pew Research Center, 2019).

Increasingly, we are using our phones rather than devoted computers to get online and communicate with others (Figure 2.9). Once relegated to voice-only communication, mobile telecommunication networks now allow data to be carried as well as voice. This allows our phones to send audio, text, video, and other forms of messages, increasing their utility and the means by which we can communicate through them. The increase in channels through which we can communicate through our mobile phones has been mirrored by the adoption of mobile phones. Ownership of smartphones—mobile telephones able to access and make use of Internet-based tools—has risen from 35% in 2011 to more than 81% in 2019. These data tell us that the means by which we go online and communicate have changed drastically in a relatively short time, so much so that smartphones and tablets are now used more frequently than desktop/laptop computers to use the Internet and CMC tools (Heisler, 2016). How individuals get online can tell us much about how they use CMC tools.

Two Approaches to Media Use

The previous section made the point that we have many devices and means by which we can now get online to communicate. Yet this could pose a problem: As we have more channels to communicate, how do we determine and make sense of which channels to use? Amid the many, many social media now available, you likely have an account at perhaps a dozen, and regularly use far fewer. And that's not counting your emails, your group chats, your messengers, and your other CMC channels. Assuming we have access to the entire cornucopia of CMC tools available, we make regular decisions—whether we're aware of making that decision or not—about which channels to use and with whom. How we select which media to use (and which to not use) to communicate with our friends, family, work associates, classmates, and myriad other connections is an important consideration. Though the nuances of which particular CMC channels an

FIGURE 2.10 Today, we have many means of CMC at our disposal, both as devices for communicating and the communication channels we use via those devices. What devices do you have and what channels do you use to communicate online? What made you select those devices and channels over alternatives?
metamorworks

individual uses may be somewhat idiosyncratic, we can consider two of the general processes that affect which CMC individuals use and how they are used to communicate: media multiplexity theory and the social information processing model.

Personal Use: Media Multiplexity

One way that we may choose media through which to communicate is based on the nature of our relationship with the person or people with whom we're communicating. In other words, our media use is governed sometimes by our relationships more than the innate characteristics of the channels themselves. The media multiplexity approach (Haythornthwaite, 2002, 2005) suggests that individuals with closer relational ties—those who have more network overlap, provide more social support, and are more interdependent—use more channels to communicate and to maintain those relationships. At first glance, this approach seems counterintuitive, as the people you're closest with, like your roommate or significant other, you likely spend the most time with face-to-face, and therefore you could think you need CMC the *least*. But media multiplexity suggests that as we grow closer to and more connected to individuals, we use more media and more often to communicate with them.

Even when we are geographically near someone—sometimes even physically next to them—we will still use CMC to supplement our FtF interactions, and more so for those with whom we're relationally close (Ledbetter, 2010). Anecdotally and experientially, this is likely true for you. Go back to those people you're closest with and consider how you've communicated with them in the past week. There's a good chance that although you've spent a lot of time together communicating face-to-face, you've also texted, direct messaged, and emailed them, in addition to communicating with them in a group chat or on a social medium. I've often messaged someone in the same room rather than verbally speaking with them, either for privacy or to not interrupt what they were doing. Data support this more broadly, as both personal relations (Miczo, Mariani, & Donahue, 2011) and classmates (Haythornthwaite, 2000) report using more channels to communicate with those to whom they feel closer.

More so than predicting whether or not we'll use a particular medium, media multiplexity theory predicts that the nature of our relationship with someone predicts how we'll use media to interact with that person. Consequently, as our relationships change, our use of CMC to develop or maintain those relationships should change as well so that as we grow closer we use more CMC and as we grow apart we use less CMC—both in quantity and in the number of channels. For example, as you start a romantic relationship and begin to not only meet for coffee but also to call, text, message, tweet, and Snap your significant other, you begin to think of your relationship as closer; whereas if you feel yourself relating less with someone, you may begin to use fewer channels to communicate (Taylor & Ledbetter, 2017). Said succinctly, media multiplexity theory supposes that strong, close ties use many media or channels to communicate and weak, distant relational ties use fewer media or channels. So although media multiplexity does not necessarily predict which medium you'll use—though it's likely the same channel as you use in your other close relationships, so that you can actually communicate—it does predict that as we communicate with those to whom we're closer we'll use more forms of CMC and mediated communication in general.

Professional Use: Social Information Processing Model

Though media multiplexity is helpful to understand how we select and use communication media in many situations, it does become more limited to explain our media selection and use in contexts and environments where we may have less control over what media are accessible or acceptable. For example, though many means exist to manage our classes and coursework online, your school has likely selected a particular learning management system (LMS) (e.g., Blackboard, D2L, Canvas, Moodle) that all faculty and students use to post readings, submit assignments, send messages, and communicate grades. How, then, do we make sense of our media use and selection in contexts where we may have a narrower decision set of available media? One way to consider how media are selected and used is through the social information processing model (SIPM) (Fulk et al., 1987). It is important here to caution readers not to confuse the social information processing *model* with social information processing *theory* (SIPT) (Walther, 1992), which we address in Chapter 4.

The social information processing model was designed to explain how individuals in organizations select among and use the various communication channels they have available to them. Broadly speaking, the SIPM suggests that how we perceive and use various communication media is influenced by two factors: the *objective traits* of the medium and the *subjective perceptions* of relevant peers of the medium. We use media to communicate—particularly in organizations or groups where we may not have as much choice over which media channels are available—based on a combination of these properties.

Every communication medium has objective traits—fixed characteristics and abilities of what users can do with the medium. Returning to your LMS, that system likely has a set of fixed properties, both in terms of things it can do and how it does them. For example, LMSs typically have spaces built in to upload assignments and for instructors to post (and students to check) course grades, and perhaps built-in email or social network-type systems for intraclass communication (Dougiamas & Taylor, 2003). Additionally, LMSs may have built-in functionality about which you're less aware, like logging to keep track of when users access a course site and the time they spent opening or reading files. Your particular class may or may not use these features, but they are there nonetheless as means of communicating. The SIPM argues that more useful or usable features can increase our desire to use a medium.

Our use of a medium is also influenced by how we hear others talk about the channel. Relevant peers likely have an opinion or attitude about each communication medium available in an organization, and often these perspectives are shared either directly (e.g., specific statements about the channel) or subtly (e.g., observable usage patterns). How others perceive a communication medium can influence our own views on that medium, provided these "others" are perceived as relevant or similar to us. Importantly in the SIPM, it is the attitudes of *relevant* peers that guides your ultimate perceptions and use of that channel. Your boss's perceptions of a channel would likely not matter to you as much as those of a coworker or friend in another department who would use the technology similarly to how you would. But how these relevant peers view the medium influences how we view it beyond the actual properties of the channel. Returning to your LMS, for example, it doesn't matter if your school's administration and/or faculty like it—if other students whose opinions you value think the LMS sucks, your attitude about the system and using it will become more negative.

Sometimes the objective properties of a medium and others' subjective opinions of that medium are in line. When both are perceived as beneficial, we are likely to adopt and regularly use the medium, and when both are perceived as detrimental, we are likely to reject the medium or to use alternate channels when possible. More complicated is when the objective properties of a medium and relevant others'

Research in Brief

You Can Call Me On My Cell Phone

Let's be honest: most of us are probably guilty of using technology during work or class. Whether it's a quick look at Facebook or a long time spent online shopping, the practice is common. This ubiquity of media use during work occurs even though we know doing so can be distracting—both to ourselves and to our coworkers—and reduce our engagement and comprehension in a meeting. So why do we do this? One explanation of our media use is the social influence model, which suggests that our media use is guided by both what we think we can do with a channel and how our relevant peers talk about that channel. To understand why employees may distract themselves during work meetings, Stephens and Davis (2009) explored workers' motivations for what they call *electronic multitasking*, or using information communication technologies during FtF meetings.

One hundred nineteen employees from across 20 organizations were recruited to complete a survey about their use of media during work meetings. Those with more experience using communication technologies were more likely to use them during work meetings. But more so than individual predictors, the biggest predictor of using communication

technologies during work meetings was organizational norms: seeing coworkers use media during meetings made individuals think the practice was acceptable and motivated them to do it themselves.

These findings provide strong support for the social influence model. Electronic multitasking was influenced by individuals' personal uses and preferences, but the effect of peers was even stronger. Whether it's to check an online sale or to check in on social media timelines, people are increasingly tempted to be online while in FtF situations. But what Stephens and Davis (2009) show is that even more important than the siren call of being online or whether you go online during a work meeting, class, or perhaps even a date is whether you think those around you are okay with you doing so or would do the same. So do yourself and your officemates a favor: turn off your phone before entering that meeting or the movie theater.

Reference

Stephens, K. K., & Davis, J. (2009). The social influences on electronic multitasking in organizational meetings. *Management Communication Quarterly, 23*(1), 63–83. https://doi.org/10.1177/0893318909335417

attitudes toward it differ. The SIPM predicts that we weight the relative costs and benefits to guide our media use. For example, if a medium is objectively good (helpful, intuitive, well structured, and with good communication tools) and yet viewed poorly and unused by others with whom you work, you may be less likely to use it because your coworkers avoid that channel. Alternately, all of the good vibes and love your coworkers send to a new time-reporting system may not outweigh how much you struggle to make it work for you, so you keep using the old system. And of course all of this is compounded by the structural properties of the organization: whether you love or hate your school's LMS, you *must* use it to get class information, submit assignments, and other communication functions critical to your class experience.

The SIPM can help explain why functional media sometimes are not adopted or used as intended, particularly in organizations. One of my first research studies explored how a local school district used its instant messenger system to communicate. The intention of the messenger system was that it could be used for classroom teachers to communicate with each other and their building's administration and support staff in a way that was more private and less disruptive than calling on a telephone or leaving their classrooms. Teachers, counselors, clerical staff, and administrators all agreed when interviewed that the messenger system actually had these benefits. Teachers reported liking being able to type a quick note to the office staff—"I'm sending Jeff down"—to let the staff know to expect his imminent arrival and without having to verbalize this in front of other students on the class's landline phone. Staff and administrators reported liking the quicker, asynchronous, and more informal way to communicate throughout the school. And yet teachers reported avoiding the system. The interviews revealed some teachers had developed the concern that administrators could use the messenger program to discreetly track teachers' time use: as the messenger program displayed to the network when the computer was being actively used or not (intended as a here/away indicator), teachers became worried about excessive computer use, which administrators may interpret as not being more mindful of classroom management. Principals assured me this was not the case, but as that perception spread among teachers, the schools saw the messenger system used less. Ultimately, and consistent with the SIPM, how teachers used the instant messenger system was informed by both the objective properties of the messenger system (which teachers identified as useful) and how relevant others talked about the messenger system (as a means for administrators to monitor and snoop on classrooms). Both of these factors influenced whether and how much teachers used the messenger system, which was ultimately used but used less than anyone in the district had expected or desired.

Though the SIPM was developed as a way to explain the use of technologies within an organizational or professional context, it can also be used to explain media use more broadly, including personal use. How individuals adopt and use new communication technologies is often a function of both what the technology can do and how their personal networks perceive that channel (Papacharissi & Rubin, 2000). Particularly as several similar communication technologies emerge around the same time, the SIPM can therefore help explain and even predict which channels will be used and how. To use a concrete example of this, consider MySpace and Facebook—two social network sites that emerged within a 12-month period between 2003 and 2004, offering a means of developing personal profiles and connecting with others. With about 7.5 million active users, MySpace remains a vibrant community and social network site, particularly among those in the music community and from urban areas (Armstrong, 2019), but not remotely close to Facebook's almost 2.4 billion users (Hutchinson, 2019). In many ways, MySpace was (and remains) superior to Facebook: it allows users more customizability of their profiles, greater protections of user privacy and data, and more channels for self-expression and interaction. What allowed Facebook to surpass the once-dominant MySpace as the preferred social network site? Think about why you use Facebook: it is likely not your favorite channel to use, or even the best medium through which to interact with your family, friends, or coworkers. More likely, you use Facebook primarily because that's what everyone else you know uses, and so its utility is in connecting you with the largest number of people you know, even if it's not the best channel to use. That is the SIPM at work, where—whether you know it or not—your media choices and use are a function of both (a) the innate and functional characteristics and **affordances**—potential opportunities enabled by the channel's properties—of the medium, and (b) how the people you want to communicate with talk about and use the medium.

Using and Adapting Communication Technologies

Finally, it is important to go beyond considering which medium or media we use to communicate, and think about *how* we use those particular channels to communicate. Computer-mediated communication systems are designed and programmed by humans who anticipate and intend the channel, program, or application to be used in specific ways. Email was designed to send data and information between sender and receiver, Tumblr to share personal experiences and ideas, and DeviantArt to showcase and seek feedback on artists' talents. How, then, do we explain when users go beyond how a channel was intended, adapting it to their own needs and wants? One way is to consider CMC adoption and adaptation through the lens of adaptive structuration theory.

Adaptive structuration theory (AST) (DeSanctis & Poole, 1994) proposes that communication media can be used in two ways: either faithfully or ironically. When a channel is used consistent with its intended purpose and function, it can be said to be used **faithfully**, commensurate with its design. Alternately, when the affordances of a channel are used in ways its designers or its primary function never intended, it can be said to be used **ironically**, as users adapt the channel to be used in new ways. For example, Twitter says its purpose is to "give everyone the power to create and share ideas and information instantly without barriers" to improve—not detract from—a free and global conversation (Twitter, 2020). As such, sharing ideas or generating information is a faithful use of Twitter. But we have seen Twitter used for purposes beyond "creating and sharing ideas and information," and when Twitter is used as a means of political manipulation, to threaten and harass users, or to collaboratively play a game of *Pokémon*, it is being used ironically.

Sometimes users can use CMC channels to communicate in ways not initially designed or intended. Examining how Rwandans used mobile phones, Donner (2007) found college students had developed complex systems of "beeping," using patterns of calling and hanging up to signal messages to each other. Similar to Morse code, these prearranged cues allowed complex messages to be transmitted without incurring charges on their mobile plans for a connected phone call or text messages/SMSs. Other times, just part of the channel can be adapted and used ironically, such as the use of Facebook's "Like" button to denote emotions other than enjoyment or appreciation of a post (Hayes, Carr, & Wohn, 2016). When you "Like" someone's status update showing them in the hospital with a new cast on their leg, you're not faithfully liking their injury. Instead, you are using the Like function ironically, in that your Like may communicate sympathy, sadness, or social support for the recently injured.

Concluding Users and Uses of CMC

Computer-mediated communication continues to become prevalent, increasingly integrated into our societal fabric and our personal lives (Figure 2.11). As we move forward in the following sections to discuss how CMC is integrated into and affecting our lives—including self-perceptions, relations with others, and the functioning of our groups and organizations—it is therefore important to keep in mind the physical systems that underlie and enable our CMC tools and influence how we use them. For many of us, it may seem that *everyone* has multiple computers and abundant connectivity as radio waves and microwave signals now allow us to have Internet access in our pockets wherever we go. And yet this experience is not universal, even in developed countries. For those without the financial capital, social status, geography, skills, or self-efficacy to use CMC tools, the modern world is a very different place.

For those who have the opportunity and means to access the Internet and use digital devices to communicate with others, CMC provides opportunities to communicate across time and distance, in new (and old) ways and with new (and old) connections, and to take advantage of educational, commercial, and personal opportunities not accessed offline. In a relatively short time, CMC has become a dominant means of human interaction. But it can be difficult to meaningfully talk about the role and effects of technology in human communication in any given moment due to how fast technologies and systems can emerge and change. Consequently, it can be more effective and beneficial in the long term to address what digital communication tools can do and the processes by which those effects occur rather than to discuss the specific channels themselves. The next section therefore explores some of the key theories of CMC, which are just as applicable to the newest CMC channels

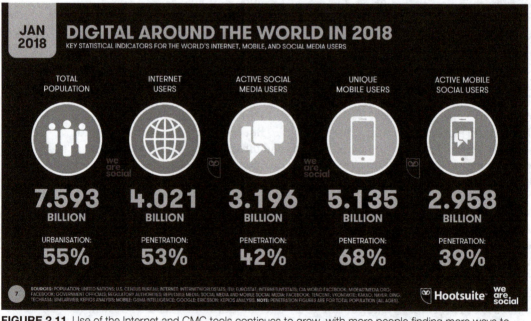

FIGURE 2.11 Use of the Internet and CMC tools continues to grow, with more people finding more ways to access CMC tools. These figures will continue to increase as CMC becomes more accessible and integrated into human communication processes. Given the meteoric rise and adoption of CMC tools, we will take a theoretical approach to understand how these tools are used in and influence human communication.
Credit: DataReportal. Hootsuite & We Are Social. "Digital 2018 Global Digital Overview" https://blog-assets.hootsuite.com/wp-content/uploads/2028/01/DIGITAL-IN-2018-001-GLOBAL-OVERVIEW-V1.00.png

as they were to the bulletin board systems (BBSs) and Usenet groups of the 1980s and 1990s. By knowing the theories that underpin and guide human interaction online, you can then apply that knowledge to make sense of the communication you see in whatever medium did not yet exist at the time this book was written.

Key Terms

Internet, 15
Internet of Things, 15
decentralized, 16
distributed, 16
last-mile problem, 18

digital divide, 22
affordances, 33
faithful adoption, 33
ironic adoption, 33

Review Questions

1. What are three ways the Internet is different from other forms of mediated communication? Consider differences in infrastructure as well as social differences.

2. What parts of your life would be most impacted were you to not have access to CMC tools? One way to answer this would be to go through either yesterday or the past week and identify all the things you have done online, and then consider if or how those could have been done either offline or through analog media (e.g., written and mailed letters, telephone) and how those processes would have been different.

3. What are the two sources of influence that guide how we choose communication media according to the SIPM? Which of those do you think has a stronger influence on the specific computer-mediated channels you frequently use to communicate?

4. Identify one way you have used a computer-mediated channel ironically, adapting the medium for a new use. How did you realize that other use was possible? How have other people with whom you communicate via that channel also begun to use that ironic adaptation to communicate?

PART II

Theories

CHAPTER 3

Impersonal Communication Theories of Computer-Mediated Communication

LEARNING OBJECTIVES

After reading this chapter, you should be able to ...

- Define impersonal communication.
- Explain how impersonal communication could result using perspectives that consider CMC as limiting the number of cues communicators can exchange.

- Apply media richness theory to predict and explain media choice when communicating.
- Discuss how the anthropomorphization of technologies guides how we perceive and interact with CMC tools.

Although the ARPANET had existed since the mid-1940s, scholarship looking into how communicating online alters communicative processes did not begin in earnest until the mid-1970s. One reason scholarship lagged was the limited opportunities for online communication. The Internet was relegated to the military and large research universities until the early 1970s, and only slowly was integrated into large organizations in the mid-1970s (primarily for email and throughout the financial industry). Because home and personal computers were not yet available, there was little opportunity to interact online and therefore little interest in studying the phenomenon of CMC because the limited technology diffusion did not allow users to talk to friends, family, or colleagues at an interpersonal level.

A second reason scholarship into CMC lagged until the 1970s was reflected in nearly two decades of research once CMC did finally emerge as a field of study. In addition to being limited in availability, CMC was perceived as a poor medium for communication. Central to one of the first CMC paradigms was that communicating online did not facilitate the rich, interpersonal, and social exchange of face-to-face (FtF) communication or even telephone calls, and was therefore a poor medium for interpersonal exchange. Reigning as the dominant paradigm for nearly two decades, this mindset can still be seen in some CMC scholarship today. This paradigm focused instead on the impersonal communication facilitated by CMC.

Impersonal communication is an exchange between participants that facilitates communication but does not allow the participants to form a meaningful relationship based on their individual selves, traits, and personalities (Walther, Anderson, & Park, 1994). Impersonal communications are not inherently bad. Indeed, we likely have impersonal communication every time we are guided by social scripts rather than individuating interactions. For example, when you make a store purchase, you and the cashier likely engage in communication as the cashier thanks you for your purchase and you say goodbye to the cashier as you leave. Although you and the cashier have exchanged messages, this exchange has not created or enhanced the relationship between you, and therefore it exemplifies impersonal communication. In fact, many of the social scripts we follow in our everyday conversations are impersonal, simply serving to facilitate common exchanges and social experiences from making purchases to acknowledging others in social settings (Langer, Blank, & Chanowitz, 1978). In the early paradigms of CMC research, the Internet was not expected to facilitate the types of messages and content required for relational development, and it was therefore largely ignored as a viable medium for communication worthy of investigation early in its development.

Though CMC was perceived as an inadequate tool for interpersonal communication, scholars began looking at it in the mid-1970s as organizations and work groups began to use CMC in decision-making. The grandparent to current workplace collaboration tools like Slack, Flock, and Microsoft Teams, group decision support systems (GDSSs) were popular tools among organizations and were used to help members exchange relevant information and make decisions through computer systems. Rather than call all members together for an FtF meeting, an online discussion or "yes/no" indicator could help group members coordinate ideas and make decisions. As they propagated in both the business and academic worlds with the adoption of these early GDSSs used in the wild (rather than in a military or academic lab), CMC tools finally provided the nudge academics needed to take note of the growing field and begin to conduct research on how communicating online may influence the communication process.

Cues Filtered Out (CFO)

The earliest paradigm of online communication, the **cues filtered out** (CFO) approach (Culnan & Markus, 1987), simply suggested CMC could not facilitate the cues necessary for interpersonal communication. A cluster of more specific theories and models, the CFO paradigm argues that because CMC inherently limits the amount of information and the number of cues that can be conveyed, computer-based communication is incapable of carrying the emotion and socialness necessary for a meaningful interpersonal exchange or relationship. More simply put, the CFO approach suggests that since you cannot see me smile or frown, hear my laugh or frustration, or shake my hand through those early text-based CMC tools, we are unable to feel the social, personal connection that we could face-to-face, and consequently our communications are limited to whatever we can put into a written message. The CFO paradigm has lost a lot of its luster (more on this in a moment), with more recent theories, studies, and experiences clearly demonstrating that we *can* have meaningful interactions and relationships online. However, we still need to discuss the CFO approach for two reasons. First, the CFO approach was critical both to establishing CMC as a field of research and to beginning the careful, scientific examination of the processes and effects of interacting online. Second, although the CFO approach may not broadly describe many contemporary CMC tools, it still accurately describes some online contexts and experiences. We still can appreciate the approach because of its historical role and its limited current applicability. Broadly, the CFO paradigm predicts CMC is limited to impersonal interactions, though this conclusion is informed by aggregating two theories that are more defined and precise than the overarching CFO paradigm: social presence theory and the lack of social context cues hypothesis.

Social Presence Theory

One reason CFO theories predicted impersonal communication was its inability to facilitate socioemotional cues, as posited by social presence theory (Short, Williams, & Christie, 1976). **Socioemotional cues** are the nonverbal and verbal indicators we have of a communicative partner's social presence, emotion, and interpersonal closeness. These signals often take the form of gestures, facial expressions, tone of voice, stance, eye contact, and proximity. However, online, technological limitations strip away most of these cues. Because FtF communication allows a lot of modes through which a message can be sent, social presence theory suggests it is inherently a better way to communicate socioemotional cues than CMC, which is limited to a basic text-based message. When you pass your old roommate in the quad and invite him to a party that night and he replies, "No," you don't take it personally because he looks frazzled and sounds tired—he must have a paper due tomorrow. However, via email you cannot tell if your roommate is angry, sad, antagonistic, or apathetic toward you and your invite based just on the text of the email. As much of our perceptions of relational closeness and socioemotional communication are derived from small cues, when such cues are minimized or absent due to technology's mediation of our interaction, CFO suggests we do not feel an interpersonal relationship when communicating online because CMC lacks the ability to facilitate the perception of social presence.

Lack of Social Context Cues Hypothesis

A second reason CFO theories predicted impersonal communication was the inability of CMC channels to transmit social cues. Social cues are those contextual tips based on our environment, our relationship, and our communication partner that we use to navigate a relationship and anticipate the course of the communicative exchange. Early studies in CMC (Siegel et al., 1986; Sproull & Kiesler, 1986) found communicating via email did not allow participants to observe or assess these norms as they weren't speaking with another person—just some name in the digital ether. Because individuals were communicating with faceless others, and without the benefit of social rules or expectations to guide the exchanges, they were expected to be unable to assess the individual characteristics of others, leading individuals to become belligerent, selfish, resistant to influence, and disinterested in interpersonal affiliation. (As we'll see in the SIDE model studied in Chapter 5, this notion of deindividuation would crop up again in CMC studies.) Due to CMC's inability to transmit social cues and its lack of affordances to transmit social norms, the CFO paradigm predicted that individuals could at best maintain an impersonal relationship.

The State of CFO

The CFO paradigm was the dominant ideology of CMC research for many years, and it guided many studies on the impacts of online communication on interpersonal and group communication processes. However, since the mid-1980s, this CFO paradigm has been mostly shelved (Walther & Parks, 2002). In large part, it has been set aside not only because of the changes in the technologies available for communication, but also due to how individuals have learned and adapted to online communication—more on that in the next chapter. In the meantime, it is important to note that though the CFO paradigm is not regularly used in modern studies and discussions of CMC, its echoes are still felt today. We still see elements of CFO applied to studies addressing the absence of some cues, most commonly cues related to identity, such as demographic information or nuanced interpersonal characteristics. For example, Walther and colleagues (2011) found individuals assimilate feedback to a personal disclosure differently when they think the feedback comes from a computer or from another person—even when all feedback is identical and computer-generated with just a few statements altered to indicate the feedback had been provided by a graduate student rather than a program. Because participants were not able to actually see the entity providing feedback, they were forced to rely on the few explicitly stated cues available to them regarding the identity of their interactant.

Consequently, although CFO is rarely used as a dominant perspective to guide our understanding of CMC, applications of elements of CFO still can help us understand how we necessarily alter the way we relate online. This is also true as CMC channels can let us deliberately select which cues we use to communicate online and which we choose to limit (Burgoon et al., 1999). As we can choose to have a text-only, voice-only, or audiovisual communication session with someone all via the same device or on the same system, how we choose to communicate via CMC can influence the cues filtered in and out. As such, CFO can still be brought to bear meaningfully in situations in which impersonal CMC processes occur.

Impersonal communication still has much value, and we use it in our daily lives: taking and placing a dinner order, getting information about a client from a colleague, and even making small talk with strangers about the weather. Importantly, impersonal communication is not inherently bad or functionless. Impersonal communication is simply not helpful when rich, emotional, interpersonal interactions with others is required or appropriate. Elements of CFO are still used to explain some kinds of communication, such as when individuals communicate in online spaces and cannot see each other or interpret messages beyond verbal (even when typed out) messages, or in cases where interpersonal communication may not be expected or necessary, such as during one-time interactions between unacquainted lab participants (Figure 3.2). But more often than not, these

FIGURE 3.1 Perhaps because early computer applications were focused on the transfer of data (including stocks, business reports, and news updates), initially computers were thought of as only good for impersonal information exchanges and could not facilitate interpersonal relationships.
Federigo Federighi

FIGURE 3.2 Socioemotional cues are not needed for—and may even distract from—our communication on many occasions. For example, when we are very task-focused (e.g., completing monthly budget reports) or when we may not want to be known interpersonally (e.g., anonymous online support groups), cue-lean channels may be desirable.
simpson33

elements of CFO are combined with other theories and models to examine how CMC works in a world now more saturated with the Internet and the means to communicate with others online.

Media Richness Theory (MRT)

As computer and Internet use began to expand in the 1980s, it did so primarily in organizations, largely because of the Internet's "killer app": electronic mail (email). Email allowed organizations to send messages and large amounts of data instantaneously around the globe (or down the hall) at no cost. As more businesses and organizations used email, GDSSs, and file sharing, it became clear there were times certain media choices made more sense than others. In 1986, Daft and Lengel forwarded media richness theory (MRT) to explain why and when individuals select specific media tools to send specific messages and make (particularly organizational) decisions.

Media richness theory suggests every medium can be considered either lean or rich based on its ability to carry socioemotional message content. **Lean media** are those channels that are effective at transmitting raw data, but not a sense of the personality or presence of the message sender. In contrast, **rich media** are those media that provide many social cues and transmit a sense of social presence. When, then, is it best to use a lean versus a rich medium? To answer this question, MRT considers the type of information to be communicated.

Uncertainty and Equivocality

Media richness theory divides information needs based on the two types of information a medium can convey: reducing uncertainty and reducing equivocality. **Uncertainty** in MRT refers to concrete information and individuals' needs to access specific information to make decisions. For example, when considering how to price next year's product, managers often look at the previous year's sales records—concrete data often found in a spreadsheet. Consequently, and according to MRT, the manager could send the supply chain director an email or a message in Slack (both leaner channels) requesting previous sales figures. **Equivocality** in MRT refers to the ability of information to identify one option among several, all of which may be vague or nebulous. Often, equivocality is referred to as ambiguity or lack of clarity in MRT. Returning to the manager's pricing problems, knowing how a market demographic will react to a change in price point may be a more complex and high-equivocality problem than simply identifying historical pricing data, requiring longer conversations with marketing researchers to understand how buyers will respond to various price points given cultural and financial factors. This conversation may require a telephone call or Zoom video chat—both richer media according to MRT, as they allow greater opportunity for rapid feedback and clarity in negotiating meaning—to discuss with the marketing researcher potential price points and outcomes. In this case, answering one question may create another—a frequent sign of high-equivocality information needs.

Applying Media Richness Theory

Given the considerations of uncertainty and ambiguity, MRT proposes a spectrum of media from which an individual may select to strategically send a message. Moreover, MRT proposes that when sending messages, there is a medium that best fits our communicative needs. As more equivocality is needed, such as when you want to get feedback from your coworkers about a new policy, a richer medium is used,

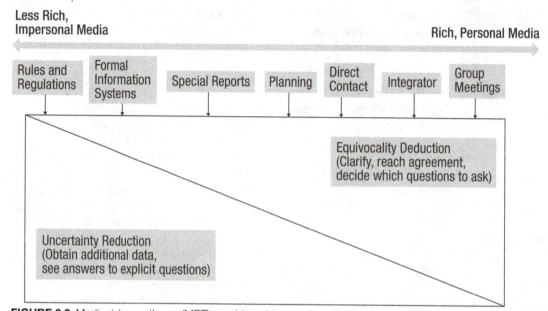

FIGURE 3.3 Media richness theory (MRT) considers richer media to be those that facilitate greater reduction of equivocality (i.e., ambiguity), and suggests media should be selected to achieve communicative goals.

such as asking them face-to-face or Facetiming them. Contrarily, as more uncertainty exists, such as trying to find a coworker who is available to pick up one of your shifts, a leaner medium like a bulletin board or group text becomes more appropriate for communicating. This is why your instructor provides a syllabus at the beginning of the semester and then may be frustrated when you ask a question proactively answered in the syllabus: it is a lean channel to provide direct, concrete information. Face-to-face questions are better for follow-up questions that are novel, going beyond information in the syllabus and requiring additional clarification or feedback. This spectrum of media richness is often illustrated visually (see Figure 3.3), helping account for and placing popular media along a continuum from lean to rich channels.

The demand for uncertainty-reducing information and equivocality-reducing information helps guide media selection. As more uncertainty reduction is needed from a message, leaner media suffice. As the need for equivocality reduction increases, richer media are needed. Note that uncertainty and equivocality are often different ends of a spectrum, so that as one increases, the other decreases. Media are often used based on uncertainty and equivocality needs and on how different media fulfill different richness needs. You'll note the medium selected often includes channels not considered CMC: memos, telephone calls, and bulletin boards and even non-mediated FtF communication. And within CMC, this illustration may also be considered dated, as originally Daft and Lengel did not (and could not have) account for social media like group chats or Snapchat stories.

Media richness theory remains popular in CMC, likely because its tenets have high face validity and are easy to digest and interpret. However, MRT does have some limitations a careful scholar should consider when applying MRT to understand human communication. First, MRT does not account well for people using traditional media in unconventional ways. For example, a friend may send you a handwritten birthday card. Although MRT should classify this written note as a leaner medium, you actually get a lot out of it (more like a richer medium) because of the thought and time put into it. Second, MRT does not deal well with media that may have multiple uses, like WhatsApp. You may create a lean WhatsApp group for an upcoming party, listing its time and location, but when people start posting personal messages or videos to the group chat, the same medium becomes simultaneously rich. In all, though MRT has received some inconsistent support in scientific tests of its applications (Walther, 2011), it remains a prominent theory in understanding how individuals communicate online.

Given Daft and Lengel's (1986) emphasis on MRT for decision-making, CMC channels—particularly channels that use primarily text to communicate, including listserves (i.e., group emails) and online bulletin boards like Craigslist—are often considered leaner and therefore impersonal. Media richness theory has historically lumped CMC channels into lean media because of the emphasis put on information dissemination as a primary communicative use of lean media, with little emphasis placed on the interpersonal significance and ability of rich media. Discussions and applications of MRT have therefore been frequently limited to or focused on information exchange via lean media channels, leaving richer channels within the domain of FtF or interpersonal communication. Though some notable examples exist using

Research in Brief

That's great it starts with an earthquake, birds, snakes, and an airplane.

Imagine you've just learned that a tornado is headed toward your city. Where do you turn for more information? After initially learning about potential public risks, people often seek out additional information before taking action, but with an increasingly broad and fragmented media environment, we have more choices than ever before as to where and to whom to turn for more information and guidance. To explore this problem, Liu, Fraustino, and Jin (2016) conducted an experiment exposing 2,015 Americans to 1 of 12 hypothetical crisis announcements and then asking where they'd seek additional information, who they would listen to, and what they would do in the event of an actual crisis.

Participants were exposed to an initial announcement of a hypothetical coordinated terrorist attack. Participants were told of the disaster either via social media (Facebook or Twitter) or traditional media (website article). Researchers also manipulated the source of the message, so participants saw information from a national (FEMA) or local (city) government source or a national (*USA Today*) or local (*San Francisco Chronical*) media source. After seeing this initial announcement, participants were asked what media they'd next use to learn more about the disaster and what they'd do. Findings revealed people would most likely use traditional (starting with television and interpersonal conversations) media to learn more

about the situation. In an even more nuanced response, national television was more likely to be used if the initial announcement was national rather than local, but local sources (e.g., local paper, local government websites) were more likely to be used if the initial announcement was local rather than national.

As interesting as these findings may be, at first glance they may seem to be more about mass media than CMC, as they deal primarily with media selection. But the researchers were careful to note that although mass or online media may initially be used to learn more, participants were more likely to initially communicate with others via direct interpersonal communication (either face-to-face or mediated via phone/email/text), even before they liked, shared, or commented on governmental posts. Keeping with MRT, when faced with a very uncertain situation, people were more likely to want to learn more and reduce uncertainty through lean media (e.g., television, newspapers), but to talk about the situation and its implications to reduce equivocality through rich media. Even in the face of disaster, people innately selected communication channels based on the information they needed.

Reference

Liu, B. F., Fraustino, J. D., & Jin, Y. (2016). Social media use during disasters: How information form and source influence intended behavioral responses. *Communication Research, 43*(5), 626–646. https://doi.org/10.1177/0093650214565917

MRT to understand media choice in a multimedia environment, such as how managers' choice of email or FtF messaging affects supervisor-subordinate relationships (Braun et al., 2019), MRT research retains its emphasis on lean media. This is perhaps due to subsequent theories focusing more on richer media and their ability to convey socioemotional meaning to guide interpersonal relationships. Before discussing interpersonal relationships and how rich (and lean) channels facilitate them, let's look at a final theory of CMC as communication that is impersonal, not due to the nature of communication, but due to with whom you are communicating—a computer.

Computers as Social Actors (CASA)

Another theory of impersonal communication is the computers as social actors (CASA) paradigm, which addresses how people interact with computers (Reeves & Nass, 1996). Calling CASA an impersonal theory may be a bit of a misnomer, though. The computers as social actors approach addresses how people relate to and with computers and technology as if they were real beings, treating their interaction with technology almost as an interpersonal relationship. This paradigm is labeled an impersonal theory here because computers are inanimate beings...so far. At least for now, computers are limited to the constraints of their programming. Although at first, **anthropomorphizing**—thinking of an inanimate object as something person-like—a computer seems odd, there's a good chance you do this in your daily life. Do you have a pet name for your car or tablet? When your phone isn't doing what you want it to, do you say it's "acting up," as if it was consciously choosing to be persnickety? Have you ever yelled at your calculator for not providing the output you expected? These all are examples of you treating an inanimate object as a social actor.

The computers as social actors theory comes from researchers at Stanford University (Nass, Steuer, & Tauber, 1994) who argued that people's interactions with computers are inherently social. This is not to say that people think computers are living things. Indeed, studies repeatedly acknowledge people are aware that computers are not living, animate things with which they can interact (e.g., Banks & Carr, 2019; Cathcart & Gumpert, 1985; Spence et al., 2014). However, even as series of programs and algorithms, computers and technologies can fulfill the functions necessary for fundamental models of communication.

As you probably (...hopefully? ...should!) remember from your introduction to communication course, Berlo's (1960) transactional model of communication posits that a sender encodes a message to be transmitted to the receiver via a medium. Upon receiving the message, the receiver decodes it and provides feedback to the sender (see Figure 3.4). Within the CASA perspective, a computer can be considered as part of the transactional model. When you provide a computer a command, it accepts that command

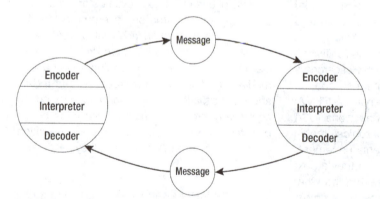

FIGURE 3.4 The transactional model of communication suggests communicators serve as both message senders and receivers at the same time. How do the components of the transactional model overlay onto the CASA paradigm to make it work?

according to script programming algorithms and provides you feedback. Sometimes that feedback confirms the computer successfully "decoded" your intention, such as to post a status message to your Facebook wall. Other times that feedback indicates a problem, such as when you cannot post the status update because your wireless Internet connection has dropped. Humans are increasingly accustomed to interacting with computers, even though a computer certainly is not a living, social being. As we grow more comfortable anthropomorphizing technology, the CASA perspective aids in understanding the intrapersonal communication that occurs with technological devices and tools.

FIGURE 3.5 Talking with a robot can be just as meaningful and impactful as talking to another human. But talking to a computer or robot may also feel safer. Children, assault survivors, and those worried about judgment from others may actually feel more comfortable disclosing to and receiving feedback from automated machines.
Sean Gallup / Staff

Experiments have generally supported the CASA paradigm, noting that individuals often treat computers as living things complete with cognition and feelings. Von der Pütten, Krämer, and Eimler (2011) even found senior citizens' health may be improved through interacting with computers or computer programs, which are sometimes less stressful to talk to than human medical professionals (see Figure 3.5). The feeling of simply talking to *something*—even if it's a machine—can make us feel less isolated and alone, even if it is by definition not an episode of interpersonal communication. Just how we interact with and receive benefit from computers can be influenced by several communicative elements, including how the computer talks, looks, or acts.

Textual CASA

Sometimes, just the words a computer program uses can influence the way a person thinks, feels, or behaves, simply based on how human the program is perceived. One way to increase the perceived humanity of a computer is to alter the language it uses. In a study of the effect of feedback, Walther and colleagues (2011) had participants input a story of a time they had been outgoing into a text box. Participants then received feedback assessing their extroversion either from a computer analysis of their word choice or a graduate psychology student who read and considered their story. Although participants who received feedback from a graduate student reported viewing themselves as more extroverted in the post-test survey, in reality there was no graduate psychology student. Both forms of feedback were created by a computer and provided identical feedback—all that changed was how natural and conversational the feedback read to the participant. When the text feedback *sounded* more human and natural, there were stronger effects on how participants saw themselves. Therefore, one way to increase the anthropomorphism of a computer program is to make talking to a computer more similar to the word choice and tone of talking with a friend. Beyond word choice, *how* a computer interacts with you can also alter your perceptions of the computer as a social actor.

Increasingly, computers are able to not just present text, but to read/speak text, sometimes even going so far as to appear to have a dialogue or interaction with you. How a computer talks to you can further influence you through the perception of the computer as a social actor. Lee (2010) had a computer program provide verbal feedback on a trivia game to subjects using either a voice-synthesizer program or recorded human speech. Again, even though the same feedback was given in both conditions (the computer program simply randomly played one of two prerecorded sound files), individuals evaluated the human-voiced feedback more positively and conformed more to its suggestions. In short, Lee's study indicates individuals' actions and thoughts can be persuaded by a computer just like they can be by a friend. More recently, virtual assistants provide us verbal cues and even means of perceived interaction, and how that text is verbalized can affect our perceptions of the computer and the effects of the interaction. If you have a GPS give you verbal directions as you drive, you've likely seen this idea at work. You are more likely to "turn left in 100 m" when your GPS sounds like a person (like Mr. T or the default female voice) rather than when it sounds like a robot (like C-3PO or *Knight Rider*'s K.I.T.T.). Text interactions with computers as social actors are also becoming common in virtual assistants, including the virtual assistants on our phones (e.g., Siri, Google Voice) and in our homes' smart speakers (e.g., Alexa, Google Home). As interacting with these tools becomes more humanlike, users can attribute the voice they're talking with to either the assistant role or the device itself, changing where the locus of interaction is (Guzman, 2019). When a computer sounds more like a human in the words and voice it uses and the way it interacts with you, the more you perceive the computer as a social actor and can be influenced by it.

Visual CASA

Another line of research has looked at how individuals interact with digital **agents**. Agents are visual online representations of computer programs intended to look and interact like humans (Guadagno, Swinth, & Blascovich, 2011). You likely interact with agents often without even being aware you are doing so, such as when you chat with a purported customer service representative on a company's website (Figure 3.6). In reality, there is likely not a person on the other end of the conversation: it's really a complex program designed to answer common and prepopulated questions. Often these agents are presented physically in some way, either through a picture of a real person or as a cartoonish avatar. As an agent's graphic representation becomes more humanlike, it is generally perceived as more credible and attractive (Nowak & Rauh, 2005). This interesting finding addresses how more visually anthropomorphic agents are perceived more as social actors. Practically, it also explains why the agent you may interact with on a company's support web page appears to be a human (to increase your acceptance of the agent and willingness to interact with it), and why

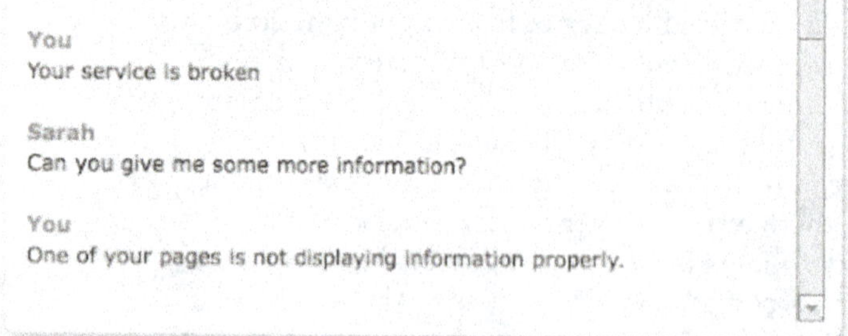

Sarah – *Our virtual agent*

AA Resize text

Sarah
Hello, Jamin. I'm Sarah and I'm here to answer your questions about PayPal. Even though I'd like to be a real person, I'm not. I'm programmed to answer your questions. How can I help you?

You
Your service is broken

Sarah
Can you give me some more information?

You
One of your pages is not displaying information properly.

FIGURE 3.6 A virtual agent, such as PayPal's Sarah, can help you resolve routine and common concerns through complex algorithms and preprogrammed scripts. However, if you saw a link on a website that said, "Click here to speak with a representative" and saw this window without the "Our virtual agent" included, would you assume that you were talking with a human represented by the name and image of "Sarah," or would you start by thinking you were talking with a digital agent?

Microsoft's Office assistant "Clippit" was such a flop: nobody wants to talk to a paperclip because it *looks* like you are talking to a computer.

Behavioral CASA

Finally, how the computer actually behaves can influence your perceptions of the computer as a social actor, and in turn your interactions with the computer. Remember Mr. Data, the android from the sci-fi television series *Star Trek: The Next Generation* and *Star Trek: Picard*? It was easier for the crew to interact with Mr. Data because he *acted* like a human, even if his brain was just a positronic network computer. Though he existed in a fictional universe (you can tell it was fictional because it was possible to afford rent in San Francisco), Mr. Data helps us understand that individuals are also able to anthropomorphize computers as social actors based on their appearance and behavior. We see similar depictions of humanlike robots in other media, from Alita in *Alita: Battle Angel* to *Battlestar: Galacta*'s Cylons to T-800s in the *Terminator* series—all are easier to interact with and personalize because they look humanlike, even when you know they are just robots. As the ability of computers to look, act, and interact corporeally with us improves, it is likely that we will increasingly perceive computers as social actors.

Krämer, Von der Pütten, and Eimler (2012) have suggested it is worth investigating how computers physically behave and how individuals interpret those behaviors. Though not as natural as Mr. Data, early results suggest people can interpret some verbal and facial expressions from agents and robots. For example, Astrid von der Pütten, a researcher at Germany's Aachen University, has explored how individuals interpret the facial expressions of a robotic rabbit and how such interpretations influence perceptions of that rabbit (see Figure 3.7). If the rabbit appears to be sad (e.g., its eyebrows are arched and its mouth is turned down in a frown), are you more likely to treat errors in your interaction with the rabbit empathically? Beyond taking cues simply from what's written or said, nonverbal cues such as space, touch, and kinesics will play increasingly large roles in how we perceive, interact with, and are affected by computers as social actors.

Whether cues are textual, visual, or behavioral, the CASA paradigm suggests individuals can have meaningful interactions and relationships with computer devices and programs. We may well recognize we are not interacting with a *person*, but we will still treat the technology as a social actor (Guzman, 2018; Spence et al., 2014). As interpersonal as the interaction may be, though, it is considered and addressed here as a means of impersonal communication because ultimately there is only one person interacting: you. Though algorithms become more advanced, and programs become more adaptive, anthropomorphic, and humanlike when they interact with you, ultimately they interact with you using predetermined logic. The concept of interpersonal interactions requires another person with whom you may actually interact. Although the CFO paradigms initially considered CMC as unable to offer opportunities for interpersonal interaction, as

FIGURE 3.7 Increasingly we find that we can read emotion from even robots and machines. What emotions or attitudes do you see these three robot dogs expressing?
PhonlamaiPhoto

we will examine in the next chapter, online communication is rife with opportunities for interpersonal communication that is equal to and sometimes even transcends FtF interaction.

Concluding Impersonal Communication

Before moving on to the interpersonal models of CMC, it is important to note that the impersonal approaches do not inherently mean CMC cannot facilitate a socioemotional, friendly relationship. (Well, the CFO perspective tried to, but this is why it has fallen into disuse.) Rather, these models and theories articulate that under certain conditions, online communication may not be the rich, vibrant exchanges we can get face-to-face with friends and family. Media richness theory, for example, is often used to determine when selecting a specific online tool for strategic goals, and sometimes suggests using a lean medium because there is not a communicative need to feel close to a relational partner—you don't need to have a heart-to-heart with your boss every time she releases a memo about casual Fridays. Although early research into CMC operated under the assumption that online communication could inherently never facilitate interpersonal communication, CFO has been generally discredited and subsequent theories have developed using assumptions like when specific media tools are used (MRT) or when communicating with a computer system (CASA), then the resultant communication may be impersonal. An important thing to remember as you finish this chapter is that communication is not inherently impersonal or interpersonal, it all depends on how you and the people you interact with communicate—online just as much as offline.

Key Terms

impersonal communication, 40

cues filtered out, 40

socioemotional cues, 41

lean media, 43

rich media, 43

equivocality, 43

uncertainty, 43

anthropomorphize, 46

agent, 48

Review Questions

1. Why did early CMC theories and models assume that CMC innately leads to impersonal interactions? Under what conditions or situations could these impersonal CMC episodes occur today?

2. Per MRT, why is your class syllabus a static document while your class discussions are either verbal or exchanged via discussion boards?

How does channel choice affect how messages (both in quantity and attributes) are exchanged?

3. Using CASA, explain how digital assistants (consider *2001: A Space Oddyssey*'s HAL, *Her*'s Samantha, *Interstellar*'s TARS, or your Alexa) are treated as communicators. Are these digital assistants really interacting with you?

CHAPTER 4

Interpersonal Communication Theories of Computer-Mediated Communication

---

LEARNING OBJECTIVES

After reading this chapter, you should be able to ...

- Define interpersonal communication.
- Distinguish between digital natives and immigrants, both conceptually and based on how they approach CMC.
- Contrast SIPT with the CFO perspective to explain how people can form interpersonal relationships via CMC.

- Discuss how CMC tools can create both equal and inflated perceptions of communication partners.
- Apply the concept of masspersonal communication to explain the communication that occurs on social media and how it may differ from interpersonal communication.

My childhood was quiet. Except for the occasional ring of the house's phone, communication didn't interrupt or punctuate my day. The postal mail arrived daily but silently. Friends coming over often called first. Years later, my days now often seem a cacophony of alerts: email chimes, text alerts, social media dings, and even the audible vibration of my phone when its ringer is off mark a frequent flow of communication from family, friends, and colleagues. It's not always an interruption; it is often a chance to catch up with friends worldwide. My email and social media help me stay connected with my friend in Taipei, my family in New Jersey, and my dear colleagues in Bielefeld (if such a place actually exists). Very counter to early assumptions of CMC as unable to facilitate interpersonal relationships, we now experience daily the ability of digital technologies to facilitate communication between relational partners.

How do *you* feel when you get a Snap from your friend? How about when your phone chirps, letting you know you got a text from home? What about when you see your name on the school website for making the dean's list? Likely, your experiences—and those of most of your classmates—made you a little resistant to the idea of impersonal communication online discussed in the previous chapter. In your daily online interaction with family, friends, and others, you likely interact with many people beyond impersonal communication, exchanging socioemotional information and helping establish, maintain, or end relationships. One reason for this experience (and perhaps resistance to impersonal CMC) is that you are likely a digital native.

FIGURE 4.1 Because they grew up alongside CMC channels, digital natives are familiar with and comfortable using digital communication tools. Messages, video chats, and social media are all natural parts of the way digital natives conduct their relationships, often not even aware of the norms and behaviors implicit in their media use.
wundervisuals

Digital natives are those individuals who have grown up with technology—video games, personal computers, and the Internet have always been part of their lives (Figure 4.1). Various people will claim different cutoff ages to be considered a digital native, but most seem to circle being born around or after the year 1992 as the cutoff point for categorizing someone as a digital native. As a digital native, you likely have turned your entire life to increasingly specialized websites to access information, you feel confident in your skills to use technology to achieve specific goals, and you are familiar with technical terms (even more so after Chapters 1 and 2) and jargon.

Alternately, **digital immigrants** are those who may be adopting and integrating technology into their daily habits later in life. Your parents and grandparents likely did not have ubiquitous access to interactive digital technologies growing up, and as a result do not learn new media and tools as quickly as you do, do not always understand how or why a technology works, and may not even feel the need to adopt a new technology. It took about two years to convince my grandmother of the advantages of email over postal mail: she just didn't see what was wrong with spending money to send a letter to arrive two weeks later.

Prensky (2001, 2009) has worked to identify differences in attitudes and use of technology between digital natives and immigrants. Among the differences, dig-

FIGURE 4.2 We now recognize that many CMC tools can facilitate interpersonal communication. From text-based emails and messages to rich audiovisual conferences, we increasingly use digital technologies to complement and replace offline communication.
FilippoBacci

ital natives typically speak a different language (or at least have a significantly different vocabulary including technology terms, lol-speak, and text pidgin), are used to processing information much faster and from multiple sources, and are more comfortable communicating online than digital immigrants because online communication has always been part of their lives.

Understanding the difference between digital natives and digital immigrants helps explain the shift in CMC research that led to theories of interpersonal communication. The earliest Internet researchers were, by nature, digital immigrants. As the Internet grew, researchers in the late 1970s could only learn about it and access it as they entered college or job fields. By the early 1990s, personal computers had emerged and started diffusing into the household. As a result, scholars began to study CMC from the mindset of a digital native. Rather than approaching the computer as a new and clandestine tool, in the early 1990s, scholars began approaching CMC based on the way they used it in their labs and personal lives: interpersonally.

Interpersonal communication includes exchanges between at least two inter-actants that allow the participants to form a meaningful understanding and/or relationship based on their individual selves, traits, and personalities (Bochner, 1989). Typically, we consider interpersonal communication socioemotionally rich as it allows perceptions of relational closeness, understanding, and empathy. Interpersonal communication has one of the longest traditions of study within the field of communication (Knapp & Daly, 2011). Many of the relationships we value and that help us in our daily lives are interpersonal relationships. We maintain close relationships with friends, family, and coworkers through interpersonal communication. As we learn about them and they learn about us, our relationships become not only broader (in that we know more about each other) but also deeper as we become more integrated into each other's lives. Indeed, it is our interpersonal ties that provide us social support, guidance, and a sense of belonging.

Scholars (as well as most individuals who regularly use the Internet) have long recognized the ability of CMC to allow us to feel closer. Interpersonal communication has also been valued as a building block of education, community, political engagement, and relational development. Though early theories denied CMC as a viable alternative to FtF communication for interpersonal communication, by the mid-1990s, scholars had begun to predict how and when online communication could be just as socioemotionally rich and interpersonal as FtF interaction. However, note that the idea that technology could facilitate interpersonal communication actually developed soon after the CFO paradigm. Like yet another reboot of the *Batman* movie franchise, the resurgence of the theory of electronic propinquity was inevitable; and as it lay in wait, the theory helped guide scholars to a better understanding of interpersonal communication online.

Electronic Propinquity

How do you feel when you talk with a parent? How about when you talk to a stranger? Likely the answers to those questions are different, even if the discussion is the same. The reason they are different is because you have a different level of propinquity with each of them. **Propinquity** is the perception of relational or psychological closeness felt toward another person. This psychological closeness can take many forms and has been measured using the variables of interpersonal attraction, homophily, and social attraction (Burgoon et al., 2002; Korzenny, 1978; Walther & Bazarova, 2008); but all these variables consider the same thing: how psychologically close we feel to another person.

Initially, propinquity was addressed offline using analog channels like telephones and closed-circuit audiovisual conferences. Studies have long showed that physical closeness—proximity—is related with interpersonal communication. We tend to spend more time geographically near family and friends than we do less close acquaintances, and we more often play softball and go out for coffee with good friends than with mere classmates. Even professionally, we tend to like and work with those we see more often (Allport, 1979). Although this makes sense to us when we are colocated—existing in the same space—with someone, when we are together geographically, it may be more difficult to think about propinquity online. Interestingly, the notion that people could feel emotionally close to others online was put forward about the same time as the CFO perspectives were being articulated. However, due to some unfortunate early experiments and slower developments in technological affordances, the idea of online propinquity was shelved for almost three decades (Walther, 2011). Fortunately, the idea of propinquity has recently seen a resurgence and empirical support.

In 1978, psychologist Philepe Korzenny, then at Michigan State University, forwarded the idea of **electronic propinquity**. Electronic media, claimed Korzenny, could allow communicators to experience a sense of closeness just like an FtF exchange. In 1978, "electronic media" were much more limited than they are now. Recall from Chapter 2 the state of the Internet in the late 1970s: few beyond the military and a very few academics working on computer systems had access to tools such as email and file transfers, limiting the available "electronic media" to television, radio, telephones, and written letters. However, even in this limited media environment, Korzenny still put forward his notion that mediated interactions could allow individuals to *feel* close to each other. Importantly, the theory of electronic propinquity (TEP) was put forward relying on factors that, although precise, were not medium-specific and were therefore able to guide research even as technologies such as the WWW, Instagram, and Zoom appeared—some as much as forty years later.

Theory of Electronic Propinquity

The theory of electronic propinquity predicts the psychological perception of closeness with a communication partner is influenced by several specific factors: channel bandwidth, mutual directionality, task complexity, communication skills, communication rules, and available media choices (see Figure 4.3). Some of these factors are guided primarily by the features and affordances of the medium itself, including bandwidth and mutual directionality. Although it has another, more specific meaning when discussing Internet infrastructure, in TEP *bandwidth* refers to a channel's capacity to carry multiple cues, similar to MRT's notion of a "rich" channel. A Skype audiovisual conference—a channel with high bandwidth—allows you to verbally talk with a communicative partner who can also see your nonverbal messages (including facial expressions) and any text you type to supplement

these channels. Alternately, an instant messenger chat program may have less bandwidth when it facilitates only text communication. The more bandwidth a channel has, the greater its ability to create a sense of psychological closeness—*electronic propinquity*—because we can use that greater number of channels to convey meaning and socioemotional content. Again, bandwidth in TEP does not mean the rate of data transfer enabled by an Internet connection by the

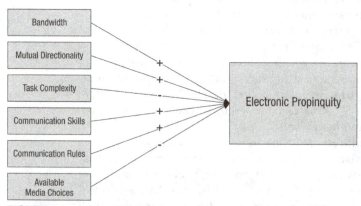

FIGURE 4.3 Korzenny's (1978) theory of electronic propinquity (TEP) makes predictions about the perceived closeness interactants feel when their communication is mediated, based on the characteristics of the medium and their use of it.

medium. Even a phone call offers a moderate amount of bandwidth through vocal and temporal cues, as conceptualized in TEP.

Next, *mutual directionality* addresses the quickness of feedback facilitated by the channel. Mutual directionality is positively associated with electronic propinquity, so that channels allowing quicker interactions among communicators should enable higher levels of electronic propinquity. Radio and Periscope streams, media that limit or make feedback difficult, facilitate lower propinquity because communicators experience limited abilities to react to each other. Alternately, instant messenger, a medium facilitating rapid and mutually directed messages, facilitates high propinquity by enabling more real-time reactions and discussions among participants.

Other factors of electronic propinquity are influenced by the needs and experiences of the individual users and their interactions, including task complexity, communication, skills, and communication rules. Propinquity may be decreased as users attempt to engage in greater *task complexity*, which refers to how complicated, intricate, or involved the focus of the interaction is. As users attempt to complete a more complex or difficult task, they will spend more time processing procedural and task communication than they do processing socioemotional cues, and consequently they experience lower levels of propinquity. For example, while on a videoconference to set out the new quarter's budget by the end of the day, members on the call will likely be focused on spreadsheets and previous earnings and budgets, and as such not have much time to socialize (and thereby increase propinquity). In a less complex task, like picking the date for the next budget review, members may have more time to communicate about relational or off-task topics, thereby increasing propinquity.

Additionally, individuals with greater *communication skills* can increase the propinquity of a medium. Those able to make more sense of cues, particularly nonverbal cues, in the messages the medium facilitates will be more able to use the available bandwidth to both encode and make sense of messages, thereby feeling closer to their communication partner. For example, while your grandmother may not like (or even be able) to use Instagram because she just doesn't understand it and doesn't feel like she's talking to a person, you likely are able to meaningfully talk with friends via Insta as you've developed your own set of cues to send and interpret, like emoticons and ingroup language.

The final two factors in TEP—communication rules and available media—deal with some of the structural and technical options communicators face. *Communication rules* refers to the technological and social guidelines that govern interactions. As the number of communication rules increases, communicators are constrained by these rules as they interact, and therefore they may feel less close because of a more structured or formal interaction. For example, the social norms of Facebook dictate with whom you can interact (those who you already know and will accept a Friend request) while the system of the Bumble dating app instructs that only females can initiate communication. In media and online environments with fewer rules, users who understand the communication rules can send a wider array of messages that allow for greater diversity in the messages sent and received.

Finally, propinquity decreases as there are more *available media choices*. This may sound contradictory at first: Shouldn't *more* options mean you can pick the best medium for communication? Korzenny proposed that as users have fewer media at their disposal available for communication, they are forced to make the best of the limited tools or media at their disposal for interpersonal communication. In practice, we see this happen quite often. When you have a seemingly unlimited number of media at your disposal to communicate with your family, you don't spend as much

FIGURE 4.4 One group that often experiences electronic propinquity and its tenets is deployed servicemembers. With their available channels limited, even limited-bandwidth messages like letters or texts from home are carefully read and reread, as both the deployed and family on the homefront seek to maintain emotional closeness through mediated communication. How do the other tenets of electronic propinquity theory help (or hinder) a sense of psychological closeness with geographically distant relational partners?
Mark Edward Atkinson/Tracey Lee

time carefully constructing your messages and interpreting your family's response. However, when you are only allowed to use postal mail to communicate with them, you begin to more carefully construct and interpret messages, resulting in an increased sense of propinquity. We see some evidence of this in instances where family members are communicating with military servicemembers deployed abroad (Figure 4.4): when the opportunities and channels for interaction are limited, communicators spend more time considering and structuring their messages to maximize their communicative and relational potential (Merolla, 2010).

Tests of Electronic Propinquity

After proposing TEP in 1978, Korzenny went on to test the theory (Korzenny & Bauer, 1981) with disappointing results: none of the proposed relationships were supported. Why this initial test of TEP failed is not quite clear. It could have been because of the relatively limited media available at the time (the 1978 manuscript only got as far as closed-circuit camera feeds), or (more likely) due to the public's limited interaction with and use of electronic interpersonal media. But for whatever reason, TEP was generally ignored by media and CMC researchers and lay dormant for thirty years.

It wasn't until Walther and Bazarova revisited TEP in 2008 that it received renewed attention. Their test of TEP used experimental groups and controlled for media alternatives to communication, and they found strong support for the propositions of TEP. Specifically, results supported the effect of bandwidth and media choice on electronic propinquity. When small groups used multiple modes of mediated interaction (i.e., text, audio, and video channels) to communicate, the increase in available bandwidth was related to stronger perception of social closeness with fellow group members, and groups asked to undertake difficult, complex tasks reported lower levels of propinquity afterward.

Walther and Bazarova's (2008) support of TEP also shed light on previous findings regarding CMC that challenged the dominant early paradigm of CMC as a lean medium. Since 1992, research had begun to assert that CMC can be used to facilitate rich, interpersonal communicative exchanges. The ultimate support of TEP helped explain why online communication, even without the nonverbal cues of FtF communication, allowed individuals to feel relationally and psychologically close to those with whom they communicated online. Constraints within computer-mediated channels *forced* users to make better use of the more limited channels they had available via CMC tools to feel psychologically close to those with whom they communicated.

When the sun go down, ain't nobody else around. That's when I need you baby.

Long-distance relationships are a challenge. Whether the couple are deployed in the military, away at school, or simply live in different states or nations, CMC increasingly helps couples span time and distance, communicating to maintain their relationship. But with all the CMC channels now available, what's really needed to make us feel psychologically close with our loved ones?

Kaye and colleagues (2005) sought to answer this question through the lens of the theory of electronic propinquity (TEP) (Korzenny, 1978). Interestingly, one of propinquity's tenets is that lower-bandwidth channels should increase perceptions of psychological closeness: as individuals are able to convey *less* information, the information they *can* convey takes on greater meaning. To test this proposition, Kaye and colleagues (2005) built an app and distributed it to five couples in long-distance relationships. The app was quite simple: a little transparent circle was placed on their Windows taskbar (near the time and calendar). When one partner clicked the circle, their partner's corresponding circle glowed red and then faded over a 12-hour period. That's it. That's really all there was to it. Clicking the button only conveyed one bit of information—that the button had been clicked. And yet, comparing couples'

self-reported closeness after a one-week period of using this one-bit cue revealed couples felt closer and more intimate, from nothing more than that little glowing dot.

So what happened? Well, consistent with TEP (and perhaps SIPT [Walther, 1992]), couples made the most out of the limited information they had and learned to use that simple cue. Participants reported that it provided just a little extra way of communicating, so that coming back to one's computer to see the button glowing was interpreted as the other person thinking of them. In fact, although the button glowed for half a day, couples used their button an average of 35 times each day. Unlike a phone call or text that requires a response or some form of processing, the one-bit dot simply acknowledged the romantic partner and relationship, and partners used it as a means of remaining close even when apart. What, then, does it take to feel close to someone even when separated by oceans? Perhaps a simple click is all that's needed to remind them you're thinking of them.

Reference

Kaye, J. J., Levitt, M. K., Nevins, J., Golden, J., & Schmidt, V. (2005, April 2–7). *Communicating intimacy one bit at a time* [Paper presentation]. Human Factors in Computing Systems (CHI'05), Portland, OR.

Social Information Processing Theory

When you talk with people online, you often likely do so because it makes you feel close to them. You email a former professor who wrote a letter of recommendation to let her know you got that summer internship, you post "Happy Birthday" on a friend's Facebook wall to let him know you care, and you Skype your parent because you felt a little homesick this weekend. You probably know inherently (You! The digital native!) that CMC allows rich communication that facilitates electronic propinquity. However, you may not exactly know *why* CMC allows you to feel close to someone you email, Facebook, or video chat, how it allows you to form and maintain relationships, and how what was once considered a lean set of media can now allow us to exchange emotions, feelings, and other socioemotionally rich messages. Social information processing theory was one of the first theories to explain the mechanisms and processes that allow CMC to facilitate interpersonal exchanges and relationships. Early support of this central CMC theory helped extend scholarship exploring the processes and effects of CMC beyond the CFO paradigm.

Joseph B. Walther's (1992) social information processing theory (SIPT) of communication has been widely used to predict and explain how individuals can form rich, socioemotional, interpersonal relationships via CMC. Indeed, SIPT was formed under the premise that media users are motivated to form rich, deep impressions of each other, regardless of medium. After all, why would people try to communicate with others online if such communications didn't foster the perceptions of propinquity that drives much of human interaction? From the CFO perspective, people were wasting time on the Internet talking with others impersonally and without the benefit of relational development, which by 1992, had started to seem as naïve as it does to digital natives now. The articulation of SIPT explained how and why individuals may be using CMC interpersonally, even if they weren't aware of doing so.

Time in SIP

At its core, SIPT predicts that, given enough time, users adapt the limited cues available via CMC and ultimately use the available cues to form rich, interpersonal connections. The primary mechanism of SIPT is *time*: Given enough time, users can adapt their verbal communication online to accommodate for the loss in nonverbal cues and to facilitate perceptions of electronic propinquity. Importantly here, "time" does not just mean the passage of minutes or days, but rather should be interpreted to mean the sequential exchange of interactive messages. Just because you email someone now and then again in a year does not mean you have built a relationship. Instead, SIPT requires you and your communication partner to exchange multiple messages that reply to each other and advance the communication between you. Time—and the number of exchanges and the amount of information exchanged—is a critical component of SIPT because online communication takes longer than other forms of communication. In fact, it takes more time to communicate online than it does face-to-face simply because typing and reading is about three times slower than talking and listening (Burgoon & Hale, 1987; Tidwell & Walther, 2002). Not only does it take more time to compose and reply to messages online than it does face-to-face, it may take more time to learn how to encode and decode simple text-based messages, and consequently more time spent interacting with someone gives you both more time to learn, create, and adapt to the types of cues and messages the CMC channel allows you to send.

Adapting Cues in SIP

Face-to-face, we know how to use cues to tell friends we are happy: we smile, raise the pitch of our voice, and have a spring in our step. But how do we communicate happiness without facial expressions? Recalling the CFO paradigm, one reason CMC was originally predicted to not facilitate interpersonal communication was because it inherently lacked the cues necessary to provide a sense of closeness: appearance, proximity, and facial gestures. Social information processing theory does not try to argue these cues are directly facilitated by online media. Instead, SIPT points out that, given enough time, users will find or develop new ways to transmit, receive, and make sense of new verbal cues that replace missing nonverbal cues. If we cannot smile at a friend online, how do we tell them we are happy, beyond simply typing "I'm happy"?

Language

One way we may communicate our happiness is through subtly altering our language and word choice. We may use more positive expression words, even when not describing our current emotional state. Texting your friend one night, you comment that the coffee this afternoon was "great," your morning class was "excellent," and the evening class was "better than usual." Using this many positives may not be typical in your daily exchanges, and having received so many to establish a norm your friend replies that you seem happy today. Even though you did not directly state this, your friend has learned your linguistic habits and can interpret cues even from a limited amount of text. Walther, Loh, and Granka (2005) experimentally validated these linguistic adaptations to our conversational partners by having lab assistants assigned to a dyad increase or decrease their friendliness in either an online or FtF session. Consistent with SIPT, verbal behaviors counted for a significant amount of variance in participants' perceptions of the confederates' friendliness, but more so in the online rather than condition. Restated, the specific words used online counted for more online when they were the only cues available rather than offline where alternate cues could also be used.

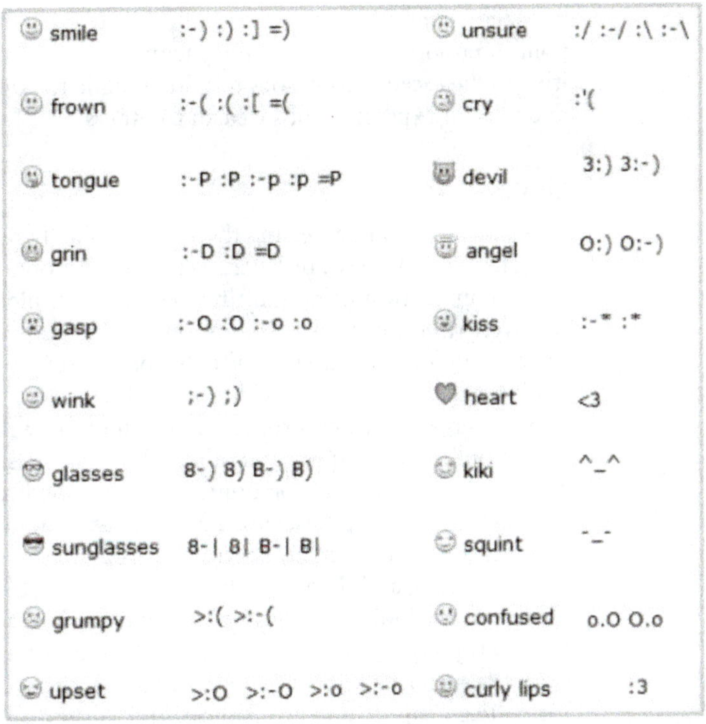

☺ smile	:-) :) :] =)	☺ unsure	:/ :-/ :\ :-\
☺ frown	:-(:(:[=(	☺ cry	:'(
☺ tongue	:-P :P :-p :p =P	😈 devil	3:) 3:-)
☺ grin	:-D :D =D	😇 angel	O:) O:-)
☺ gasp	:-O :O :-o :o	☺ kiss	:-* :*
☺ wink	;-) ;)	♥ heart	<3
😎 glasses	8-) 8) B-) B)	☺ kiki	^_^
😎 sunglasses	8-| 8| B-| B|	☺ squint	-_-
☺ grumpy	>:(>:-(	☺ confused	o.0 0.o
☺ upset	>:O >:-O >:o >:-o	☺ curly lips	:3

FIGURE 4.5 Common emoticons created through alphanumeric characters and their emoji equivalents. These emoticons can be used to send nonverbal cues through text, letting users communicate emotions beyond the body of their message's text.

Emoticons Besides altering the words we use, we may also use text characters to create images to replace our missing nonverbal cues in CMC. Sometimes, we may simply type out a nonverbal gesture or expression, such as when we type "lol" or "::grin::" into an email to indicate the humor has us laughing out loud. Another option is to use alphanumeric characters that have become commonly accepted as nonverbal expressions. Commonly called **emoticons**, these icons are configurations of characters meant to represent facial expressions or physical action (see Figure 4.5). Configurations of letters, numbers, and punctuations have evolved and been integrated into the online lexicons of communicators to compensate for not being able to physically express a smile online. Indeed, these icons now commonly appear online and have even been codified for digital communication (typing ":)" into new versions of Microsoft Word will automatically replace the two characters with a ☺). However, what effect do emoticons actually have on how we interpret a message?

While emoticons have certainly pervaded the lexicon of digital natives, the use of emoticons is often a means of adapting to channels where nonverbal cues are limited. Though it's better than nothing, it is notable that emoticons may not have the same effect as the nonverbal cue they textually represent. Walther and D'Addario (2001) studied the relationship between verbal messages and emoticons online. The verbal content (i.e., the words used to communicate) in an interaction affected the way the message was interpreted more than the emoticon that followed. But interestingly, not all emoticons seem to have the same communicative weight. Walther and D'Addario found that although negative emoticons (e.g., ":(") negatively impacted the positivity of a message, positive emoticons (e.g., ":)") did not have an equal effect to make the message seem more positive to receivers. Although posting a smiley emoticon after a message may make you feel better about the message's contents, it is more important to make sure the contents carefully encode your intention because the smiley icon does not make the message's meaning as positive to the receiver as it does to you. A smile to you as a means of happiness may be taken as an expression of sarcasm to others.

Beyond Text

Social information processing theory was initially developed and posited only considering text-based CMC. In 1992, there was no Zoom, YouTube, or Instagram as means of communication using multimedia and multiple cues. However, SIPT has proven relatively robust even as CMC moves beyond text-only forms of communication, and remains useful as we are increasingly able to communicate via CMC tools that include audiovisual cues. Now, individuals may learn about each other online through photographic representations, profile photos and other imagery common in websites, social media, online newspapers, and other CMC channels. Pictures of an interaction partner can instill a feeling of propinquity, even if the picture is shown before interacting with the person (Tanis & Postmes, 2003). That a picture can increase feelings of closeness explains why so many services for online communication—including Instagram, Twitter, Zoom, and discussion boards—ask users to upload a profile photo. Having a picture of the individual with whom you are interacting can increase perceived electronic propinquity before you actually directly communicate. Even in high-bandwidth channels such as Zoom, users do not have access to all forms of nonverbal communication (e.g., haptics and proxemics), and so they can read into the cues available (like a picture) to compensate for missing cues. Consequently, though SIPT may have been put forth to deal with text-based CMC, critical tests of the theory have found its basic tenets apply even in richer media that afford users more than just text-based channels for interaction (e.g., Sprecher, 2014; Tates et al., 2017).

Equaling Face-to-Face Closeness Online

Ultimately, SIPT predicts that, given enough time and interactions, users can develop and maintain relationships online that are on par with offline relationships. Though the Internet had not quite diffused to the general population when it was put forth, Walther's SIPT represented a change in almost 20 years of theorizing by predicting that online communication could facilitate interpersonal exchanges equal to the gold standard of offline exchanges. However, some SIPT findings soon began to suggest perceptions actually *exceeded* the perceptions hypothesized. These findings, as well as the increasing prevalence of CMC tools, led Walther to put forth another model that went a step further and predicted CMC could actually *surpass* FtF perceptions.

The Hyperpersonal Model

If you have ever had a pen pal, a summer camp romance, or a long-distance relationship, you likely have already experienced a hyperpersonal relationship without even knowing it, or at least without the words to explain your experience. The hyperpersonal model of communication (Walther, 1996) was put forward to explain how individuals may use limited-cues channels to develop relationships that may go beyond what is possible face-to-face. The hyperpersonal model is particularly helpful to understand how people communicate and relate when interactions occur entirely online. In stark contrast to the CFO paradigm, the hyperpersonal model of communication argues individuals can actually form enhanced impressions when their communication is computer-mediated.

One of the more heavily utilized models of CMC, Walther's hyperpersonal model is also one of the most complex, even though it closely parallels the familiar transactional model of communication. Before getting into the model, it may be helpful to begin by defining and discussing the model's outcome. A **hyperpersonal relationship** is one in which the sender is idealized, so that impressions of the individual are stronger and more salient than similar impressions that may be developed face-to-face, and are thus *hyper*personal (rather than *inter*personal) because they surpass the levels of perceptions and attributions typically occurring in an interpersonal relationship. These enhanced impressions and effects are the result of four elements of the hyperpersonal model of communication.

The Hyperpersonal Model

The hyperpersonal model explains how CMC may facilitate interpersonal impressions "that exceed the desirability and intimacy that occur in parallel offline interactions" (Walther, 2011, 460). When individuals communicate through mediated channels, they have opportunities not afforded in FtF interactions that can be used to develop enhanced perceptions. Note that the model, while developed specifically for CMC, should still hold when not interacting specifically online (Carr, 2020a), as the elements discussed in what follows also manifest in other forms of mediated interactions, including written communication.

A shift happens when we are able to selectively present ourselves and the traditional transactional model of communication is modified. While the elements of the transactional model remain the same, individuals interacting through some intermediary channel are able to take advantage of features of the medium and the changes in how messages are encoded and decoded that result in idealized perceptions of each other (see Figure 4.4). These idealized perceptions can occur when using online tools like dating websites (Hancock & Toma, 2009) or blogs (Walther et al., 2011), but they should also occur when interacting with a pen pal via written, postal mail (Shulman, Seiffge-Krenke, & Dimitrovsky, 1994). The hyperpersonal model therefore becomes an effective explanatory tool for idealized impressions in

online dating, summer camp romances, and long-distance relationships. The hyperpersonal model predicts individuals derive enhanced, idealized impressions of others when interacting via mediated communication due to four factors: senders' selective self-presentation, receivers' idealization, the disentrainment of mediated channels, and a feedback loop reinforcing idealiziation.

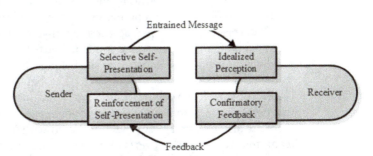

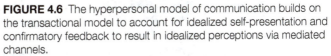

FIGURE 4.6 The hyperpersonal model of communication builds on the transactional model to account for idealized self-presentation and confirmatory feedback to result in idealized perceptions via mediated channels.

Sender's Selective Self-Presentation When you present yourself in a mediated context, you have an advantage you do not have offline: you get to pick how you are represented, unlike in FtF interactions where you can't always choose what messages you send off to make sure they are the "you" you are trying to represent. Remember that yearbook photo of you with the goofy smile? When you build your Facebook profile, you pick the information that makes up who you are: the pictures, the interests, and the status updates. And because most of us want to put our best selves forward, we pick the information that presents the ideal, best version of ourselves (Schouten et al., 2014). We populate our Facebook profile with pictures in which *we* look good (Who cares if our friend has a goofy expression?!) and are doing fun things, we list only our favorite cool bands while not acknowledging our hidden and shameful love of Nickelback, and we create and share posts we hope reflect our thoughtful musings rather than the ridiculous conversation we had last night at 3 a.m. In short, we are generally careful to purposefully and carefully present the self we think our intended audience will most value.

Senders' ability to selectively present themselves is the first and most powerful aspect of the hyperpersonal model. Because they have control and time to manipulate what they present, individuals can carefully construct the persona they want others to perceive (Turkle, 1995), and in doing so often present a more favorable version of themselves than may exist offline. These selective self-presentations can range from simple omission of negatives to explicit deception. For example, when you interact with a stranger in a Chatroulette chat room, you may decide not to disclose you are a smoker (a perceived negative) while emphasizing the work you did for Habitat for Humanity. (In fact, you were just lending a friend your pickup truck to move plywood and didn't even personally enter the build site. But online, who's to know otherwise?) As we want others to like us, we typically present an idealized version of ourselves to maximize our strategic and relational goals, and these idealized self-presentations lead to hyperpersonal perceptions. In the words of Brad Paisley (2007), "I'm so much cooler online."

Receiver's Idealization of Sender Like the sender, the receiver in a communicative exchange plays a role in the hyperpersonal model by attempting to idealize the sender. Just as senders want others to like them, receivers want to like those with whom they interact (Berschied & Walster, 1969). (Why would you want to interact with someone you *don't* like?) Consequently, we often seek out interaction with those we like or try to justify why we like those with whom we interact. Consequently, when interacting online, we use the available cues to support our affinity toward one another.

When interacting with someone with whom we have not previously interacted, we try to find cues we recognize from our previous experiences. These cues are then interpreted by comparing them against known group identities and stereotypes. Sometimes, we recognize someone's attributes based on the social categories to which the sender indicates he or she belongs. For example, were you to encounter someone online with the username YankeesFan14, you may assume your partner is a fan of the New York baseball team, which provides you some clues as to what this person values and how you may interact (Heisler & Crabill, 2006). However, there are also certain stereotypes associated with Yankees fans (beyond their sports team) that you may also overlay onto your chat partner: Assumptions of geographic location, extroversion, political leanings, and even dialect and vocal patterns may be made simply based on the user name and the social group it suggests (more on this in the next chapter).

As receivers selectively focus on positive group identities and stereotypes, they engage in the second process of the hyperpersonal model: *idealized perceptions*. Forming impressions of others based on limited cues, often selected by senders to increase receivers' impressions, receivers often form impressions that are more positive or of traits more desirable for communicative partners, viewing them in the best possible way. Receivers play an important role in hyperpersonal communication by exaggerating perceptions of message senders, compensating for missing information or cues with the receivers' own assumptions and desires. In this way, receivers often aid senders by filling in mental blanks in senders' self-presentation in ways that benefit the senders.

Channel Entrainment When individuals have the time and ability to carefully construct their message, it allows them additional opportunities for careful and strategic message creation. **Channel entrainment** refers to a sender's ability to synchronize a message with a strategic self-presentation as facilitated by a particular communication channel. A channel that enables the deliberate construction of a message can be considered a highly entrained channel, while a channel that does not allow careful management of self may be considered a less entrained channel. By taking advantage of the limitations and abilities of a channel to carefully construct a message, a sender can more carefully present an idealized self.

Channel entrainment is often closely tied to the synchronicity and richness of the channel. The faster information and feedback are exchanged, and the more cues that can be exchanged simultaneously, the less opportunity for individuals to carefully construct their messages to achieve strategic relational goals. When using asynchronous media such as email or a word processor, most of us naturally find ourselves using the property of channel entrainment as we delete errant phrases, replace words with synonyms to better our rhetoric, and carefully proofread our composition before finally submitting it to its reader (Walther, 2007). In less entrained channels, such as FtF communication, our ability to self-edit is reduced by communicating in real time. Thus, we are forced to interact with people through our verbal faux pas, while wearing the "Bieber Fever" shirt we donned at the gym, or with our uncontrollable bad breath due to the garlic bagel that was our lunch. This may not be how we would present ourselves through CMC; but face-to-face these cues *must* be given off to communicators. Clearly, the differences in channel entrainment in these two scenarios have huge implications for how we would be perceived by our receiver due to our ability to control and edit the cues to which the receiver has access via our channel of communication. Put briefly, the ability to selectively edit a message enhances a sender's ability to selectively present her or his idealized self.

Idealizing Feedback Loop The final element of the hyperpersonal model is idealized feedback from the receiver regarding the sender, which the sender can then use to guide even more favorable self-disclosures. This reinforcing process between presentation and feedback is known as the *feedback loop*. The hyperpersonal model predicts that when receivers obtain selectively presented information from the sender and idealize the sender itself, they respond to the message source that reciprocates and reinforces that idealized information. For example, if we are texting and you mention that you love dogs and fishing, and I reply that I also love dogs, you take this as confirmation the topic or trait of "dog lover" is valued (at least over fishing), and therefore talk more about dogs rather than fishing, which ultimately makes me like you even more. One explanation suggested as to why receivers may provide idealized feedback to a sender was Snyder, Tanke, amd Berscheid's (1977) **behavioral confirmation**, which describes how individuals' impressions of a partner affect response patterns. For example, when we are led to believe the person with whom we are talking is physically attractive, we treat them as more attractive. This feedback loop, whereby selective self-presentations lead to reinforcing idealized feedback, completes the hyperpersonal model and helps explain how the hyperpersonal model may result in more positive impressions: good presentations beget good impressions and feedback, which beget more good presentations.

Hyper-positive and Hyper-negative Impressions

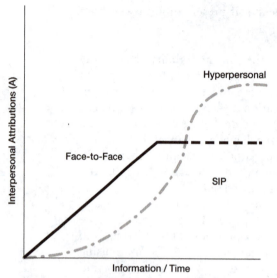

FIGURE 4.7 Three processes—FtF interaction, SIPT, and the hyperpersonal model—predict different processes and outcomes related to how we form impressions of others. Typical perceptions via offline interaction grow steadily until stabilizing. Social information processing theory proposes that perceptions start slowly and then develop more quickly, until also leveling off at levels comparable to perceptions formed offline. The hyperpersonal model suggests that perceptions may also start off slowly, but ultimately achieve higher levels than their offline counterparts.

Mostly, hyperpersonal impressions are addressed as positive: we feel greater attraction, better intimacy, and closer to people we only meet online than those we meet face-to-face. Indeed, this is one of the problems of online dating: we find a suitor whom we begin to idealize as we talk with him or her. However, research has slowly begun to acknowledge the potential to develop a hyper-negative impression of another (High & Solomon, 2014; Walther, 2007). For example, should an online dating partner display attributes you do not find desirable, you may actually hyper-dislike the suitor rather than just feel apathetic about a bad match and move on, as you would face-to-face.

Although the hyperpersonal model is often discussed with respect to positively valenced impressions, it is important to remember that, at its core, the hyperpersonal model discusses the *intensification* of impressions. Negative impressions can be intensified just as readily as positive impressions. The reason researchers typically study positive hyperpersonal effects is simply because it is more natural to have prolonged interactions (thereby giving time for the feedback loop to close and engage) with those whom we view positively than with people we dislike. But the principles of the hyperpersonal model should work to guide both overly positive and overly negative impressions of others, surpassing what could be created face-to-face.

Masspersonal Communication

Social information processing theory and the hyperpersonal model were developed to explain interpersonal communication in its purest sense: one-to-one communication. These theories generally left explaining large-scale interaction to theories of mass communication or group communication. Historically, mass communication theory (e.g., Shannon and Weaver's [1949] SMCR model) described one-to-many communication with minimal feedback while interpersonal communication theory described one-to-one communication with greater opportunities for feedback. These theories rarely were cross-applied, leading to divisions in the communication discipline (Rogers, 1999), especially at the intersection of mediated communication. However, new media have emerged that have blended mass communication and interpersonal communication, blurring the lines that previously delineated subdisciplines in communication science. Online tools like TikTok, YouTube, Twitter, and blogs now allow individuals to broadcast a mass message simultaneously to many recipients (a la mass communication), but are unique in that many of the recipients are friends and acquaintances of the sender and readily able to provide feedback regarding the mass message (similar to interpersonal communication). In this way, these new media facilitate masspersonal communication.

O'Sullivan and Carr (2018) articulated masspersonal communication as a means of describing and conceptualizing communication that conflates mass and interpersonal channels and interactions. **Masspersonal communication** occurs when an individual uses a mass channel for interpersonal communication, interpersonal channels for mass communication, or both simultaneously (see Figure 4.8). When graffitiing your message of love on an overpass, you are using a mass medium to target an individual receiver who you hope notices and replies, and are thereby using masspersonal communication. Perhaps the cleanest example of masspersonal communication is a status message on Facebook. When you post a status, you do so in a mass medium—the status is immediately transmitted to all of your Facebook Friends. However, Facebook allows friends and family with whom you have interpersonal ties to reply personally and individually to you about your status. Online tools—especially social media—have made it easier to engage in masspersonal communication.

FIGURE 4.8 When Sarah Michelle Gellar tweeted to and about Ben Platt on Twitter, that message was also accessible to anyone else on Twitter, and was therefore masspersonal. She could have used interpersonal communication by direct messaging him to privately congratulate him on slaying his musical performance that night, but instead she tweeted masspersonally. What motivation would she have had to communicate this way, and how would it affect (a) Ms. Gellar, (b) Mr. Platt, and (c) their respective followers?

Although masspersonal communication seems to be emerging quickly as a context for communication, research has been slower to understand its effects. Some research suggests masspersonal communication alters the way we see ourselves, so that creating a message in a masspersonal context effects the way we think about ourselves and our relationships. Walther and colleagues (2011) found that individuals told to create a blog post recalling a time they were extroverted viewed themselves as more extroverted than individuals who posted similar recollections to a Word document. Similarly, masspersonal effects have begun to be explored in online education, as college courses move to collaborative learning environments. Carr and colleagues (2013) found students learned better watching an online lecture when they perceived they were among strangers rather than classmates, suggesting learners can take advantage of a masspersonal communication environment to help them learn. Another way masspersonal communication has been explored is through looking at when masspersonal channels rather than interpersonal channels are used for relational communication. Exploring how couples in relationships used various channels available to them, Tong and Westerman (2016) found that couples used masspersonal channels (e.g., Facebook wall posts) to communicate more mundane relational maintenance messages and more positive communication, but used interpersonal channels (i.e., private/direct messages) to communicate intimate, private, or negative information. Finally, research is beginning to emerge indicating that social network sites afford meaningful social support, as support can be drawn from a mass audience comprised of meaningful interpersonal ties (e.g., Krämer et al., 2014; Rozzell et al., 2014). Much work remains to be done to understand the complexities of interacting in a mass medium on an interpersonal level and vice versa, but the emergence of social media as masspersonal channels has challenged research to quickly catch up. This challenge is compounded by the rapid advancement of media tools that change the nature of relationships and interactions.

Concluding Interpersonal CMC

It likely did not come as a surprise to learn that interpersonal communication can occur online. The world may be a little louder; but that's because it's also filled with more opportunities to connect with friends and family. Perhaps even while reading this chapter, you heard your device chime and you took a break to check Instagram, reply to a friend's WeChat message, or Skype home to check in with the family. Though they took almost 20 years of CMC scholarship to develop, theories of interpersonal communication are now a dominant force in the field and greatly help us understand how people relate online. And while the idea that we can have computer-mediated interpersonal interactions is likely not awe-inspiring to you, hopefully the SIPT, hyperpersonal, and masspersonal perspectives can shed light on how you view your online communication and help you understand both the technological and communicative processes that enable that interpersonal communication. Many digital natives have grown up with CMC and therefore find processing CMC very natural, but unpacking and understanding the mechanisms that differ between online and FtF communication greatly aids in understanding how communication works when mediated, and in ways that transcend specific devices to be applied when the next apps emerge.

Key Terms

digital natives, 52

digital immigrants, 52

interpersonal communication, 53

[electronic] propinquity, 54

emoticon, 59

hyperpersonal relationship, 61

channel entrainment, 63

behavioral confirmation, 63

masspersonal communication, 65

Review Questions

1. Think of the last message you sent or received through a computer-mediated channel. Applying TEP, how did the characteristics of the channel, the communicators, and the communication goal influence how close you felt to that sender/receiver?

2. Using SIPT or the hyperpersonal model, explain how CMC can foster a sense of presence or closeness. What changed from the impersonal CFO perspective of the previous chapter to enable this process?

3. What is meant by "time" in SIPT? What role—if any—does this "time" play in the hyperpersonal model?

4. How may masspersonal messages be different than interpersonal messages? Think about the composition of the messages, but also the relational implications and effects. How does the audience (both intended and accidental) of a masspersonal message influence how it is constructed and interpreted?

CHAPTER 5

Group Communication Theories of Computer-Mediated Communication

History demonstrates the importance of configurations of individuals grouped together for various purposes. Without groups, we would never have engineered cars, developed the Internet, or been able to experience the Beatles' *Abbey Road* album. Groups, such as those waiting for a bus, sometimes simply share a goal or purpose, but achievement of that goal does not rely on interdependence among group members. If you don't board the bus, I'm certainly not going to stop it and drag you aboard: we're commuters, not Marines. But more often, as the initial examples illustrate, without people using communication to facilitate collaboration, they could not achieve the group's greater goal. Offline, we see groups all around us, from religious groups to campus groups to volunteer groups. Online, groups may be less initially visible, but certainly have a dominant presence within CMC. Early CMC tools were often bulletin board systems (BBSs) for social support and GDSSs (essentially chat boxes and discussion boards with voting systems to help groups collaborate). We now see tools like Discord servers, Reddit subreddits, and group chats used to bring members together based on common interests (e.g., fans of baseball, science fiction, or Korean pop group BTS) or common associations (e.g., coworkers, old roommates, families) and facilitate communication among them.

Increasingly, CMC occurs through both text and audio-visual channels to help group members interact, coordinate behaviors, and engage with other group members (see Figure 5.1). Now groups can communicate globally in real time through Zoom or Skype, or asynchronously through file shares, websites, or shared documents.

FIGURE 5.1 Group members have many digital channels available to help them communicate, coordinate activities, and maintain their groups. Recently, videoconferencing has been popularized for groups, facilitating discussions among members that occur in real time and affording greater cue bandwidth (including both verbal and nonverbal cues) for richer interactions.
fizkes

Group communication can be complex, as it involves the interactions among a network of many individuals brought together by a common interest. Groups help governments function (or dysfunction, if *Last Week Tonight with John Oliver* is to be believed), organizations conduct business, and nonprofits provide society with security and fun. Though group communication has long been studied, only since the late 1980s have scholars studied group communication as it is mediated by online tools. In only a few decades, our online group communication has grown from a novelty and oddity of the academic and business worlds to a common occurrence for many online denizens. As Tapscott and Williams (2008) pointed out in their book *Wikinomincs*, online tools have enabled huge numbers of individuals from diverse cultures, backgrounds, and time zones to collaborate on substantive projects ranging from Wikipedia to designing airplanes. As more tools facilitate online collaboration and communication, we will increasingly engage in CMC to identify and collaborate with others in groups.

Group communication refers to the interactions of multiple individuals who are associated through some shared attribute or commonality. Groups can exist and function in many contexts, "including work teams and family, religious, educational, and recreational activities; in activist and social movement contexts; and in virtual environments, public meetings, or laboratory settings" (National Communication Association, 2020). Groups are often distinguished from organizations (see next chapter) by their lack of formal hierarchical structures, means to control members' behaviors, and more permeable boundaries so that it is easier—or at least less formal—to enter and leave a group than an organization (Miller, 2009).

How we define and conceptualize groups can vary widely within the communication discipline, with membership ranging from three to infinite. Though we must have at least 3 people to form a group (any less is just a relationship or you talking to yourself) and most groups commonly tend to lose some functionality and ability to communicate with memberships beyond about 20 people (Thomas & Fink, 1963), groups have no theoretical maximum. As group size increases offline, it becomes more challenging to maintain functional communication patterns, preventing all members from communicating and feeling engaged. However, tools in many CMC systems, including threaded conversations and a searchable archive of prior messages, make online group interaction at massive scales more structured and manageable than their offline group counterparts (Baltes et al., 2002). Most often, *small groups* are understood to be networks of between 3 and 20 members, with groups comprised of more than 20 members considered *large groups*, but this distinction is fuzzy and imprecise. Within the communication discipline, communication within groups can serve several functions, including maintaining social ties and relationships, information sharing and analysis, decision-making, developing and executing plans, coordination of activities of both group members and those not in the group, and attempting to persuade or influence others (Poole & Hirokawa, 1996).

Groups can communicate differently online than they do offline, in both processes and outcomes. Even the way individuals and members of a group identify with a group (and, inherently, with those *not* in that group) alters when members are not face-to-face. The interactions of groups are often guided by an aggregation of interpersonal relationships, which admittedly sometimes makes it challenging to label an interaction as either interpersonal or group communication. Several scholars may correctly argue both of the theories addressed in this chapter are interpersonal in nature, as they specifically relate to one-on-one relationships. The two theories presented here—the SIDE model and the contact hypothesis—are covered in this chapter as both theories fundamentally address how individuals affiliate with a given group, distance themselves from other groups, and perceive others based on their affiliation with groups online. Even the intrapersonal and interpersonal effects described here are ultimately guided by how individuals perceive themselves with respect to a broader group. While some of the communication that occurs in groups may be dyadic or interpersonal in nature, even those interactions will be guided by group processes.

Social Identity Model of Deindividuation Effects

The *social identity model of deindividuation effects*, more commonly known as the SIDE model, was developed to explain how individuals may act and group with others when individuals are deindividuated (Reicher, Spears, & Postme, 1995). The SIDE model helps explains how individuals identify online, both identifying themselves and identifying with others. It is a helpful tool for those seeking to understand not only how individuals form groups online but also how individuals may distance themselves from others online. Though one of the most popular and commonly researched theories of CMC (Walther, 2011), the model's roots trace back more than two centuries to its grounding in the concept of deindividuation.

Deindividuation

Deindividuation refers to the loss of awareness of one's self as an individual, often while in a group. The notion of deindividuation was initially put forward by LeBon (1895) and related to the idea of mob mentality. Observing the rioting and looting taking place by otherwise upstanding Parisians in the tumult of the French Revolution, LeBon sought to explain why a typically honest person could suddenly engage in criminal and unethical acts. What came of this was the idea of a loss of self: when situational effects occur that reduce one's sense of self and individuality. LeBon postulated this loss of personal identity reduces one's concern for self, so that social pressures, norms, and expectation of consequences of one's actions are all reduced. In short, when you are deindividuated, you feel that you are not really you, and therefore you are not concerned about consequences (both positive *and* negative) of your actions.

There are many situations where your awareness of yourself as a unique individual is reduced. When attending sporting events, concerts, and political rallies, you rarely consider yourself in terms of all the individual traits that uniquely comprise *you*.

FIGURE 5.2 Fans at a sporting event may become deindividuated as their personal identities are suppressed in favor of the salient social group. Though an individual may not walk down the street yelling, "E-A-G-L-E-S," it would be normal to do so at an Eagles football game surrounded by fellow fans of Philadelphia's National Football League team.
Doug Pensinger / Staff

Think of the last time you went to a sporting event: you likely wore clothes that demonstrated your fandom for your home team, chanted and cheered when other fans of your team did the same, and may have even jeered at and harassed the other team or the referees in a way that you wouldn't holler at someone outside of the arena (see Figure 5.2). Clearly, numerous times in daily life, we find ourselves in a deindividuated situation.

One of the seminal studies of deindividuation used participants with whom you are very familiar: college students. In 1969, Philip Zimbardo used female students at New York University (NYU) to test the effects of deindividuation. Zimbardo (1969) asked female participants to either wear large coats and hoods to conceal their identity or remain in their street clothes as they administered electric shocks to another NYU female who was behind a one-way mirror in another room. (Unbeknownst to participants, the student in the other room was actually a lab assistant acting as if they were about to be electrocuted, and the electrodes did not administer the supposed shocks.) Consistent with the notion of deindividuation, participants administered more shocks and of a longer duration when their identities were concealed behind large coats and hoods. Because they were less concerned with retribution or the consequences of their actions (due to their lessened personal identification under the coat and hood), deindividuated participants were more aggressive toward their peer. Online, this suppression of a sense of personal self and the consequences of our actions is even easier to facilitate. As soon as we're hidden behind a user name or feel untouchable in the ether of the Internet, deindividuation can quickly suppress our sense of self.

Without this clear sense of personal identity, you are forced to rely on alternate means of identification with which to relate. In these situations, the SIDE model predicts that when we are deindividuated, we seek out and rely on a social identity to guide our interactions. A **social identity** is the characteristics and traits associated with a group of individuals (Tajfel et al., 1971; Tajfel & Turner, 1986) that often govern the group's behaviors and interactions. When a person's individual identity is reduced or suppressed, a social identity may be elevated to guide interactions and behaviors. Recall that sporting event from a few paragraphs ago: your behaviors were guided by your social identification with fans of that sports team and their social identity. The color of your clothes, the ritualized cheers, the mutual celebrations and commiserations were all guided by the social identity of being a fan of that sports team rather than by your individual habits. As we go through our days, we often give off clues to our social identities as we wear clothing or tattoos, use specific language and jargon, and even perhaps affect a particular way to act or carry ourselves. All of these behaviors may display our affiliation and identification with a college, athletic team, musical genre or performer, social groups or status, and/or hobbies. Online, where we have more control over which of these social cues may be salient or which we give off, the processes and effects stemming from our social identities may be different than they are offline.

Ingroup and Outgroups

You likely belong to many social groups: your college or university, an extracurricular organization, a religious group, a band or choir, or even a group for this class. But how do you know you are a member of these groups, and how do these groups actually define themselves as a group? What is the boundary between being *in* the group and *not* being in the group? Research into SIDE has helped us understand how we identify and affiliate with our own groups while disaffiliating from other groups.

The SIDE model is guided by the effects of social identification, or how we identify with other groups and their members. By realizing who we *are*, we also recognize who we

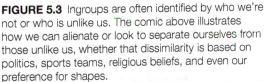

FIGURE 5.3 Ingroups are often identified by who we're not or who is unlike us. The comic above illustrates how we can alienate or look to separate ourselves from those unlike us, whether that dissimilarity is based on politics, sports teams, religious beliefs, and even our preference for shapes.
Credit: Poorly Drawn Lines by Reza Farazmand. (CC BY-NC 3.0) https://creativecommons.org/licenses/by-nc/3.0/

are not. Perhaps surprisingly, studies of how we form associations with groups have noted that we identify who is like us based on who is *unlike* us (Allport, 1979). This categorization based on "like us" and "unlike us" then creates social ingroups and outgroups, respectively. **Ingroups** are associations of individuals whom you favor and identify with. Allport (1979) states that ingroups are "psychologically primary," as the familiarity with, attachment to, and preference for associating with members of the ingroup takes precedence over associating with members of other groups. **Outgroups**, then, are associations of individuals unlike your social group. How we distinguish our ingroup guides how we identify with our ingroup and its members, and also informs how we relate with members of an outgroup (see Figure 5.3).

The process of identifying and defining an ingroup—those like you—actually seems to begin by identifying the outgroup—those unlike you (Mullen et al., 1992). As we begin to see individuals with whom we are dissimilar—those we are not like—we begin to formulate an understanding of our own self and identity. In other words, by understanding who we are *not*, we develop an idea of who we *are*. You can probably remember this process from your first day of college. As you walked around campus, you saw hundreds of new faces. As you started to seek out people to talk to and make friends with, you most likely took note of people who looked, acted, or thought totally different from you and initially dismissed them from consideration as potential close friends. As this process continued, you slowly winnowed down a list of unlikely friends, eventually leaving you with a remaining group of people you felt were enough like you to seek out as friends, thereby defining your ingroup.

For some, the explication of this process may come off as shallow, and critiques of ingroup/outgroup formation have likewise criticized the nature of ingroup identification and association. The favoritism and preferential treatment of ingroups and their members was initially likened to ethnocentrism, as we seek to associate with those like us while disassociating from those unlike us based on demographics or ideologies (Sumner, 1906). However, subsequent significant research has indicated ingroup/outgroup processes more closely resemble discrimination *in favor* of ingroups rather than discrimination *against* outgroup members (Brewer, 1999).

Common-Bond and Common-Identity Groups

One key way we identify an ingroup with which to associate is through a common bond or a common identity. The differences in common-bond groups and a common-identity group are defined by how group members come together and associate. A **common-bond** group is an association of individuals brought together and who maintain relationships due to close interpersonal ties. Individuals remain part of a common-bond group because they like individuals in the group as individuals. Relation-based groups (including fraternities/sororities, live-action role players [LARPers], and families) have often been cited as examples of common-bond groups. Your friend group reflects a common-bond group: you do not need to associate with them, but as you begin to meet and like fellow students you begin to hang out and eat dinner with that group. Online, common-bond groups may take some time to form as they require interpersonal interaction, but many individuals have formed strong ties with groups online, including members of support groups and fantasy sports leagues. Unlike members of common-identity groups who just feel attached to the group, members of common-bond groups often feel attached to the group holistically and feel strong attachments to the groups' members (Prentice, Miller, & Lightdale, 1994).

Alternately, individuals may be part of a **common-identity** group: an association of individuals brought together and who maintain relationships due to their affinity toward the group as a whole. Individuals like the aggregate of group members, even if they don't have a particularly close interpersonal tie, and remain in the group to continue to associate with a group of people with a common interest or attribute. Examples of common-identity groups abound online, as individuals can use Web tools like discussion groups and social media to find and affiliate with like-minded individuals (see Figure 5.4). For example, as a fan of the New York Yankees, you can go online to find other Yankee fans even if you live in California. You don't necessarily have to like the individuals who participate in the online discussion, and it's enough that you're able to talk with others who share your love of baseball players A-Rod and Jeeter. You continue to talk with them because you share the common identity of a Yankees fan, rather than because you interpersonally like the other members of the online group.

Interestingly, studies have shown your ingroup identification does not have to be a large part of your personal identity, merely made salient or important in the moment for it to actively drive intergroup processes. The **minimal group paradigm** supposes that, under certain conditions, simply being assigned or categorized to an arbitrary group is enough to activate intergroup processes (Billig & Tajfel, 1973), even

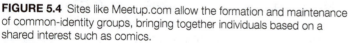

FIGURE 5.4 Sites like Meetup.com allow the formation and maintenance of common-identity groups, bringing together individuals based on a shared interest such as comics.

if the groups are fabricated or irrelevant. For example, arbitrarily labeling people as "under-estimators" and "over-estimators" is enough to activate a common identity and guide interactions when "estimation" is made a salient attribute in group discussions (Tajfel et al., 1971). Online, such minimal groups are common: games typically utilize groups to activate arbitrary divides among players (e.g., *World of Warcraft*'s Alliance and Horde factions; *Pokémon Go*'s Team Mystic, Valor, and Instinct), travel websites will appeal to vague identities to spur purchases (e.g., "Three travelers like you have booked this hotel in the last hour"), and even social media may foster minimal groups through the random assignment of default background colors or avatars until users enter a more individuating cue. You may not care about your astrological sign, but if the dating site you use makes a big deal about the role of the Zodiac in finding love, suddenly your Capricorn designation is critical to how you interact with others and how they interact with you. The minimal group paradigm is particularly critical when associating with others based on a common identity—even these small distinctions are enough for individuals to cling to and can lead to a strong sense of affiliation with that group and its members.

Once you know that we like to associate with members of the ingroup and disassociate with members of the outgroup, and that these associations can be guided by either common bonds or common identities, it becomes much easier to understand how and why SIDE may guide group communication. Though it probably doesn't come as much of a shock, we like to associate with those we like and those who are similar to us. Meanwhile, we don't spend much time interacting with people unlike us or with whom we have little in common. This idea is a fundamental component of SIDE. However, we certainly don't go through life basing our friendships and interactions on who is like us in some arbitrary group identification. Just because you're a fan of *The Hunger Games* series does not mean you'll yell at anyone you walk by who is wearing a *Divergent* sweatshirt. Likewise, as you encounter someone online who likes Harry Styles, that may not affect your perception of them (or yourself) until your attention is drawn to the importance of "music" as a social category or your own love of Juice WRLD. So when do these social group distinctions guide our interactions and group communication? The power of SIDE as a model is to help predict when and how social rather than personal identities guide our interactions.

Returning to SIDE

The SIDE model predicts that when they are deindividuated, individuals associate with others based on a salient social identity—a group category that is made important given the situation. The SIDE model is particularly applicable in CMC for two reasons. First, the anonymity allowed by CMC doesn't let users identify based on individual differences. Visual cues let us learn a lot quickly about who an interactant is, and the lack of cues in many CMC channels (recall the CFO perspective) eliminates much of the information necessary to guide individual, interpersonal interactions. Second, when deindividuated, users seek to orient themselves based on a salient social group, which are often readily available online (Postmes, Spears, & Lea, 1998). Individuals can draw cues to a social identity from information users post, the type of online space in which they are communicating, and even the nature of the medium itself. In sum, SIDE predicts that when someone's individual identification is minimized, their social identification is made salient and guides interactions.

Although the SIDE model is not indigenous to CMC, many of CMC's features and abilities make SIDE particularly applicable in online communication. Many of the ways we communicate online deindividuate users. For example, when communicating in online chat rooms, you often communicate in a deindividuated state as you (hopefully) do not provide identifying information (such as a name or hometown) and are visually anonymous as you communicate without a picture of yourself next to what you post. Moreover, many sites online are specifically designed as places for interaction among individuals based on common interests. Discord servers allow individuals to relate based on common identities rather than common bonds, and YouTube videos are often viewed and commented upon by individuals who have an interest in the video's topic. In both instances, your individual self is often not as critical as the social categories relevant to the online space.

Walther and Carr (2010) used the following example of a common online experience to exemplify SIDE effects: the classic *Star Wars* vs. *Star Trek* debate. Imagine you log on to a discussion forum that focuses on debating the relative merits and quality of two popular science fiction franchises: *Star Wars* and *Star Trek*. You, like all the other users on the site, post messages under a pseudonym, and your posts cannot be linked to your offline identity. Because of the nature of the group and the discussion topic, social identities are made salient and used to drive the conversation. Being the *Star Wars* fan you are, you log in under the username "Jedi1" and seek out someone with whom to interact. In the forum, you see two active users: "WarsFan" and "TrekFan." With whom do you associate? Because your own identity has been subsumed by your social identity, it does not matter who the individual is behind each username; and instead you associate based on the salient social identity, in this case sci-fi fandom. The SIDE model therefore predicts you feel greater social identification with "WarsFan," perceive more social attraction and affinity toward "WarsFan," and begin talking with "WarsFan," who undoubtedly shares your view that Star Destroyers could blow up the *Enterprise*. All the while, you distance yourself from the outgroup member "TrekFan," who exemplifies a different social identity and whom you therefore initially dislike because of your differing views on the relative cuteness of Ewoks and Tribbles. Though a little unusual, this scenario demonstrates how SIDE may be used to explain not only social identification with others online, but also communication patterns and behavioral norms.

Applying SIDE

One of the reasons the SIDE model has been such an important theory for communication over the past 30 years is its explanatory power. The SIDE model has allowed us to begin to predict and explain how individuals identify themselves with social groups and how their identities guide their interaction when they are online. Think back to your attendance at a sporting event: you throw on a jersey, wear the team hat, and know the team cheers. In the stadium, surrounded by 86,112 other fans of the same team, you no longer identify as you—your personal identity has been suppressed and your social identity has taken over. Unlike your usually reserved and thoughtful self, you start yelling insults about the opposing team, their fans, and their stupid animal of a mascot. Similarly, when playing *League of Legends* online, you may uncharacteristically insult other players over voice chat, going so far as to imply intimate relations with one of their parental figures. What happened to make an otherwise considerate scholar such as yourself throw caution to the wind and act this way? The SIDE model can help explain interactions ranging from heated team rivalries to interethnic conflict, and also begin to suggest ways to resolve differences among these groups. Once you understand and can apply SIDE, you can begin to use it strategically.

For example, if you are a social media manager for a clothing line, how do you want users to interact on your social network site? According to SIDE, if you want users to identify with the brand, you want them to interact in a deindividuated state, which should increase their identification with the brand, and most likely their spending (see Figure 5.5). The less you can get users to think about how they will look in the clothing, and the more they realize that members of social groups they associate with (perhaps similar demographics like age or geography, their friends, or other students at their university) want to wear that clothing, the more you can expect sales to rise. Because you've activated the social identity (e.g., college student) and indicated that other members of that ingroup wear this clothing line, you've begun to subtly influence users' attitudes about your clothing line.

These SIDE outcomes can occur even as individuals are more clearly identified, particularly as visual cues are apparent. For example, exploring how people make decisions online, Lee (2004) had participants discuss potential solutions to a problem in online chat rooms where the users were depicted by cute animals such as penguins, parrots, and polar bears. When participants were in rooms with many different characters representing the group members, there was more debate and less agreement about the final solution.

FIGURE 5.5 Activating social identities related to a brand can increase consumers' pro-brand behaviors, such as purchasing frequency and advocacy (i.e., word of mouth) to others on behalf of the brand, even going so far as to dress up as their favorite brand. Coast-to-Coast

However, when participants were in rooms where all participants were represented by the same character (e.g., all penguins), participants were much more likely to conform to and agree with the group's decision. Such a finding is likely to explain why the Snapchat or Instagram profile pictures of your close friends have a composition similar to your own profile picture: consciously or not, your group has converged toward a similar idea of a "good" profile picture.

The SIDE model processes and guides negative effects just as readily as positive ones. For example, workgroups whose members are partially distributed (i.e., small subgroups work at the same location, but the overall group is distributed worldwide and collaborates online) can lead to cliques guided by SIDE. Members of the subgroups located in geographic proximity begin to act as an ingroup and can seek to distance or disaffiliate themselves from outgroups: team members in other locations (Tidwell & Walther, 2006). Even physical location may guide intergroup perceptions and behaviors online. We'll get back to this in Chapter 7 discussing group CMC in practice. But for now, it's enough to note that how individuals see and configure themselves online in relation to others can guide perceptions and interactions based on group-level processes, even before individuals first communicate.

Contact Hypothesis

It may be tempting to interpret SIDE effects as "bad," as they emphasize group differences and suggest a lack of prosocial communication between even nominal social categories. However, remember that theories and models are not inherently good or bad, positive or negative: only their *applications* are "good" or "bad." Strategic use of SIDE can be applied to very positive outcomes. One way that careful manipulation of individuation and deindividuation can be used pro-socially is to reduce intergroup conflict. Many groups have a fundamental dislike of each other: Palestinian Jews and Palestinian Muslims, Democrats and Republicans, East Coast and West Coast rappers, or even Yankees fans and Red Sox fans. These feuds are not necessarily rooted in two individuals disliking each other personally, but rather in differences between social groups. A Yankees fan doesn't need to know a Red Sox fan personally to dislike him or her—that he or she is a Sox fan is enough. So how can these intergroup differences be reduced? Particularly, how can individual and group identities be highlighted or suppressed to cross intergroup boundaries? One way is the contact hypothesis.

The **contact hypothesis** predicts that if individuals first have an opportunity to communicate with each other at an interpersonal level or based on a common group identity, that interpersonal relationship may minimize subsequent interactions at an intergroup level. The contact hypothesis was put forward as a way to reduce interethnic conflict (Amir, 1969), proposing that if two individuals from opposing groups get to know each other guided by other identification before learning about their affiliation with differing social groups, once their respective social identities are revealed, their interaction should be guided by the favorable interpersonal impressions that initially developed rather than the antagonistic relationship guided by their group-level identification. Returning to the earlier example of Yankees/Sox fans, consider two individuals who get to know each other initially as "Carolyn" and "Paul" and have time to develop a friendship based on who they are as people and perhaps a shared love of baseball. When it's finally revealed that Carolyn likes the Red Sox while Paul follows the Yankees, their personal friendship should continue, guided by their like for each other rather than their dislike for each other's social categories. Because they relate on the personal and social identities they *share*, before they relate on oppositional social identities, their relationship may continue positively. After learning they both like reading, architecture, and nature walks, Carolyn and Paul can continue to relate based on those shared attributes rather than focusing just on their different preferences for baseball teams.

Several scholars have noted the ability of strategic applications of online tools to reduce the role of social identities and therefore intergroup conflict. Amichai-Hamburger and McKenna (2006) recognize Internet-based interactions, where individuals can carefully craft and control the cues they send about their social identities, represent an environment that readily facilitates contact among oppositional social groups. Because the Internet crosses geographic and social boundaries, anyone can access and use CMC tools to interact with others, including those of different social groups. Consequently, CMC can be used to allow oppositional groups to communicate without being in the same room—a benefit when close proximity may result in violence. Amichai-Hamburger and McKenna recall the challenges of Catholic and Protestant children in Northern Ireland. Because these groups have a history of antagonism and bloodshed, even children may be verbally and physically aggressive when face-to-face with someone of the opposite group. However, one cannot punch another through a TCP/IP connection, making it somewhat safer to communicate online. Consequently, CMC affords members of oppositional groups, like Irish Catholics and Protestants, to initially connect in a way that allows communication without enabling interaction.

We Come Together, Cuz Opposites Attract

Coca-Cola vs. Pepsi. Marvel vs. DC comics. The Jets vs. the Sharks. East Coast vs. West Coast rap. Rivalries and conflicts among groups are common occurrences, and though some, such as sports rivalries (think the Chelsea and Arsenal football clubs), are fun means of engaging fans, others can represent strongly held prejudices or geopolitical tensions with serious implications for the groups and people involved. Can CMC reduce intergroup conflict?

Walther and colleagues (2015) set out to answer this question by applying the contact hypothesis to three groups with centuries-old rivalries: Muslims, religious Jews, and secular Jews. Members of any one of these groups typically have very negative impressions of members of the other two groups, often going so far as to provoke hostilities or physical violence based solely on their group status. Because they are "not us," they are bad. The contact hypothesis suggests that by getting to know someone from a different group interpersonally first, any positive feelings generated toward the individual may be overlaid onto the group once that social category is revealed. Because CMC can filter out some of the visual cues to one's religion (e.g., by hiding attire or not making note of religious practices), individuals can begin their interactions as individuals.

Knowing this, Walther and colleagues (2015) assigned students in nine Israeli colleges to work together in groups comprised of two members of each of the three religious groups to complete a year-long course focusing on online learning. Students communicated entirely online (and interpersonally, without cues to group identities) for the first semester of the course, and finally met face-to-face after the end of the first semester. Findings revealed students in the study reported significantly lower prejudice toward members of other religious groups following the class compared to students enrolled in the same course, but not to groups from across multiple campuses.

Given these findings, one way to use CMC to get people from different groups to communicate effectively is to first make them interact as individuals. Use CMC's ability to filter out cues to group identity that may trigger intergroup debate and discrimination, and only once interpersonal bonds are built should they reveal their affiliations. By then, the individuals may have realized they share more commonalities than differences, reducing tensions between them and perhaps even prejudice toward the other groups as a whole.

Reference

Walther, J. B., Hoter, E., Ganayem, A., & Shonfeld, M. (2015). Computer-mediated communication and the reduction of prejudice: A controlled longitudinal field experiment among Jews and Arabs in Israel. *Computers in Human Behavior, 52,* 550–558. https://doi.org/10.1016/j.chb.2014.08.004

Interestingly, evidence for the contact hypothesis seems to support two other CMC theories: the CFO paradigm and SIPT. When we interact with members of other groups online, we can choose to use cue-lean text-based channels, audio channels, or even cue-rich audiovisual channels, and the channels we use to interact affect the perceptions that are formed. Consistent with SIPT (Walther, 1992; see Chapter 4), when we interact with someone over video chat, we get to know them personally and begin to like them more. However, consistent with SIPT, we like *them* as an individual person more, and those positive interpersonal gains don't necessarily translate into reduction of outgroup prejudice. However, when we interact with someone over text-based chat, we get to know them, but in a less personalized and individuated way. In these text-based interactions, the CFO paradigm (see Chapter 3) says the development of close interpersonal relationships is challenged. (Now aren't you glad we haven't forgotten about the CFO paradigm? I told you it wasn't useless.) However, the subsequent reduction of intergroup biases may be stronger as we are more apt to view our partner as an intergroup member. Cao and Lin (2017) demonstrated this effect by having mainland Chinese students (ingroup) chat with a confederate purporting to be a Hong Kong student (outgroup) through either video

FIGURE 5.6 The contact hypothesis argues that by first relating interpersonally and seeing each other as individuals, people may reduce and overcome intergroup bias when they meet face-to-face, even if the groups have been deeply divided historically.
Richard Baker / Contributor

or text. Findings revealed that while participants reported interpersonally liking the confederate more in the video condition, perceptions of the outgroup were more favorable following interaction in the text condition.

Even minimal interactions with stigmatized others on social media can reduce prejudices, albeit only mildly. Simply looking at profiles of individuals from stigmatized groups (e.g., people with physical disabilities, with schizophrenia, transgendered individuals) can reduce intergroup attitudes or perceived differences if individuals feel like the profile allows for an imagined conversation with the profile owner (Neubaum et al., 2020). But when interactions are primarily imagined, individuals still need to feel some degree of similarity to the stigmatized individual. If the other person is seen only as the stigmatized trait and with no connection to the observer, it is not sufficient to humanize the other person and provide the prejudice-reducing sense of interpersonal awareness and relationship.

These findings also help us remember an important caveat to the effectiveness of the contact hypothesis: individuals must view their partners as individuals while still viewing them as members of the outgroup for the positive associations with the individuals to be transferred onto perceptions of the outgroup (Figure 5.6). If you think your new friend is a great person, but not reflective of the broader group, then the reduction in intergroup prejudices will not follow. This may explain why some individuals claim not to be biased in some way because of a token association of that group. The fact that you have one minority friend may not reduce intergroup prejudice if you do not see that one acquaintance as faithfully representing or as a member of the broader group. Alternately, the fact that you are in a workgroup with that person is not enough to reduce intergroup prejudice if you simply associate but do not perceive any similarity with that person. These perceptions of interpersonal relationships and closeness are critical to the workings of the contact hypothesis.

The contact hypothesis has received compelling initial support, but researchers are still testing its boundaries. How disparate can groups get before group members are incapable of identifying outgroup members as individuals? As more tools become available online to learn about an interaction partner, is it even possible for an individual to hide a group membership? Are "ingroups" and "outgroups" even relevant online as Web-based tools like social media collapse the relational contexts in which we interact and potentially make groups less salient? The contact hypothesis has the potential to be a powerful, prosocial theory that enables previously disparate and hostile groups (and their members) to effectively communicate and relate. However, it will be up to future Internet users to determine whether long-term effects are practical and possible.

Our society certainly faces large intergroup conflicts today guided by racial, religious, geopolitical, and ideological differences. It would be naïve to think that chatting online is the cure to some very substantive problems. However, as these studies demonstrate, if you can find ways for group members to first interact as individuals, the negative perceptions that may be associated with their respective social categories may be diminished, leading to more positive interactions later on. In other words, it may be easier for us to interact with different types of people online than offline, helping to reduce our prejudices and stereotypes by interacting with others as people rather than immediately relegating them to their social groups by their appearance or other physical cues. Just as study abroad programs make students more aware and inclusive of other cultures and societies (Goldstein & Kim, 2006), used strategically, CMC tools can allow users to interact with diverse others, potentially reducing intergroup differences over time.

Concluding Group CMC

Groups increasingly use online tools to collaborate. You likely are a member of many groups, including familial, social, and organizational groups. And most likely you use CMC to interact with members of that group: you organize group events through a Facebook group, and you use Yammer or Slack to let group members know the progress you've made on a project. Although the processes of group communication are complex when we interact with our groups face-to-face, the Internet has altered some of the ways we communicate and in doing so has provided new challenges and opportunities for group communication. We can now interact with others based on common interests or shared personality traits, while minimizing the influence of interpersonal cues such as physical appearance or dress. The way we perceive ourselves not only as individuals but also as group members can change when we interact online, as we can more readily (but not uniquely) lose our sense of self and be guided by a larger social identity. The SIDE model and contact hypothesis are two theories of CMC that help explain how mediating communication can affect intergroup communication, and they can serve as helpful insights as we seek to understand how individuals, groups, and organizations interact online—they will serve us well in Chapter 7.

Concluding Theories

An old adage in academia says that "nothing is quite so practical as a good theory" (Lewin, 1945, 129). The previous three chapters have explored theories central to the study of CMC. But, like many scholars, at some point amid long discussions of theory you have likely felt frustrated at times, wondering what good these head-in-the-cloud ideas are for you. One function a good theory provides is a foundation on which to build practical application. With that application—and good CMC theories—in mind, the next section will start to put these notions to use and explore how CMC can be used and applied in the contexts in which communication occurs.

Key Terms

group communication, 68

deindividuation, 69

social identity, 70

ingroup, 71

outgroup, 71

common-bond group, 72

common-identity group, 72

minimal group paradigm, 72

contact hypothesis, 76

Review Questions

1. Deindividuation can be a powerful force both offline and online. Is there a time you may have acted in a way while in a group that you wouldn't have while alone? (Concerts, protests, sporting events, and large lecture classes are often fruitful contexts to think about for this.) Using the concept of deindividuation, explain the process by which you were more or less likely to behave in a way other than how you would have done alone.

2. Compare how you are presenting yourself offline right now (e.g., clothing, books, status items, and other artifacts) and how you present yourself in a profile on a social media site. What social identity cues do you give off in both? Are there social identities you put forward in one and not the other? Why do you think differences appear in how you present your social categories offline and online?

3. At any given time, we hold numerous social identities. Take a moment and list as many identities as you can think of for yourself on a sheet of paper. Once done, look back and identify which two or three may most commonly guide your behaviors, thoughts, and communication. Why are those identities more salient than the others? Is their relevance and impact because of the environment you are in or because of how you view yourself?

4. How can CMC be used to reduce a key outgroup perception in your life? That intergroup challenge may be based on political, religious, ideological, or geographic social categories, but consider how and where you could engage in the types of interactions necessary to activate the contact hypothesis, and articulate the process by which you could reduce intergroup differences.

PART III

Theories

CHAPTER 6

Organizational Computer-Mediated Communication

LEARNING OBJECTIVES

After reading this chapter, you should be able to …

- Identify technologies—both analog and digital—that organizational members use to communicate.

- Understand how digital media are selected to communicate both within and beyond organizations.

- Explain how the properties of communication media can affect how organizational members use them and the types of communication facilitated.

- Articulate how CMC tools can be used as knowledge management systems to store, index, and retrieve organizational information.

The 1893 World's Fair in Chicago commemorated the 400th anniversary of Columbus's first trip to the Americas. Between May and October of that year, the Windy City hosted more than 27 million visitors who came to see exhibits, amusements, and cultures from around the globe. The shores of Lake Michigan that summer presented the world with many firsts: Juicy Fruit gum, Pabst Blue Ribbon beer, and prototypes for a dishwasher and fluorescent light bulbs all debuted during the 1893 World's Fair. Among these technological wonders was Dr. Nathaniel S. Rosenau's gold medal exhibition—the world's first vertical file cabinet.

Though vertical filing cabinets are now found in virtually every office environment, they were a radical shift in how individuals and organizations in 1893 stored, organized, and accessed information. Before the vertical file, printed information was either stored in pigeonholes (i.e., rolled into small cubbies similar to post office boxes) or stacked in piles on desks, tables, and whatever surfaces had room left. Though such stacks and piles may be *your* preferred filing system in your dorm room, for organizations dealing with large volumes of information, these filing systems are space-intensive. For example, an accounting office would need a pile of ledgers and receipts for each of its clients, necessitating either huge amounts of space for storage or frantic searches through mounds of paperwork to find a specific receipt or tax return. In short, organizations were often limited by the space and manpower they could devote to organizing information.

FIGURE 6.1 An early vertical filing cabinet—a technology that allowed the easy storage and retrieval of various physical files.
Early Office Museum

FIGURE 6.2 A contemporary vertical file cabinet, which still looks similar in many ways to its early predecessor.
Our United Villages (CC BY-NC 2.0)

The development of the vertical file cabinet was a game changer for how information was stored and accessed. With the vertical file (see Figures 6.1 and 6.2), documents containing needed information could be hung vertically—thus taking up much less space than laying pages flat—and quickly removed and refiled—thus allowing information to be arranged and rearranged readily based on categorization. What we now look at as a staple of an office was initially a groundbreaking paradigm shift of how we store and access information, and allowed organizations of all sizes and purposes to grow and change as their space and manpower could be reallocated from filing to achieving the organization's strategic goals. Though the vertical file is certainly not a computer-mediated tool, its importance and revolutionary nature in organizations do illustrate the role of *technologies* in organizations.

Colloquially we often think of technologies as electronic devices such as telephones and computers, but within an organizational context, **technology** more broadly encompasses any application of knowledge for practical purposes that enhances work processes or flows, and it is not limited to things that must be plugged into an outlet or Internet connection. In this way, we can certainly consider computers and email "technologies," but we must also include pens and books as they facilitate the transmission of knowledge and information more readily than FtF knowledge exchanges. Can you imagine if all the information in this class (and in your other courses) could only come from FtF, verbal transmissions of messages that you later had to recall from memory alone? A book—bound pages containing knowledge—represents an early technology that facilitates the exchange of information between the sender (i.e., the author) and receiver (i.e., the reader).

While books and chapters could (and have) been devoted to *technology* in organizations in this broad sense, this chapter focuses a bit more on computer-mediated technologies—media and channels that are connected to the Internet and exchange messages between people. To this end, we will admittedly minimize our discussion of telephone and written communication in favor of emphasizing how email, file transfers, and social media have affected and been affected by organizations. However, it is important to keep in mind that, at a broader level, even analog technologies such as the vertical file cabinet have played a part in changing the way organizations and their members communicate.

Organizational communication refers to the flow of messages within a network of interdependent relationships (Goldhaber, 1974), often with a focus on power structures or differences among network members. Computer-mediated communication has changed the way members of these organizational networks structure themselves and relate in many critical and impactful ways. This chapter will focus on several facets of organizational communication affected by the influx of CMC, specifically changes in the structure of organizations, how organizational members communicate, and how leaders may have to adapt to these newer means of communication.

Organization and Structure

One of the biggest and most noticeable changes CMC has had on organizations is how organizations structure themselves. Even before CMC was popularized and common in organizations, the ability to communicate asynchronously and across geographic distances was noted and lauded (Short, Williams, & Christie, 1976). Whereas organizational members typically work with those geographically close to them (usually within 30 meters, or about 100 feet) (Allport, 1979), the Internet allows organizational members to expand their reach and communicate more readily with those more distant. No longer are we constrained to interacting only with other organizational members working at a desk or in a cubicle near ours—email and file transfers allow us to meaningfully interact with those on different floors, in different departments, or in different states or countries.

This change was mostly brought about by email, an asynchronous digital medium that allows messages to transcend time as well as space. Surely, organizational members could and did communicate across distance via the telephone: as early as 1856, transatlantic cables allowed telecommunication between New York and London (Schwartz & Hayes, 2008). However, telephone conversations require interactants to be on the phone at the same time. Often members of different departments of an organization do not have the same times free, as someone in accounting may have time for a conference call at 11 a.m., but her colleague in finance may be in a meeting then. An international conference call in New York City at 2 p.m. would be even more complex, requiring organizational members in London to be at work until 7 p.m. and colleagues in Singapore to wake up at 2 a.m. the next day. For extremely urgent matters, the challenges associated with scheduling synchronous interactions may be acceptable, but they make daily interactions complex and beyond the scope of when and how most people want to work.

Email affords organizations a means of asynchronous interaction, allowing someone in accounting to send a message to her or his colleague in finance that can be received and responded to when convenient, even if the finance person is away from the desk when the accountant sends the message. Even more important, the asynchronicity of email allows individuals in different time zones to send and receive messages on their schedule irrespective of the corresponding time in their colleague's country. An email sent from New York at 3 p.m. will be instantly received by a Singaporean colleague, to be read upon arrival to work the next morning, and by the time the New Yorker returns to work the next day a reply will be waiting in a virtual inbox. The affordances of email and other CMC tools have allowed many organizations to expand geographically—both domestically and multinationally—as much of an organization's daily, routine messages can be exchanged asynchronously (Sproull & Kiesler, 1986). Indeed, studies have shown that email increases the amount of communication within organizations (e.g., Sarbaugh-Thompson & Feldman, 1998), enabling members to communicate even when separated by more than 30 meters. Now, through CMC, we can communicate easily across time, geography, and culture, bringing together diverse organizational components.

As our media tool kit has expanded, the structures of organizations have expanded as well, empowered by the emergence of newer technologies. The rise of the Web in the mid-1990s enabled organizations to expand their communication beyond local media and markets by reaching out globally to stakeholders, and to potentially market their business to consumers not able to go to a single physical storefront. Local newspapers could provide news of interest to those who had moved away from home without incurring costly shipping charges to mail Michigan's *Claire County News* to former Claire residents now residing in Arizona or Azerbaijan. And individuals could cheaply and efficiently email colleagues in multiple office branches simultaneously rather than making a series of phone calls or mailing multiple letters. As tools to facilitate communication and collaboration across time and space proliferated, organizations spread from dense, single geographic locations into distributed organizations, groups, and work teams (Tapscott & Williams, 2008).

The rise of Web 2.0 tools has further expanded the media mix available to organizations. Now, tools like Etsy and Folksy provide small entrepreneurs an online storefront from which they may market and sell homemade wares, which (alongside rapid shipping by UPS, FedEx, and DHL) can allow entrepreneurs to reach out to previously untappable markets and grow a business (see Figure 6.3). Organizations are turning to social media as a means of reaching and activating members for social change (e.g., Change.org or Kickstarter.com) or as brand communities to engage further with an organization. And organizational members supplement their formal, company-sanctioned channels of communication (e.g., email, chat fora) with informal channels. For example, Best Buy employees can access complex guilds and gaming structures in *Battlefield 2* both from work and after hours as a way to encourage collaboration across multiple stores and to facilitate a repository of company and task-related knowledge (Tapscott & Williams, 2008). If a Best Buy Geek Squad member in Anchorage, Alaska, doesn't know how to fix a problem on your computer, it may be faster to jump into a game with other Geek Squad members worldwide to ask than to research the answer firsthand.

The challenge, then, is to select which of the ever-increasing number of media channels to employ for a particular communicative interaction. With the bevy of channels available via your smartphone alone, how do you choose whether to email someone, send a private chat message in the company's enterprise social medium (e.g., Chatter, Yammer), schedule an appointment via linked calendar apps, or just

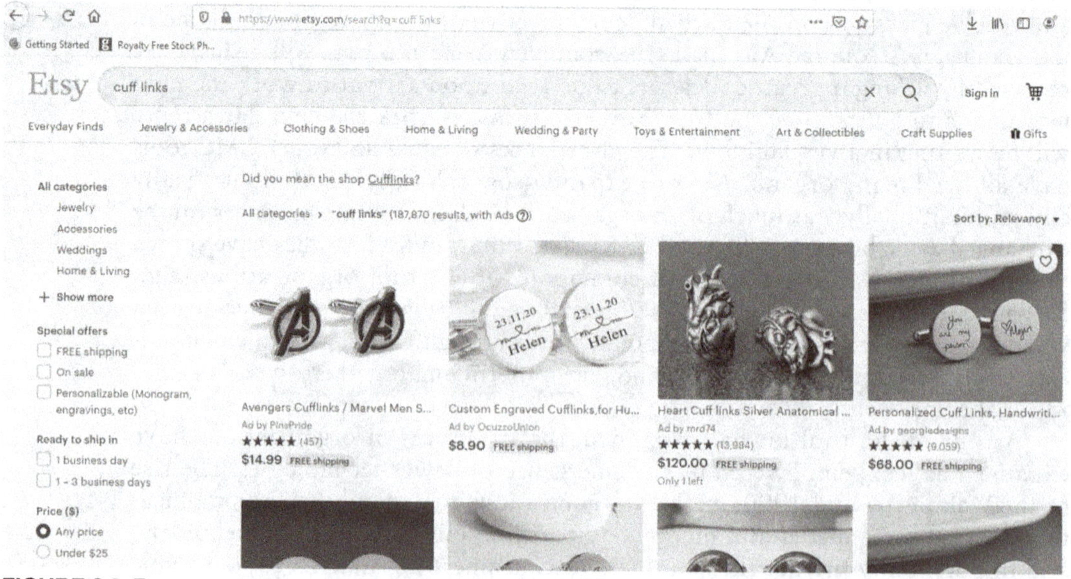

FIGURE 6.3 Etsy.com, displaying handmade cufflinks. Etsy provides an online storefront for local producers to market and sell goods globally, even if they do not have the resources to create their own custom online interface.

call them on the phone? The explosion in available media has made the concept of media choice even more salient.

Media Choice

One of the first challenges in integrating CMC into organizational communication is selecting *which* medium to use. Organizations have numerous channels from which to choose to interact within and beyond the organization. Certainly we see an increased push toward social media adoption by and within organizations. But this drive toward social media is often because they're the new hotness, and this drive often occurs without the adopting organization considering whether social media are most appropriate for the organization's communication. So how do we choose whether to send messages via social media or email, deliver them face-to-face, or simply post them on the bulletin board in the break room?

Media richness theory (MRT) (Daft & Lengel, 1986) provides an effective lens when considering what channel(s) to use to transmit a message. As you recall from Chapter 3, MRT suggests channels vary in their ability to transmit information to reduce uncertainty and equivocality, and particular media are more suited to certain situations. Just as you wouldn't break up with someone via Post-it® Note (a lean medium and an inherently bad choice for what may be a high-equivocality situation, as well as just a scumbag move), you wouldn't celebrate an employee who shattered the company's sale record via an impersonal and lean memo. As these examples indicate, we often have a feeling for what is and isn't an appropriate medium through which to transmit particular messages. However, are there more concrete guidelines or protocols than our gut feelings that we can follow? The following situations present several scenarios and desired goals. Which media may be most effective to achieve the organization's goals from the perspective of MRT (see Figure 6.4)? Remember that richer media are better at reducing equivocality (i.e., addressing unknown or idiosyncratic feedback of a message) and are typically better at facilitating socioemotionally rich messages.

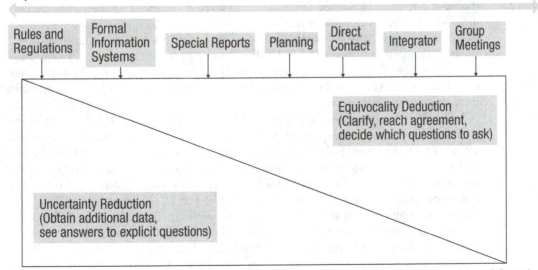

FIGURE 6.4 Media selection per media richness theory (MRT). Based on MRT, why is your syllabus delivered by a written document at the beginning of the semester rather than presented verbally by your instructor?

- You want to let your employees know about new policies for break periods at work.
- The organization wants to provide a forum for customers to interact both with other customers and with the organization.
- You have been asked to provide training in new features of the company's software to a specific department of about 25 workers.
- You want your employees to provide open, honest feedback about one of their managers.

A savvy communicator will think of both the message and the audience when considering which medium to match to each scenario, as these scenarios each present different communicative goals and intended audiences. For example, the first scenario presents a situation high in uncertainty (i.e., communicate the new break period policies) but low in complexity (i.e., the policies are pretty straightforward and clear-cut, without much need for feedback or discussion), so a lean medium would be effective at conveying such a message. You may send out a mass email to all employees or post a message on the bulletin board in the break room that employees can read at their convenience. Alternately, if you are looking for a forum to foster interactions among customers and between the organization and those customers, as in the second example, you may seek a rich medium. In this case, online discussion boards or social media (e.g., Facebook fan pages or Instagram posts) may allow large numbers of your customers to interact about topics of their choosing and provide an opportunity for feedback both from peers and from the organization itself. The richer channels commonly available in social media may therefore be preferable to leaner channels.

The appropriate medium in the third scenario, training departmental members on new software features, may depend on how complex the software is and what potential questions or challenges the departmental members being trained may have. If the new software is relatively straightforward (i.e., high uncertainty but low equivocality, such as an updated version of a previously used software package) and you just need to pass simple steps along to a larger audience, you may use YouTube videos demonstrating the new software and its use or provide screenshots or screen capture of your own use. Sharing the link to the training video(s) may be enough to get tech-savvy departmental members confident in using the new software. However, if the new software is complex or the departmental members are not as confident in their own computer skills, you may expect a lot of questions about the software, its use, or its functions (i.e., high equivocality). In this case, you may need to hold FtF meetings or a synchronous Web seminar via Zoom. The affordances of richer interactions (e.g., synchronous Web meetings or FtF training) would be better suited for the more complex and less predictable interactions that may occur.

The final scenario, seeking open and honest employee feedback about manager performance, is an important part of organizations' performance (Bernardin, 1986). Employee appraisals of managers and leaders can often identify strengths and weaknesses in organizational structure, particularly when managers' strengths and weaknesses are unknown. Initially, it may seem like, because this reflects a high-equivocality situation, a rich medium like FtF interviews or a social media discussion would be ideal. However, this scenario contains an element that goes beyond MRT: the nature of identifiability and its effect on messages. Employees are much less likely to give open, honest feedback when they are identifiable—when their name or position is attached to their comments—because of fears of the manager's retaliation (Waldman, Atwater, & Antonioni, 1998). Imagine if your manager asked you how s/he was doing as a manager and you replied, "You suck and you're mean," only to have your employee review the next week. Do you think your review may be affected, at least in part, by telling your manager s/he was terrible just a week before? Because of this fear of retaliation, though this scenario calls for a richer medium, the medium must also afford employees anonymity. Thus, rather than channels in which the communicator is identifiable (like email or social media), you hopefully selected a medium that allows richer feedback but still affords communicators the shield of anonymity. Consequently, tools like anonymous online surveys or suggestion boxes are often the best balance between rich channels (allowing for equivocality-reducing feedback from employees) and those that strategically obfuscate communicators' identities are often best for upward feedback about managers (Walker & Smither, 1999).

You can see how media choice can be complex, often taking into account the message, the communicative goals, and the communicators to decide which channels are strategically the best for each individual situation. Media choice is not a one-size-fits-all scenario. Rather, communication managers should be strategic when considering which channels—CMC and otherwise—to use to transmit a message. After all, as McLuhan (1994) told us, the medium is as much a part of what's communicated as the message itself. Whether you are fired over a phone call or by email, the result is the same ("You are no longer employed here"), but the medium through which that message is communicated carries additional meaning about how your employer viewed you and your employment. Much of the message conveyed by channel is carried by what the channel allows us to do, including managing time.

Chronemics

Digital tools are fast, allowing us to send and receive messages anywhere in the world almost instantaneously. This rapid transfer of messages may make you think that time and **chronemics**—the use of time to communicate—would be relatively meaningless. And yet, even though you can have a real-time synchronous Skype chat with someone across the globe does not mean that time is not something to consider in CMC, particularly with respect to organizations and their communication. For now, let's focus on the use of time in organizational CMC in three ways: the channels we choose, the time needed to make decisions, and the normal response times.

Channel Choice As we discussed with respect to MRT, the channels we choose have implications for our communication. Particularly amid the many mediated (and unmediated) channels available to us in many organizations, the chronemic implications of our CMC begin even before crafting a message, as we select our channel. Our days as an organizational member are usually punctuated by continued calls for our attention: phones ring, computer notifications chime, our devices vibrate with incoming texts and social media notifications, and there are always new flyers in the breakroom—sometimes from human resources making policy updates and sometimes from coworkers with more informal or social intent. If we heeded every call for our attention, we'd never get anything done! So how are we to make sense of which channels to take note of, use, and reply to? One answer is that we triage our media use in organizations based on the chronemic agency of each medium.

Chronemic agency refers to "the relative power that messages and notifications that are received through the medium have on the recipient's attention and reaction time" (Kalman, Aguilar, & Ballard, 2018, 1965). Experientially, you probably already have a sense of chronemic agency: when someone takes the time to call you, you pick up or call back as soon as possible, but you also know that you don't need to check and respond to social media content immediately. Data support this, as

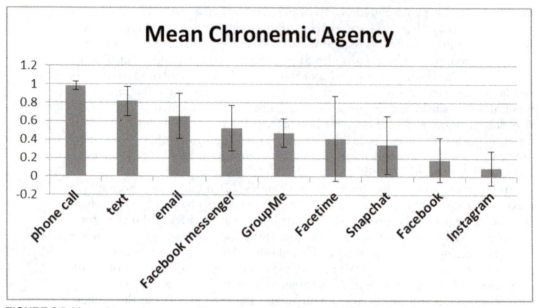

FIGURE 6.5 Mean chronemic agency of the most common communication media among organizational users. Those channels with more chronemic agency demand greater attention and more rapid response. Reproduced from Kalman, Aguilar, & Ballard (2018). HICSS

more dyadic and personal channels tend to be assigned greater chronemic agency, demanding our attention over whatever else we may be doing, rather than more asynchronous or impersonal channels (Kalman et al., 2018). Organizationally, this suggests your need for immediate action on behalf of your communication partner may be influenced by the channel you use to initiate that interaction. A phone call or email is much more likely than a message through the company's enterprise social medium to merit the receiver turning away from whatever project they are working on to attend to your message. Interestingly, knowledge of this seems implicit, as people seem to know the hierarchy of chronemic agency assigned to channels (see Figure 6.5) without having to explicitly discuss it.

Individuals monitor the various channels and will be pulled away from other duties to attend to messages in those channels. However, for more prolonged projects or sustained interactions, additional considerations must be taken. For example, though phone calls and texts may demand our attention most immediately, if every workgroup you are a member of conducts all of its business through text messaging, it can quickly become overwhelming; texting may not even be the most appropriate medium through which to make impactful decisions. As such, a second consideration of CMC in organizational communication related to chronemics is the nature of decision-making in organizations.

Decision Time Though synchronous CMC tools are common, organizations are still dominated by asynchronous communication media, particularly email. Much of this is due to convenience, as using synchronous tools—whether a Facetime video chat or a real-time Slack discussion—requires some degree of scheduling. All parties must be available and active at the same time. This can be challenging enough when organizational members have different schedules, responsibilities, and free times, and it gets even more complex when moving beyond a single location or time zone. Consider how hard it already is to find a time for all of your group members to meet. Now imagine half your group is in Madagascar. See how difficult it can be to find a time you all have free? Asynchronous tools are therefore very common means of discussion, decision-making, and interaction in organizations, especially for individuals whose desks are more than a few feet from each other. But, because of the nature of asynchronous communication and constrained cues, it takes more time to have a discussion and come to a conclusion through online channels than it does face-to-face.

It can take almost three times as long to communicate the same message online as it does face-to-face, even in synchronous text-based communication (e.g., text messages, Slack). This greater time involvement is because it takes at least three times as long to type and read than it does to speak and listen (Hiltz, Johnson, & Turoff, 1986). We speak at about 135 words a minute but type only about 40 words per minute. And even though an average reader can read about 270 words per minute, we can hear the spoken word at a rate of about 450 words per minute. These differences mean it takes more time to communicate digitally in writing than verbally, and *that* assumes the communicators are both fully engaged and communicating synchronously. When participants may be distracted or multitasking, requiring repetition or additional efforts of communication, it can take longer still to interact (Stephens, Cho, & Ballard, 2012).

Beyond the time it takes to read and write, it also takes longer to establish relationships and common meaning online via written communication. Recall social information processing theory's (SIPT) (Walther, 1992) position that online relationships begin slowly, but eventually reach the same relational outcomes as their offline counterparts. Consistent with SIPT, it takes time for communication partners to replace all the nonverbal cues (e.g., facial expression, vocalics) lost when they communicate online, but they do eventually get there. The same process occurs within organizations, so when two organizational members (or groups of organizational members) begin to communicate online, they need to form new patterns of communication and shared understanding unique to that configuration before they can really begin to effectively communicate and function. Compounding the challenge, that SIPT process and learning curve is on top of the longer time it takes to type and read. So ultimately, for organizational communication, communication sessions can take three to nine times longer to accomplish the same task, whether it's emailing your counterpart in India or having a text-based interaction with your work team via Flock. Time is such an important resource for the functioning of groups and members that if the organizational communication being mediated is a decision, meta-analysis (Baltes et al., 2002) suggests the group be given as much time as needed. By putting a time constraint on the group, you may get a quicker decision, but not necessarily a better one. Arbitrary time constraints may force the team to skip steps in group formation and socialization that are important to the decision-making process. It simply takes longer to communicate online than off, and so organizational CMC needs more time to become just as effective as FtF organizational communication and reach a comparable decision.

Response Time The other frequent chronemic concern in organizations is response time: how long it takes to respond to a message, if at all. As asynchronous communication allows messages to be sent and received any time, how long should we expect to wait to receive a response before we begin to assume the original message was not received, lost or forgotten about, or purposely ignored? As we cannot see a person to answer this question, it can be difficult and stressful to await a response via CMC.

Many organizations have policies in place stipulating the expectations of responses for their members. A common example may be seen in your syllabus, as many instructors have a specific email response policy like "I respond to all emails within 24 hours during the week and 48 hours during the weekend," or, "I observe work-free weekends and school breaks, and as such will respond to any messages received over the weekend/break on the first Monday back." Such policies and statements, whether organization-wide or specific to an individual, can help manage expectations for senders and receivers.

Beyond formal policies of response times are the norms of response time. Norms are not formal policies and often do not have penalties for noncompliance. Instead, norms are informal and generally used guidelines for behavior. Within organizational culture—and particularly that of for-profit businesses—a common norm is to respond to emails within a day. If I send something to you at noon on Monday, I should likely expect a response before the end of Tuesday's business day. This "within 24-hour" heuristic is frequently put forward and discussed in business classes, trade journals and magazines, and several books whose titles likely end in "for Dummies." Even though this 24-hour norm is unofficial, it does seem like most in the business world keep to it. Studying thousands of email exchanges among hundreds of Enron workers, Kalman and Rafaeli (2005) found that most emails were replied to in at least a day. In fact, about two-thirds of emails received a reply within six hours, and only about 5% of emails received a reply more than 48 hours after being sent.

Knowing that more than 90% of emails get a reply within 24 hours does more than serve as a fun little factoid to use at cocktail parties: it helps us set some very good expectations and norms for our own email use, especially within organizations. Even if the reply is, "Good question. I need to pull up some old files out of basement storage to find that answer. Give me a few days and I'll get back to you," it should come within about 24 hours of the initial email request, or the sender may start to worry the email was not sent, was ignored, or the receiver is mad at the sender for some reason. As organizational members, this 24-hour norm also helps us reverse engineer some timelines for discussions and deliberations. For example, if you know that you need to send in final proofs of a project by Thursday, it is not a good idea to send the proofs out for review on Wednesday afternoon: that may not give those who need to review it enough time. In fact, if you think the proofs are going to be problematic, you may want to send out the reviews on Monday or even Friday of the previous week, as each back-and-forth email may take up to 24 hours, and therefore the entire email exchange may take two or three days. This process may actually be sped up a little when the organizational members are distributed globally, as you can send an email before leaving work at 5 p.m., your counterpart across the globe will find it on her/his desk first thing and reply at the end of her/his day, and you will return to work the next day with a reasonable expectation of the response in your inbox. But again, coordination across time zones can be more difficult and take more time. For truly time-sensitive decisions, you may want to suggest or select a synchronous channel like a phone call or Skype session. Though email is convenient and instantaneous, the communication it conveys may not always be quick.

Status Equalization

A final consideration for the mediation of organizational communication is the effects communicating online can have on the structure and power differences that occur in organizations. One of the more-studied effects of being online for communication within organizations has been how using CMC tools can reduce status differences among organizational members. Think about when you talk to your boss: you probably spend less time talking and more time listening to her, are less likely to disagree, and offer fewer risky new ideas. This hesitance to talk—especially to challenge your boss—is due to the status differences that are often overt and explicit when face-to-face: you come to *her* (larger) office and stand in front of *her* (larger) desk while she sits behind it, and you may even address her more formally (e.g., "Ms. Cherson") than she addresses you (e.g., "Liz"). However, those cues to the differences in status between organizational members are often hidden or absent online, which can result in more balanced communication among interactants, both in the form of address and in the amount of "floor time" each individual holds. Such differences are especially pronounced when interactions are anonymous or pseudonymous. This leveling of status roles and subsequent communicative behaviors has been labeled the **equalization phenomenon** (or **equalization effect**), as reducing the number of social cues by moving interactions online can equalize the differences typically created by organizational hierarchy or status distinctions (DeSanctis & Gallupe, 1987). You may already see or experience this equalization phenomenon in your own organizational communication: consider when you begin an email to a professor with an overly familiar greeting (e.g., "Hey," or "Eric") that you would never dream of using in class (where he is "Dr. Dickens"). This is likely because in email you are not presented with the status cues of being in a classroom or office, sitting behind a desk, and seeing diplomas and thick academic books on a shelf. As such, you may not feel the status differences between you and your professor online as strongly as you do in his office. Though the way equalization effects manifest online generally support the phenomenon, some notes bear mentioning.

Though receiving some empirical support, the equalization phenomenon has been challenged both practically and theoretically. Equalization effects don't consistently occur, in part because people often want to be identifiable—both to themselves and to others—in online workgroups. Doing so can help interactivity by identifying to whom you're replying and being able to have some accountability for messages sent. Consequently, rather than facilitating the anonymous interactions necessary for status equalization, many management systems—especially enterprise social media—make users identifiable, presenting messages alongside names and sometimes even titles. As soon as you log on to Microsoft Teams or Slack, you see messages coming through from "Michael (District Manager)" and "Amanda (Regional Manager)," or even "Amanda (Leader)" or "Amanda (Moderator)," often with accompanying pictures. These status labels activate differences in participation and engagement based on hierarchical status (Weisband, Schneider, & Connolly, 1995), and any gains from the democratization of CMC are lost.

A second way the equalization phenomena can be stymied is when individuals are represented by avatars in online discussions, which may lead to status differences based on group effects rather than organizational status. Many conversation platforms online—from Slack to Twitter—allow users to identify themselves using avatars, sometimes providing a small set of avatars from which users may choose. The challenge in this is that avatar selection can activate social identity effects, which subsequently impact status effects and discussions. For example, when several users select the same avatar (see Figure 6.6), the similarities of avatars may create an ingroup, leading them to generate a larger proportion of the conversation and to have others follow the lead of those similar avatars or self-censor disagreements (Lee, 2004). Even if not using embodied avatars, other cues can activate social identities and influence discussions. For example, the use of gendered (i.e., feminine or masculine) user names in an online chat

FIGURE 6.6 In the discussion above, all organizational members are represented by different avatars. Imagine if two avatars were bears. How would that affect the decisions of the person communicating via the parrot? Would they be more likely to agree or disagree with them? Reproduced from Lee (2004).

can activate the social norms often associated with genders and gender roles, so that individuals communicating via a female avatar or feminine user name may actually contribute less to an online conversation, regardless of their offline selves (Wijenayake et al., 2019). Consequently, organizations and groups using online tools to democratize group processes should be mindful of the ways discussants are identified online, or the gains of CMC discussions may be lost.

As user names or avatars indicate, gender differences in language may not always equalize when online. Face-to-face, males tend to display more power behaviors in their language use, in that they write longer posts, post more frequently, and are more opinionated or polarizing in their ideas. In actual online discussions, however, different patterns seem to emerge. In CMC, men still typically write longer posts and messages than women, but women contribute more frequently than men to online discussions (Sussman & Tyson, 2000). With respect to polarizing content, there does seem to be an equalization effect, with genders writing about the same number of opinions in online discussions. Ultimately, Sussman and Tyson (2000) conclude that although CMC can equalize some communicative processes, male domination—by those identifying as or identified by others as male—may persist.

Knowledge Management

A final implication of CMC within organizations is how technology and digital tools are used for knowledge management: the ability to store, index, search, and retrieve information. Much time in organizations is spent on knowledge management, helping to identify who knows what about what topics. The ability to efficiently identify and access information within an organization can be a critical component of the organization's efficiency, so knowledge management systems are increasingly critical for organizations.

We likely see examples of online knowledge management systems in our daily lives, as wikis are one of the most common knowledge management systems. Tools like Wikipedia allow users to store and relate complex information in a way that can be accessed and used meaningfully (see Figure 6.7). If you look up "Scotland" on Wikipedia, its entry contains more than a list of information about the country. The Wikipedia entry includes

FIGURE 6.7 Wikipedia—A multipurpose knowledge management system.

cross-references for information to go more in depth about related topics. If you see that Scotland's national animal is the unicorn, you can follow links to learn about the mythical one-horned horse. The unicorn's Wikipedia entry may then lead you to another related link about the narwhal: a real one-horned whale. Though this form of wiki-roulette (i.e., following a trail of barely connected links to increasingly off-topic entries) is fun, it also demonstrates how an individual interested in learning more about a topic can use wikis to store and retrieve information in complex and interrelated patterns.

Many organizations now use their own wikis and other systems to manage their information and help members effectively retrieve information. Especially in distributed organizations—those in which all organizational members do not work in close physical proximity—it can be difficult for all members to access the same information. A file in the California (USA) office cannot be easily picked up by someone in the Catalonia (Spain) office. Likewise, for organizations in which members are often not physically in the office at the same time (e.g., parole officers, sales representatives, human services case managers), having a centralized store-house of information is beneficial. In these cases, knowledge management systems may take the form of **wikis**—an online information repository whose content and structure can be edited collaboratively. For example, an organization may set up a wiki (either publicly accessible or on the organization's internal server) to allow members to post past projects and identify any problems that were encountered and their solutions. Project teams who confront novel problems can then access the wiki to see if other teams have experienced similar problems and learn from those past experiences (Munson, 2008). This type of knowledge management has been referred to as *organizational memory*, as the wiki serves as a storehouse of the organization's collective knowledge.

A related form of wiki-based knowledge management is *collaborative learning*, as organizational members take advantage of the wiki's talk feature to interact and resolve organizational issues. Talk features are often used after identifying key information holders who may have previously experienced similar issues or have unique knowledge or perspective to help resolve new ones (Chau & Maurer, 2005; Engstrom & Jewett, 2005). In one example, a county Department of Health and Human Services may have many case managers, each assigned to individual cases of families with distinct needs. Because the case managers spend so much time doing home and site visits, they rarely get to interact in the office, where they may be able to seek help on particularly challenging scenarios from other case managers. Using the department's wiki, one case manager who is experiencing a particular problem (e.g., a child with particular behavioral issues) may post information about the case. Other case managers, seeing the post, may contribute to the entry by sharing stories of similar experiences or resolutions, or use the talk feature to seek more information or share thoughts. The case manager can then draw from the entry to try some of the new behavioral interventions suggested by coworkers, all without returning to the office. In this case, the wiki can manage organizational information and help resolve issues by connecting organizational members and their various experiences (e.g., similar past cases) and information (e.g., community resources, new behavioral intervention standards).

Another example of CMC for knowledge management is the use of social network and networking tools to identify knowledge resources. One of the biggest challenges large organizations face is knowing who knows what. To identify and connect organizational members within proprietary networks, many large organizations use **enterprise social media** (ESM), to identify organizational members, their knowledge and competencies, and their network connections

FIGURE 6.8 IBM's Social Blue—An enterprise social medium, often used for the organizational knowledge management.

(Leonardi, Huysman, & Steinfield, 2013). Tools like ESMs, accessible only to organizational members, can be efficient means of seeking out critical organizational information (Leonardi, 2015). For example, IBM's internal enterprise social medium, Social Blue (Figure 6.8), is similar to Facebook in that it allows members (who must be IBM employees) to create profiles in which they can identify their areas of expertise. Rather than spending time calling around offices to see who may have certain information, these profiles allow IBM employees to quickly identify and reach out to the right person about a particular problem (Majchrzak, Cherbakov, & Ives, 2009).

By offloading organizational knowledge to databases, wikis, SNSs, and other CMC tools, organizations can create searchable and usable channels to manage the vast knowledge, information, and experience of their members. Rather than employees aimlessly roaming the halls hoping to encounter a coworker who may know a solution, CMC tools can manage the information itself as well as organizational

Research in Brief

Now You're Just Somebody That I Used to Know

Who would you call with a question about taxes? If you or your direct family members are not accountants, you perhaps know someone who is, or at least someone who knows an accountant. Similar problems are common in organizations: knowing who to ask for information, for advice, or for help resolving a problem is a critical part of organizational information management. But particularly as organizations and workgroups are distributed—either across offices or countries—it can be hard to know everyone you work with and their various knowledge and skill sets. One way to make organizations more accessible is through CMC, which can make users more aware of their environment while not requiring direct communication.

Leonardi (2015) explored how employees may be able to observe coworkers even when not physically present. Workers at American Financial were given access to a Facebook-like enterprise social networking tool, "A-Life," which allowed them to create profile pages, send public and private messages, and access news feeds, suggested new contacts within the organization, and provided a place for the secure storage and sharing of organizational data. Twenty groups from across the company's 15,000 employees were randomly chosen to try out the A-Life system. After several months of

using the system, A-Life users and another random sample of American Financial employees not on the A-Life pilot test were asked to identify which organizational members could answer certain types of questions (i.e., who knows what) and the relational connections among members of the workgroup (i.e., who knows whom). Users of A-Life were better able to identify which other organizational members may have the information they needed as well as how to access that person.

Particularly within large organizations, it can be a challenge to find people who have the information you need, whether it's the previous year's sales data or advice on how to deal with a particular client. Enterprise social media, just like their interpersonal counterparts, provide a way to observe others' comments and interactions and allow us passively to learn about our organization and the people in it. Whether it's a social medium, a wiki, or other collaborative tool, sometimes the ability to determine who knows something is even more important than knowing the information yourself.

Reference

Leonardi, P. M. (2015). Ambient awareness and knowledge acquisition: Using social media to learn "who knows what" and "who knows whom." *MIS Quarterly, 39*(4), 747–762. https://doi.org/10.25300/MISQ/2015.39.4.1

networks and members. In other words, CMC tools not only store what the organization collectively knows but also identify who in the organization may know what, and provide means of accessing those individuals. Reducing search costs associated with finding the information translates into large gains for organizational efficiency (Leonardi, 2015), so organizations are increasingly investing in digital tools to harness the collective knowledge of its past and present members.

Organizations' Use of CMC to Communicate

Beyond using computer-mediated tools to communicate intra-organizationally, members of organizations increasingly use digital channels to communicate extra-organizationally (i.e., outside of the organization, either with other organizations or with external stakeholders or publics). Consequently, CMC gives us reason to consider new or expanded ways organizations can communicate internally and externally.

Communicating with External Individuals (B2C)

Organizations have long used technologies to communicate with individuals who are interested in the organization but not themselves organizational members, including customers, community members, and other stakeholders. For example, catalogs and newspaper inserts helped organizations advertise sales and promotions. Radio advertisements and town hall meetings helped organizations communicate to community members about upcoming community events or impacts. And quarterly earnings reports and brochures aided organizations in messaging financial performance and company information to investors and interested others. It should therefore come as no surprise that organizations have quickly sought to use online tools to likewise communicate with individuals outside of the organization. Now, organizations use websites, podcasts, digital advertising, and a host of other CMC tools to communicate with other individuals. Often, this type of communication is referred to as business-to-consumer (B2C) communication, though many of its strategies can apply to nonprofit, religious, philanthropic, and other organizations that do not necessarily interact with "consumers," but rather with some form of external client or stakeholder. Beyond facilitating a greater reach, CMC technologies can enable new opportunities for organizational communication.

One opportunity CMC tools—particularly social media—offer B2C communication is the ability to quickly create and disseminate information. Rather than having to wait for a publisher to print and send out physical copies of a communique, emails, websites, and Twitter allow organizations to communicate with stakeholders directly and quickly. The ability to quickly get information to an audience can be particularly critical for organizations dealing with crises. For example, weather bureaus, government agencies, and municipalities can use Twitter during natural disasters to rapidly send out the latest information about infrastructure (e.g., a collapsed bridge, flooded areas), responses (e.g., evacuation orders, first responder contact information), and services (e.g., shelter locations, access to clean water). Affected individuals can then seek out this information to coordinate relief and support efforts between them and support rescue and relief organizations, including the Red Cross, the Coast Guard, and the Cajun Navy (Lachlan, Spence, & Lin, 2017; Smith et al., 2018). Even in less dire circumstances, organizations can use CMC tools to quickly promote items, provide updates on new services or policies, remind individuals about an upcoming event, or otherwise rapidly communicate with interested individuals. Again, such cases may not fit the way we typically think about B2C communication, as flood victims or citizens engaged in an uprising are not purchasing goods or services, but the information and the way it is communicated fits the models and other conceptualizations of B2C communication.

A second way digital tools have changed organizations' efforts to communicate with external stakeholders is by enabling two-way, dyadic interactions between organizations and their publics. Unlike traditional media (i.e., television, radio, newspapers, billboards), which really only facilitate one-way communication from the organization to an audience, CMC tools enable feedback and the potential for the audience to initiate an interaction with or about the organization. Going beyond simply emailing an organization, the dialogic potential of CMC tools allows the creation of perceptions of immediacy or interactivity between an organization and its stakeholders. Increasingly, organizations are using newer tools—especially social media—to provide annual reports and corporate social responsibility (CSR) statements to stakeholders (Reilly & Hynan, 2014), often doing more than simply reporting quarterly earnings or stating corporate values. Social media enable an organization to create discussions about and engagement with their company, its policies, and even its corporate philosophy. Even actions as simple as replying to a Facebook comment can help organizations from Taco Bell to Harvard feel more immediate and accessible to users. For example, small airports often lack the passenger or cargo volume of large, international airports like O'Hare, Heathrow, and Changi. Small airports, including regional airports and municipal fields, can now use social media to distinguish themselves and create a unique "voice," add value to and engage with their local and regional community, build brand loyalty, and raise awareness of the benefits of smaller airports for local populations, government, and economics (Grigsby Smith, 2019). Importantly, this social media engagement goes beyond simple self-promotion to include value generation for and interaction with the relevant stakeholders.

Online dialogue can also help organizations provide information and support to individuals. For example, during the 2014 outbreak of the Ebola virus in the United States, individuals used Twitter to contact the Centers for Disease Control (CDC) to get information related to the epidemic. The CDC used this opportunity to answer questions, redress misinformation, and keep the concerned public updated on the spread of the disease and the CDC's effort to fight it (Crook et al., 2016). When traditional channels (e.g., daily newspapers, nightly television broadcasts) are too slow to update information about a rapidly changing situation, or when those traditional channels cannot allow individuals to ask questions about their own situations, CMC tools can increase the speed and responsiveness with which the public receives information when faced with outbreaks, natural disasters, or other events that necessitate quick and personalized messaging to those affected from the organizations involved (e.g., disaster relief, health organizations, weather services, community groups).

Communicating with Other Organizations (B2B)

A second way CMC has changed organizational communication is by affecting how organizations communicate with other organizations. Organizational members and leaders—at least in Western cultures—often place a large value on being able to see and talk with a business associate. Particularly when making deals or contracts, the nonverbal communication of a handshake still has significant personal and even neurological significance for individuals, increasing trust and confidence in the others with whom we make agreements (Dolcos et al., 2012; Wesson, 1992). But how do we establish that trust when we're not able to meet with a counterpart from another organization?

Several forms of B2B CMC have long been used to exchange information and other messages between businesses. For example, some of the earliest users of the telegraph were businesses communicating with other businesses (History.com, 2009), including businesses in New York City and suppliers in Pittsburgh, transatlantic stock markets, and even the infamous first wireless SOS distress call from the doomed *Titanic* to any other oceangoing ships nearby (which were not also owned by the White Star Line). The telegraph served as the initial infrastructure for the B2B communication devices that followed, including the telephone and fax machine.

More recently, new CMC tools have complemented our offline and online communicative toolboxes, giving us new ways to do old things. Sometimes, CMC is simply used to exchange data between organizations. Stock trading firms have spent billions to erect large microwave towers to shave milliseconds off of their data transfer time (Cookson, 2013). As an extra 0.4 ms can amount to millions gained or lost when trading large volumes, faster and more automated CMC between a trader and a trading floor is critical. Organizations use similar data exchange systems in the form of file transfer protocol (FTP), which is a standardized way for users to exchange files over the Internet. You may have friends who use FTP when sharing torrents of music or movies online, and businesses use FTP to exchange files and information. For example, many businesses can use FTP to file their tax information and receive replies from the state or federal taxation agencies. Organizations in the health sector can use FTP to securely transmit patient information among various healthcare providers (e.g., primary physician, insurance company, specialist), ensuring all parties have access to the same recent and correct health history.

Another CMC tool has become so ubiquitous for organizational communication we may often forget its contribution: email. Email has often been dubbed the "killer app," as it was the feature that made computers in organizations not just desirable but necessary, making desktop computers the prominent and prevalent tool they now are in organizations (Tassabehhi & Vakola, 2005). Email helped first supplement (and increasingly replaced) the telephone call and even FtF communication as a means of interaction, especially for interorganizational communication. Its asynchronous nature allowed a user in one organization to send an email to a user in another organization without worrying whether she was at her desk, in the middle of another project, or available to talk. But email does not always have the initially predicted positive impact. Though email can enhance organizational members' productivity, increasingly organizational members face **email overload**: the perception of being overwhelmed by a constant flow of messages into one's inbox and the inability to effectively manage or keep up with the high volume of communication (McMurtry, 2014). Particularly as messages can come into or leave the organization at all times, with no gatekeepers (i.e., mailrooms) to control or limit the volume we receive, managing email overload can be critical to stem the flow of intraorganizational as well as interorganizational information. We now try to control the barrage of incoming emails in several ways: limiting the amount of email communication, using email client setups to limit the hours of a day emails can be received or sent, and being mindful of listserv and other email subscription notifications (Jackson, Dawson, & Wilson, 2001).

Many B2B technologies can minimize the time humans need to create, transmit, and receive messages, offering ways to automate and streamline B2B processes. For example, global positioning systems (GPSs) on trains, ships, and freight trucks allow suppliers and receivers to track shipments in real time without having to interact with a counterpart at another organization or to call drivers, engineers, or captains in transit. However, B2B CMC can also automate information exchange to facilitate subsequent human communication. For example, logistics programs can automatically track shipments and send notifications to receivers about the timeliness of the shipment's arrival, and then prompt the receiver if they would like additional information, options, or discussion about shipping disruptions. The system then turns such requests over to the responsible individual in the shipper's organization (O'Leary, 2011). Another way B2B communication has incorporated CMC tools is through monitoring public sentiment, trends, and demands via online spaces like Facebook, Reddit, and chat fora (Cui et al., 2018). For example, by staying attentive to how fashion groups and key individuals discuss the latest trends on Instagram (both in general and in fashion-specific groups), fashion brands can gain key insights into future markets and industry trends.

Online tools can also facilitate B2B communication in the form of marketing and advertising (Chae, McHaney, & Sheu, 2020). Directed emails, listservs, and social media can be used to communicate with potential business partners, suppliers, or customers. For example, an organization specializing in janitorial supplies could use an email directory of the local chamber of commerce to send directed messages to all local businesses to inquire about their custodial needs. As we'll discuss more in Chapter 10, social media and other digital tools have become particularly valuable for their ability to target specific potential audiences (see Figure 6.9). Rather than just broadcasting a message to any billboard reader or radio listener, social media advertising can target specific industries, types of clients, and/or geographic regions, making the organization's marketing efforts more efficient and effective.

Finally, social media have become another powerful tool for organizations to interact more overtly with other organizations via masspersonal communication. In other words, social media allow two organizations to communicate in a way that others can publicly observe. As brands seek to establish their identities with their publics and stakeholders, social media can announce new partnerships and alliances between organizations, which may make them more valuable to their communities. For example, a local temple and food pantry can announce an upcoming cosponsored event to raise money for nonperishable goods. In a more recent example, Disney promoted the release of its new streaming service via a tweet: "It's moving day! Is everyone packed up and ready to go to @DisneyPlus?" (see Figure 6.10). Disney's other properties, including Star Wars, Pixar, Marvel, National Geographic, and The Simpsons, quickly replied. It may seem

FIGURE 6.9 Business-to-business (B2B) communication can include marketing and advertising directed at other businesses. Social media platforms like LinkedIn and Xing can target users from particular industries or within specific geographic regions.

odd for the Walt Disney Company to tweet at and reply to itself. But the visibility of the interactions promoted the new streaming service by demonstrating the breadth and volume of media about to be streaming exclusively on the Disney+ service, and therefore not over Netflix, Amazon, Hulu, or other streaming services (Ridgely, 2019). This example of B2B communication therefore also serves as an instance of B2C communication, as the messaging sought to inform potential customers of the new service and entice them to subscribe.

Similarly, B2B communication via social media can help organizations establish their **voice**, or "the language styles … a company uses to express a distinctive personality or set of values that will differentiate its brands from those of competitors" (Delin, 2005, 10). Voice can be thought of as an organization's unique personality or tone. Social media, more so than static press releases or traditional media, can allow organizations to demonstrate some of their personality, particularly via the voice of their interaction with customers and clients. Does an organization seem genuinely responsive to concerned customers who reach out on Facebook, do answers seem empty and formulaic, or are customer complaints altogether ignored on social media? When and how do organizations address various holidays, special events, or even tragedies (see Hayes, Waddell, & Smudde, 2017)? Whether and how an organization interacts with other organizations online can help to create its voice. Commonplace brands like fast food brands Taco Bell and Wendy's and the UK's Waterstones bookseller have used Twitter to create distinct voices and raise awareness of their respective brands. Wendy's has been particularly notable through its use of B2B communication, using Twitter to antagonize, denigrate, and otherwise give a hard time to competitor organizations (Figure 6.11). In doing so, the previously commonplace burger chain has managed to remake its image as a contemporary, sassy, and youthful organization.

As another notable example of organizations using social media to find their voices, early into the global COVID-19 pandemic, the Steak-umms brand of frozen steaks began to tweet focused statements regarding media literacy and science (see Figure 6.12). Prior to the pandemic, the brand had been posting the bland, self-promotional messages that were endemic to many brands online. In short, the brand's Twitter account

FIGURE 6.10 Disney+ leveraged its various other Twitter accounts to publicize the release of its streaming services.

Oh, hey Snake 'n Fake! Think you dropped this: @

Maybe leave the subs to the sandwich shops and leave the cheeseburgers to us.

FIGURE 6.11 Sometimes, organizations can engage in B2B communication masspersonally. The quick-service restaurant chain Wendy's has developed brand awareness and a voice for the brand through its use of Twitter to interact with—and often roast—other quick-service and fast-casual restaurants, such as Steak 'n Shake.

Steak-umm @ @steak_umm · Sep 11
conspiracy theories are baked into human nature. they don't only affect the fringes. all people naturally seek patterns to make sense of the world's randomness and develop in-group, out-group prejudices that make them vulnerable to all kinds of propaganda, paranoia, and extremism
181   2.1K   5.7K

Steak-umm @ @steak_umm · Sep 11
people are often led by intuitions and overestimate their ability to understand complex subjects. that, coupled with the natural tendencies to seek truth, meaning, and belonging, can make even the most well-intentioned and intelligent people vulnerable to conspiracy theories
11   178   1.5K

Steak-umm @ @steak_umm · Sep 11
research shows that social conditions and personal traits can make someone more vulnerable to conspiratorial thinking, such as lower analytical thinking, lower education, schizotypy, being distrustful, narcissistic, powerless, alienated, anxious, or prone to projection
21   162   1.3K

Steak-umm @ @steak_umm · Sep 11
it's important to understand the potential factors that make people extra vulnerable to conspiracy theories, but just as important to recognize that everyone is vulnerable to some degree. studies and polls repeatedly show millions of people believe in them in some form or another
6   89   1K

Steak-umm @ @steak_umm · Sep 11
there's a difference between "conspiracies" which are covert plots that can be uncovered through reasoning and evidence, versus "conspiracy theories" which are metaphysical narratives based on prejudices, sensationalism, and flawed assumptions made outside material reality
10   197   1.2K

Steak-umm @ @steak_umm · Sep 11
there's no universal way to bring people back from conspiracy theories. some take months, some take years, others stay gone. some need relationships. some need isolation. and some need an "aha" moment. some prefer debate. some conversation. and others need to find their own way

FIGURE 6.12 At a time when the honesty and objectivity of several government agencies were questioned, the Steak-umm brand rose to global prominence amid the COVID-19 pandemic due to its surprising voice as it provided thoughtful, cogent discussions of science, media literacy, moral reasoning, and social media use that belied a brand of frozen sliced beef.

was boring, with little to inspire any but the most devoted brand loyalists. However, Steak-umms began to violate expectations by changing its voice to be one of leadership, objectivity, and calm in the cacophony of Twitter. Of course the brand occasionally pointed out its product and suggested others may purchase it, and many Twitter followers tweeted their buying behaviors had changed and they were purchasing the brand's products more due to the honest and frank nature of its tweets. But ultimately, Steak-umms was able to capitalize on its newfound voice during 2020 by going beyond what many organizations were doing and providing authentic, transparent, and considered commentary to Twitter users that was more public service than hocking a product (Bogomoletc & Lee, 2021).

Communicating with Future Organizational Members

A final way CMC has changed organizational communication is by affecting how organizations communicate with *potential* organizational members—those who are not yet part of the organization but may eventually become members. Organizations spend significant time, effort, and money communicating with individuals who could join the organization. From membership drives to rush week for sororities and fraternities to job fairs, organizational efforts to communicate with prospective members are all around us. Much of this communication is centered on creating a positive, favorable impression of the organization. Online, these efforts are perhaps most evident in the form of organizational websites used to provide information about employment opportunities and benefits. In addition to listing current job openings, websites can also provide prospective job applicants information about salary, health benefits, internal promotions or management training programs, organizational culture, and other features that may present the organization as a good employer (Barber, 1998). Though these websites are still heavily used, organizations have also turned to social media to present themselves to prospective applicants and provide information about the organization as an employer (Sivertzen, Nilsen, & Olafsen, 2013). Through many online tools, from static websites to dynamic and interactive social media, organizations use CMC to communicate their identities to prospective members. The image of an organization that appears online is important for personnel recruitment, as it can influence prospective applicants' perception of the organization and intent to apply (Carr, 2019; Kissel & Büttgen, 2015). But increasingly CMC tools are more than just another channel by which organizations can present themselves, as newer channels pose problems for the strategic self-presentation organizations often seek in their own online presence.

Organizations decreasingly have control over their online identities. Unlike recruitment brochures or organizational websites in which an organization has complete control over its self-presentation (do you notice elements of the hyperpersonal model here?), online channels increasingly give others—from current and past employees to customers to simple Internet users—the ability to present information about an organization. Employer review sites like GlassDoor, Indeed, and Vault now allow individuals to evaluate their employer and provide information about work conditions, organizational climate, and other organizational characteristics that may be of interest to potential employees. Unlike information created and presented by an employer itself, the reviews on these websites are generated by other users, and therefore less likely the selective self-presentation of an organization trying to make itself sound good to potential applicants. Because the information on employer review sites is perceived as generated by current (or recent) employees who may be *just like us*, the peer-generated review can have a significant impact on our perceptions of the organization and our decision to pursue a job with that employer (Carr, 2019; Edwards et al., 2007).

Similar challenges occur for organizations beyond the context of employment. For example, when you find yourself in a new town, you may use Yelp or TripAdvisor to find a good local restaurant. Often, peer-generated reviews strongly influence what restaurants we try or how we perceive the experience. A good restaurant can be harmed by negative Yelp reviews just as an underwhelming one can be raised up by good Yelp reviews (Luca, 2016). Good online reviews and word of mouth can even allow restaurants and other business to charge more for the same product/service by increasing demand (Taylor & Aday, 2016). As a result of this challenge, organizations often benefit when they monitor and manage what others—including their own organizational members—say about them online. No longer is CMC limited to what an organization says about itself; it is now subject to the evaluations of the crowd.

It is worth noting that as complaints or concerns about an organization become more accessible and public, the articulations of organizational shortcomings also provide opportunities for managers. Organizations may find opportunities in online concerns by helping to diagnose problematic processes or find helpful advice in nontraditional ways. In person, employees may not be as comfortable raising concerns about organizational shortcomings or presenting suggestions to improve organizational processes if the organizational communicative structures do not encourage such feedback. Thin-skinned bosses not open to constructive criticism, leadership's lack of desire to hear critical feedback, and an absence of channels to provide such feedback to where it can actually be implemented often limit the willingness of current organizational members to raise issues affecting the organization. Consequently, organizational criticisms and dissent are often restrained offline, particularly with direct supervisors (Bisel & Arterburn, 2012). Online, without the fear of repercussions, individuals can raise whatever concerns they want about the organization and its members. Such was the case of electronics store Radio Shack, many of whose members (both current and former) created the site RadioShackSucks.com as a forum in which to raise grievances and provide support and outlets for ideas that were not provided in formal organizational systems at Radio Shack (Gossett & Kilker, 2006). When made aware of it, Radio Shack executives sought to close the site, worried about the misinformation and image management issues the discussion board provided. However, executives could have also used the site as an opportunity to learn from its front-line employees about the way managerial policies were implemented (or not) at the store level, cultivate new leaders within the organization from those whose initiative brought forth the site, take ideas provided by workers to improve Radio Shack's failing business plan, or even simply understand the general sentiment of its workforce. Employers can use even more mainstream online tools like GlassDoor and Indeed to gain insights into their organization, from individual reviews that may provide particular points of improvement for the organization to overarching and persistent themes among reviews that may indicate problematic organizational cultures (Das Swain et al., 2020). Ultimately, although organizations—like people—may not like to see negative information about themselves online, user-generated information presents new opportunities for organizations to learn about themselves informally, potentially resolving minor problems and gaining novel insights.

Concluding Organizational CMC

Computer-mediated communication has done a lot to affect the way organizations communicate, both internally and externally, even beyond what the vertical filing cabinet did for organizations. Internally, both "old" and "new" media have changed how coworkers, employers, and employees communicate, offering new channels and new opportunities for strategic communication. Externally, media have opened up new markets, allowed organizations to expand across geography and cultures, and created new opportunities for interactivity with stakeholders. Many organizations now face the paradoxical challenge of having to choose from amongst too many channels, mindfully selecting their communicative media mix to match the needs of the organization. What media does your organization use, and for what purposes? As the number of CMC channels propagate, it is important for organizations to select, use, and manage their media strategically to achieve communicative goals, rather than just to simply adopt the newest and hottest social medium.

Key Terms

technology, 84

organizational communication, 84

chronemics, 89

equalization effect, 93

wikis, 95

enterprise social media, 96

email overload, 99

voice, 101

Review Questions

1. How has the evolution of technologies (especially from analog to digital) altered organizations and the communication patterns within them? How have organizations altered how we see and use technologies?

2. Why is media selection so important in organizational communication? How does the same message change

in meaning based on the channel used to transmit it?

3. Computer-mediated communication can let us do many things differently than we can face-to-face, including being anonymous, not exchanging visual or vocal cues, and selecting how we are represented to others. How can organizations harness these affordances to make

the interactions of their members more effective? As you answer, pay attention to how you consider "effectiveness" and what it may mean (or how it manifests) to the organization.

4. How should an organization go about finding its "voice" online? How does that "voice" change when communicating internally or externally, and what causes that change?

CHAPTER 7

Group Computer-Mediated Communication

LEARNING OBJECTIVES

After reading this chapter, you should be able to ...

- Distinguish group communication from other types.
- Identify social groups and categories and how they can be activated online to guide group processes in CMC.
- Apply a typology of online groups to describe the characteristics of groups you encounter online.
- Discuss how offline factors like physical configuration can influence online group processes.
- Understand several forms of social support and how they can be facilitated online.

Quick: What's today's date? This may seem like a simple enough question with a single correct answer. You look at your calendar and state the date. But the date is somewhat subjective, as depending on where you are in the world, it may be tomorrow or yesterday as you read this paragraph, at least relative to other students in different countries (and perhaps even your own). Time zones and date lines are like that, and they make the answer to this question more subjective than you may initially think. I bring up this issue because a few years ago in one of my classes, we partnered students from our campus with students in a similar class at a foreign university. Groups were made up of two members from Singapore and two members from Illinois, USA. One of the first challenges the undergraduate group members discovered was that when they said, "Let's have our first draft ready at 8 a.m. on Wednesday," they had to specify *whose* Wednesday at 8 a.m. Wednesday morning in Singapore would be Tuesday evening for the group members in the United States, and Wednesday morning in the United States meant the Singaporean members had until after dinner on Wednesday night. To make matters more complicated, the course instructors specified all deadlines in Coordinated Universal Time (i.e., Greenwich Mean or Zulu, based in London, UK).

Coordinating their schedules was the first of many challenges groups had to overcome for the class project. Groups also had to deal with differences in culture, workflows, and the inability to see each other and each other's whiteboards or notebooks. However, groups did gain some advantages by working online. The students quickly recognized their groups worked on their project around the clock: US group members worked on it during *their* daytime, and then before quitting for the day sent the project off to their Singaporean group members with a list of "to dos." The Singaporeans woke up to the project having been moved forward and began to tackle their next tasks. At the end of *their* day, the Singaporeans passed it back to the US students and went to bed themselves. This cycle continued for weeks, with the occasional synchronous discussion to coordinate and strategize, and the groups found the project was constantly being worked on thanks to online word processors and email. As this experience illustrates, groups can take advantage of CMC to span space and time as well as cultures. But CMC can also provide new opportunities (and challenges) for common tasks or goals. In this chapter, we'll explore how the nature of being online may cause or break down group processes, whether those processes are task-oriented or social in nature. We'll also explore how the types and structures of groups influence how groups and their members interact online, both within a group and with members of other groups.

Groups Online

Group communication involves at least semi-frequent interactions among at least two or more people who share a common social identity. This definition can incorporate sports teams, support networks, and political cabals as groups to consider how they communicate, both internally and with other groups. Helpfully, this definition also puts boundaries around what may be considered a group, as riders at a bus stop do not share a common social identity, and thus may not be governed by group processes.

The term *group communication* has often been used interchangeably with *small group communication* (Frey, 1999) suggesting between 3 and 20 individuals in a group, with a focus on decision-making or problem-solving (Gouran, 2016). Group communication can certainly involve more than 20 participants, but it becomes harder for group members to communicate face-to-face once the size of the group is greater than about 20 (Thomas & Fink, 1963), as too many communicators can reduce the flow and directness of communication. Digital tools make it easier for groups to communicate with vastly larger numbers of participants (DeSanctis & Gallupe, 1987), as evidenced by the thousands of users collaborating on sites like Wikipedia, Slashdot, and GitHub (Matei & Britt, 2017). Though people discussing *group communication* often mean "communication among and by small groups of less than twenty individuals," it is important to remember that CMC has made group interactions scalable and manageable through threaded chats, subgroups, and other technological features (Baltes et al., 2002; Lowry et al., 2006), and as such "group communication" via CMC is not inevitably "small."

There are many places and means of CMC in which group communication and its corresponding processes are almost self-evident. Chat forums, Google Groups, and Discord servers may host social support groups, bringing together groups of individuals who share a common ailment, concern, or need. Slack or Microsoft Teams may be channels for members of workgroups who are collaborating on a shared task to communicate. Mommy blogs may bring together mothers (and other caregivers) to share parenting tips and techniques. And fan pages and social media groups may serve to connect those who share a common interest in a movie, musician, or athlete (see Figure 7.1). Because such groups can allow users to focus on the common or shared characteristics of group members, rather than the distinct and individuating elements of each individual member, CMC can help individuals interact based on their shared social category.

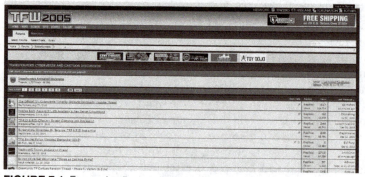

FIGURE 7.1 Fan and collector sites provide a place where people with shared interests or social identities can gather online. Transformers World 2005 provides a place for individuals interested in Hasbro's line of robots turning into other things to come together virtually and discuss Transformers-related toy lines, cartoons, movies, and comics.

Social categories are the salient demographics or psychographics shared by a group of individuals that guide a sense of social collectivity. Even offline, social categories often drive communication. When you identify yourself as a member of the Alpha Gamma Delta sorority, a fan of *Stranger Things*, a Millennial, or even the [Your School Here] Class of 2027, your social category is active. And when a CMC tool primes that social category—when it makes the social category relevant and important—those social categories can guide how we see others and interact. In short, when social identities are activated—especially by the CMC tools we use—group communication processes and theories (including SIDE and the contact hypothesis) can be brought to bear to help understand, explain, and predict human communication.

Activating Group (Rather than Individual) Processes

It doesn't take much to activate a group process. In a now-classic experiment, as campers got off of their buses to attend the first week of a day camp, the kids were handed shirts of different colors at random. Although shirts were randomly distributed and counselors made no comments and gave no instructions about the shirts, campers naturally clustered based on their shirt color (Sherif et al., 1961). Yellow shirts sat with yellow shirts at lunch and played with them in team games, all while making fun of kids in blue shirts. And sure enough, the second week, the campers were again randomly assigned shirts, only to group up based on the new shirt color, going so far as to disassociate with their friends from the prior week who now wore a different color. Consistent with the *minimal group paradigm*, in the absence of something more substantive, even something as minor as a shirt color was enough to facilitate clustering of groups.

Similar small and seemingly trivial cues can be enough to activate similar group processes and perceptions online. Email addresses and screen names (e.g., "PackerFan" or "BTSluvr" can quickly be used to make assumptions about someone's social identity (Heisler & Crabill, 2006). Even someone's school affiliation can tell us about their social categories (Tanis & Postmes, 2003). Given how many of these small cues there may be within a single profile or interaction, how do such cues guide group-based communication, particularly when the site or tool itself does not activate those social categories?

A growing challenge to group CMC is identifying when an interaction online is better predicted or explained by group theories and processes, rather than by interpersonal, organizational, or cultural theories or processes. When you interact with a classmate on your course management system, is your interaction interpersonal, so that you consider and interact with your classmate as a unique, distinct individual, based on the relationship with that person? Or is your interaction intergroup, and you understand and communicate with that classmate primarily as just a member of the same class, potentially interchangeable with many other classmates? Think about your profile within a social medium. Most likely, your profile contains numerous elements that help to both (a) create a distinct and individuating profile, and (b) identify several social groups or categories to which you belong (see Figure 7.2). Many CMC tools, from social media to email, can provide us opportunities to construct and present ourselves through a mix of personalizing and social identities. As Carr, Varney, and Blesse (2016) note, some cues that we intend to personalize us (e.g., our favorite band or movie, preferred politician, or astrological sign) can also be used to lump us into some social category or group (e.g., a Little Monster [Lady Gaga fan], Republican, or Capricorn).

If online tools give us so many ways to signal both our individual and our social identities, how then do we know if others are seeing us as the individuals we are or as members of some social group? Do others like me (or not) because I'm interesting or just because I'm a member of the right social categories?

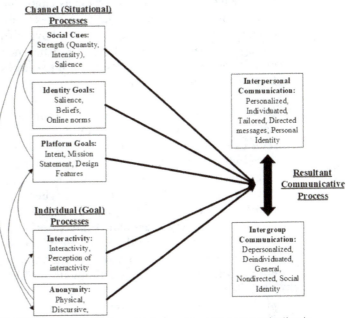

FIGURE 7.2 Many elements in social media may guide either interpersonal or intergroup communication phenomena. What elements of this SNS profile would lead you to interact with Chris Mayburn as a unique individual? Which elements would lead you to interact with this person based on some social group or categorization?

One way to answer this question is to apply the dual factor model of interpersonal-intergroup communication (Figure 7.3; Hinck & Carr, in press). The dual factor model considers the relationship between intergroup and interpersonal communication as a continuum and posits that both human and program factors influence whether CMC will be more or less guided by group communication processes. The model contends both personal factors (i.e., user) and system factors can activate intergroup over interpersonal processes.

Personal Factors Several factors that can drive intergroup communication online are guided by user choice. Anonymity can lead to group (over interpersonal) processes online by deindividuating and depersonalizing users, forcing them to rely on social identities for interactions. When you don't know much about the person behind "BlackPinkRules" online, all you have to guide your interaction is that s/he must be a fan of the Korean pop band BLACKPINK, and you relate with

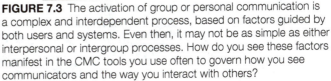

FIGURE 7.3 The activation of group or personal communication is a complex and interdependent process, based on factors guided by both users and systems. Even then, it may not be as simple as either interpersonal or intergroup processes. How do you see these factors manifest in the CMC tools you use often to govern how you see communicators and the way you interact with others?

"BlackPinkRules" based on your musical preferences. Alternately, when communicators are identifiable or known, people can interact with them interpersonally as individuals.

Next, **social cues** signal one's social identity and can guide group rather than interpersonal processes. A profile photo of the user proudly holding a "Vote for Bernie" sign can be a strong cue to signal the individual's social identity as a Bernie Sanders supporter, politically left-leaning, or even just politically active, even overriding the ostensibly individuating nature of the individual's face. When cues to a social identity or category are either numerous or prominent online—more so than cues to the individual's unique sense of self—intergroup processes may be activated.

Finally, **interactivity** refers to the "extent to which messages in a sequence relate to each other, and especially the extent to which later messages recount the relatedness of earlier messages" (Rafaeli & Sudweeks, 1997). Message interactivity is typically pretty high in FtF communication, as communicators naturally listen and respond to verbal communication. However, the disentrainment of online communication can make it so that conversations do not always flow naturally and quickly, and later messages may not always relate back to earlier ones. Take a look at any random text message exchange you've had and see how far you have to scroll back before you realize you're discussing an entirely different topic than was addressed in the last message sent. Or consider the Snapchat streak you've been on, but no longer remember the initial (or even recent) exchanges that led to your present Snaps. When messages are disconnected from later messages or are simply one-way transmissions, communicators do not establish the sense of social presence and conversational coherence necessary for interpersonal communication and relationships, leading them to fall back on social identities and group categories to guide interactions. As interactivity decreases, like in that Snapchat streak, you may be less likely to see or interact with the person as a unique friend. Instead, you may be more likely to interact with them as a more general Snapchatter or school friend, relying on those broader social categories to guide your communication. Because communicators often have great control over their identifiability (disclosing individuating information or not), the quantity and magnitude of identity cues they display and the degree to which messages are tied to earlier messages in an exchange are factors over which individual communicators have direct influence.

System Factors Communication technologies themselves can also influence whether online interactions are guided by group or interpersonal processes. Systems can influence communicative processes by the intent or design of the system itself, which can focus on either personalizing interactions or highlighting the role of groups and their members. Social network sites, from Facebook to Tumblr, have users create detailed personal profiles when logging on and then use these profiles to interact with others. Such sites are designed—and these designs are articulated to users—to encourage personal connections among users. It is therefore a common experience to interact with others on Facebook and Tumblr as individuals rather than group members. A notable exception to this, of course, is when interacting in a group within that SNS. Interacting with members of a band or movie fan group or a "College Libertarians" group on Facebook, social identities (and therefore group processes) likely drive the interaction, because the nature of those groups has made those social categories salient. But in these cases, it is a feature of the subsystem—the group feature—rather than the superordinate system (in this case, Facebook). Alternately, many CMC tools are established with the intent of fostering interactions based on group characteristics. Sites like Meetup, OpenSports, and Debate Politics are driven more by social identities (shared interest in a given social activity, sport, or political interest, respectively) and therefore driven by the relevant social category. Sites focusing more on individual-level interactions, including social network sites and dating sites, are driven more by personal identities, and therefore focus more on interpersonal dynamics. The purpose of a CMC tool can have a large impact on the way communication within that tool emerges.

A second system factor influencing intergroup-interpersonal processes is the identity or interaction goals users have when using a communication channel. Consider an online support group for recovering alcoholics or a fan group for the television series *Star Trek*. In these environments, users' social categories are more important and relevant to the interactions they would expect to have. In the alcoholics' support group, it is not as relevant that you like *The Next Generation* more than the original *Star Trek* or are looking for others to watch new episodes of *Picard*. Instead, it is more relevant that you are an individual struggling and seeking help with recovery. Likewise, your drinking habits or support needs may not be as relevant when talking with other Trekkies about the latest season of *Discovery*. In these scenarios, your social category is the identity that's relevant to the conversation, and should therefore guide more group processes. Alternately, in a group chat with your close friends, your personal identity is typically more salient to the conversation, and your friends interact with you interpersonally as "Caleb" (or whatever *your* name is) rather than as a fan of a certain sports team, reader of a certain genre of books, or member of a particular political party. The types of communication in which we engage and our goals for that interaction—and therefore the places we go online to engage in those interactions—can prime either our social or personal identities to guide our interactions.

Types of Online Groups

There are several ways to categorize groups, which we will get to shortly. Before we begin, though, it may be helpful to think about online groups not in terms of what they discuss, but in terms of how they associate. Beginning by considering groups based on the nature and duration of their association gives us an effective way to think about many of the groups online, from task to social groups, to formal and informal groups, and across multiple media. Though there are several ways to do this, I suggest one way to initially think about the types of groups you may experience online is to consider (a) what brings the group members together, and (b) the permanence or persistence of that group (see Figure 7.4). We will dig deeper into these two axes in the following subsection.

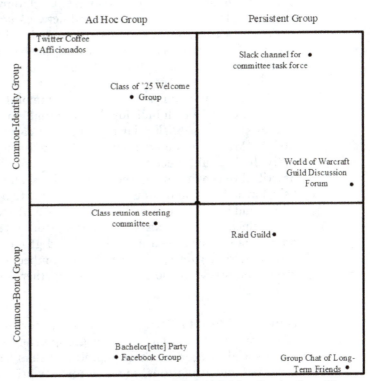

FIGURE 7.4 A typology of online groups based on the duration of the group (ad hoc v. permanent) and the nature of members' connection (common-identity v. common-bond).

What Binds Us: Common-Bond and Common-Identity Groups

One way to think about the nature of a group is to examine what makes it a group in the first place—what bind those individuals together. Think about the connections that bind individuals together, distinguishing groups as either common-bond or common-identity (Prentice, Miller, & Lightdale, 1994), which we previously discussed in Chapter 5. The common-bond/common-identity distinction is a helpful way to consider online groups because it focuses on the nature of what connects members to the group (often communicatively) rather than trying to categorize groups based on the topic or persistence of the group.

Common-Bond Groups Again, **common-bond** groups are those whose members are attached to the group primarily because of the interpersonal connections or relationships (or *bonds*) among group members. In common-bond groups, members are drawn to the group more because of their personal attraction to the group's members rather than an innate attraction to the group holistically. One example of a common-bond group is your friends: you and your friends likely like each other not just because you are all part of a Tuesday night softball league or play laser tag on Saturdays, but because you simply like each other and being together, almost regardless of what you're actually doing together. Online, common-bond groups likewise are held together by personal attachments among group members rather than innate qualities of the site itself.

Many digital communication tools, from text messaging to social media like Instagram and Cyworld, are heavily used to facilitate common-bond groups. These channels provide means by which groups can stay associated and communicate: friend groups can use emails, chat systems, file shares, and group texts to communicate about shared interests, maintain relational ties, and generally interact. In fact, exploring the composition and network structures of many of our social media ties will quickly reveal that your individual connections actually represent clusters of disparate social groups in your life: high school friends with whom you remain connected, college friends, friends from work or summer camp, and so on (see Figure 7.5).

Another way online tools facilitate common-bond groups is through providing channels by which individuals can initially meet (sometimes around shared interests) and establish relationships. The early text-based virtual worlds of MOOs (Multi-user domain, Object-Oriented) simply provided a space wherein individuals could meet and interact. In these virtual worlds, in which users described the environment, their interactions, and themselves, users often established meaningful and enduring relationships that tied MOO-ers together, both within and beyond the virtual world (Parks & Roberts, 1998). Within the gaming environment, many guilds—formal groups of players who come together to take on end-game content—are common-bond groups, as the guilds play in groups across multiple games, with the association between guild members keeping the group cohesive more so than coming together for a particular game (Nardi & Harris, 2006).

Common Identity Groups Alternately, **common-identity** groups are those whose members are attached to the group primarily because of the shared interest or social category of the group. Rather than being held together by interpersonal ties and affiliations, common-identity groups are formed and maintained by a shared interest in a topic, goal, or identification. These groups are similar to clubs or conferences offline, bringing together individuals based on a shared

Facebook Network

Mapping a single person's social graph on Facebook and analysing the underlying clusters

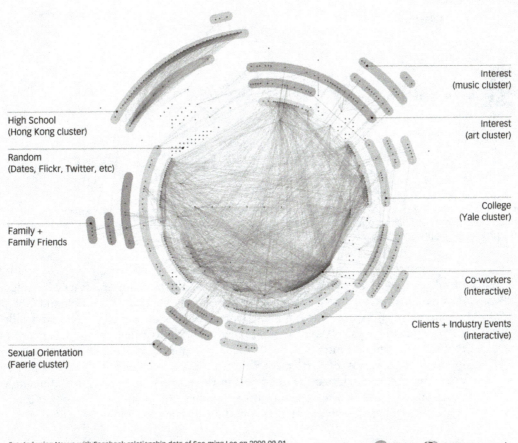

High School
(Hong Kong cluster)

Random
(Dates, Flickr, Twitter, etc)

Family +
Family Friends

Sexual Orientation
(Faerie cluster)

Interest
(music cluster)

Interest
(art cluster)

College
(Yale cluster)

Co-workers
(interactive)

Clients + Industry Events
(interactive)

Created using Nexus with Facebook relationship data of See-ming Lee on 2009-09-01
Nexus is a Facebook network visualizer and can be found at http://nexus.ludios.net/

⊙ SML   SML Network

FIGURE 7.5 If you map out the individual connections you have on a social medium, you will likely find those ties actually comprise several groups—individuals who cluster together. Those groups are often common-bond groups. See-Ming Lee, 2009. Flickr

interest (e.g., fishing club, San Diego Comic-Con) rather than based on personal associations. Online, sites focusing on connection to the broad group or site itself (rather than deep connections between users) represent common-identity groups.

If you find yourself on a chat forum for gearheads to diagnose and resolve car problems, bookmarking the image-sharing site Imgur, or taking part in an online course, you likely find yourself in a common-identity group (Nicholls & Rice, 2017). In such cases, you likely join or engage in the online site for a particular purpose (e.g., fix a car, find a funny cat photo, learn about a concept) without developing close personal bonds with other site users, and may even leave once your purpose is fulfilled (e.g., car fixed, cat lolled, class completed). Perhaps one of the most obvious places online to observe common-identity groups is Facebook or Google groups (see Figure 7.6), which provide fora for individuals to associate based on a

FIGURE 7.6 Facebook groups are often common-identity groups, as group members associate around a common interest (e.g., coffee, alma mater, travel) or identity (e.g., fandom, political affiliation, graduating class of 2025) rather than the relationships with other group members.

shared interest (e.g., travel, motivational quotations) or affiliation (e.g., association members, fans of the same book or movie series), but without necessarily seeking to develop interpersonal connections among group members. In Facebook and Google groups, as well as in groups in chat board, websites, Discord servers, and video games, grouping systems are often designed to bring together individuals based on that shared, common interest, and the focus of communication is around that unifying topic (Ren, Kraut, & Kiesler, 2007). Off-topic conversations—those about issues other than that which may be identified in the group's title, mission statement, or description—are often relegated to other channels or away from the main discussion, so as to keep the group and its communication centered on the unifying issue.

Categorizing Groups as Common-Bond or Common-Identity It is probably incorrect to declaratively say an online group is entirely a common-identity or common-bond group, as different members may have different experiences. While we may generalize and broadly categorize a "New York Yankees" discussion forum as a common-identity group, surely some members joined or remained not because of their collective love for Major League Baseball's Bronx Bombers, but rather because they have developed legitimate, meaningful interpersonal associations that keep them tied to the group. Similarly, though Imgur is likely a common-identity site as its members are unified by the collective experience of image-sharing, heavy users of the site can begin to develop relationships and strong interpersonal networks within the community, and may subsequently see and treat Imgur—or at least a subset of superusers—as a common-bond group (Mikal et al., 2016). And tools like Meetup. com or a running club Facebook group may initially bring together individuals based on a shared interest like fly fishing, running (i.e., a common-identity group), but after several meetings, those individuals form close personal ties that extend beyond that singular group and represent a common-bond group. Ultimately, while we can sometimes quickly label a group as either common-bond or common-identity to help us understand the nature of the group, its communication, and its members, it is also important to consider how the members of the group may see themselves and their relation to the group.

Temporally Bounded Groups: Ad Hoc and Persistent Groups

Another way to consider groups is by the formality and duration of their association. Just as they are offline, groups online can be a simple coming together of random people with a similar goal (e.g., morning commuters) or long-term associations of the same individuals working toward a shared goal (e.g., a group for a class project). Online, groups can be as secretive and difficult to access as hidden websites and password-protected logins, or groups can be as public and easily accessed as a public chat forum. Consequently, another effective way to consider the nature of the online group is by thinking about what challenges one faces when joining or leaving the group and the permanence of the group itself.

Ad Hoc Groups Online, low barriers to access make groups very permeable—easy to enter or leave. Because joining can be as easy as clicking a "Join" or "Subscribe" button and leaving as simple as clicking "Unsubscribe" or simply not logging back into the group, membership can be very irregular, with individuals frequently joining and leaving the group. The group itself may even come together for a short time and then dissolve as members leave or a catalyzing event (e.g., political rally) passes. These groups can be considered ad hoc groups, established or maintained only when needed or on a temporary basis. Just as a protest rally brings a group together for a single event only to disperse once the rally is over, online groups can come together for a specific purpose (e.g., event planning with friends, watching the Twitch stream of an esport event) or due to a shared social identity (e.g., housing coordination for the incoming class of 2025). After the given task or event is completed, the group quickly disassembles or loses members.

Even more challenging are the groups that do not even realize they are a group at all. Sometimes groups naturally emerge online from the networks and structures of interactions among individuals. In one example, Zappavigna (2014) documented the natural emergence of groups of coffee lovers on Twitter based on their use of common hashtags and their interactions. Analyzing how individuals interacted, Zappavigna identified distinct clusters (i.e., groups) of coffee lovers: aficionados who discussed the nuances of roasts and brews, those who focused on the experience of coffee, those who wanted or perceived they needed coffee as a stimulant, and so on (Figure 7.7). When talking about #coffee on Twitter, these individuals likely did not realize they were having conversations within subgroups of like-minded coffee drinkers. These subgroups represent ad hoc groups, as there are no formal boundaries to membership, yet the members can be distinctly identified: coffee connoisseurs view and treat coffee differently than coffee addicts, and

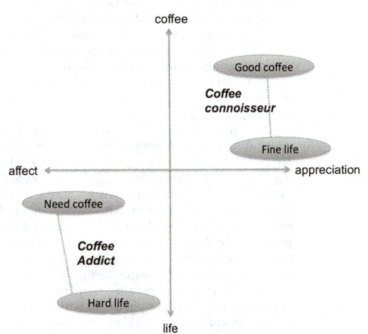

FIGURE 7.7 Connections among ad hoc Twitter coffee discussants based on hashtags and interaction networks.

both likely perceive their caffeinated beverage differently than fans of tea or caffeinated soft drinks. Consequently, the casual discussions or topics discussed online in social media, discussion boards, or in MOOs can comprise ad hoc common-identity groups, as individuals come together around a shared topic or identity.

Similarly, ad hoc common-bond groups can emerge as friends create online groups to facilitate discussions or event planning, only to abandon that channel once the problem is solved. For example, the WeChat messaging system provides groups of friends a platform to form groups for chats, allowing members to invite others to the group as they become desired or relevant to the conversation and to leave the group when they are no longer interested in the topic (Qiu et al., 2016). Even though the chat group persists in WeChat, its membership is a constant ebb and flow of friends, brought together within WeChat as a common-bond group, much as friends may congregate in a coffee shop and enter or leave as their schedules allow. When a group of people form a temporary group simply because of preexisting relational ties, they are engaging in an ad hoc common-bond group, whether that's a text chain to plan an outing or a Snapchat group during fantasy football season.

Persistent Groups At the same time, online tools can facilitate long-term groups, connecting individuals with shared interests or social attraction in enduring and persistent groups. These groups may still be relatively easy to join or leave, but often as groups are more specialized, they are less permeable. Entry to the group needs either an invitation or permission, knowing an existing member, or a password. However, the more distinguishing factor of these groups is that the group persists and its membership is relatively stable. As new members enter the group, they are likely to remain group members for a longer period, and though group members can exit, such exits tend to be infrequent. These groups can be considered persistent groups, as the group is an enduring, ongoing phenomenon. Just as an Alcoholics Anonymous group meets every Monday ad infinitum, online groups can come together for a specific purpose (e.g., social support) or due to a shared social identity (e.g., fan group) without any assumption of the group ending.

Persistent groups can reflect common-identity groups, bringing together individuals based on a common interest or social categorization. Perhaps some of the most common persistent common-identity groups are online fandom communities that bring together individuals around a shared love of a particular cultural staple (Baym, 2000; Booth, 2010; Gray, Sandvoss, & Harrington, 2017; Hinck, 2012), including television programs (e.g., *Buffy the Vampire Slayer*, *Star Trek*), music (e.g., the Beatles, Jimmy Buffett), characters (e.g., Harry Potter, Dr. Who), soap operas (e.g., *General Hospital*, *Grey's Anatomy*), and celebrities (e.g., Beyoncé, Kim Kardashian). In these online groups, individuals sharing the common interest come together to discuss trivia, write fan fiction, trade files, and discuss nuances of the subject. Though relationships between members of these fan sites do emerge (Baym, 2000), most members are drawn to the group not because of the individuals in it, but because of the shared interest. One of the oldest of these common-identity fan groups evolved into the Internet Movie Database (imdb.org): before becoming a repository of all information regarding the silver screen, IMDB was a Usenet newsgroup (rec.arts.movies) whose members discussed movies, actresses, and actors. Similarly, the image hosting site Flickr enables photographers to form communities of practice, reaching out to other shutterbugs to discuss camera equipment, lighting techniques, and other photo skills (Smock, 2012). These groups within Flickr represent persistent common-identity groups, coming together based on their shared identity as photographers. Individuals in the same residential area can use social media like NextDoor to schedule annual neighborhood events like yard sales or holiday parades, coordinate neighborhood watches, and conduct neighborhood business like contracting collective lawn services among neighbors. And listserv email groups can distribute information to selected receivers based on their common association with an organization, job role, or community. Ultimately, many kinds of persistent common-identity groups use CMC to share information and activities, but based on their shared social identity rather than personal connections.

Common-bond groups can also persist online, maintaining associations of individuals based on mutual personal appreciation and ties. Facebook and Instagram are used to keep families and friends connected, and email chains, and text message and Snapchat groups help keep old roommates connected long after they move away. No matter what channel, these groups use CMC to stay connected with specific people, meaning the topics and focal points of communication may ebb and flow over time. Think about a closed Snapchat group that you and your friends have: one week you may be discussing your upcoming midterms, and the next spring break plans. Particularly as individuals are more likely to move away from home before and after college, CMC tools can maintain relationships not just between individuals, but among complex social structures (Cabalquinto, 2018; Ellison, Steinfield, & Lampe, 2007) and therefore keep groups connected.

Physical and Geographic Configuration of Online Groups

Another concern that influences how online groups function is how groups are configured—both communicatively and physically, as CMC is altering both types of configurations. It may seem odd to discuss where individuals are physically located, as previous sections have already discussed how CMC has allowed groups and organizations to be more geographically distributed. However, when you're using a computer to communicate, you're still somewhere, and where individuals, groups, and organizations are physically located has some bearing on the processes that emerge in CMC. Consequently, this section will focus on how a group's configuration affects its communication. Just because you can distribute your group, splitting its members across multiple offices or worksites, should you? And what form should that distribution take?

Tapscott and Williams (2008) note that 80% of businesses now utilize some form of distributed groups, relying on technology for organizational members collaborating in groups toward a shared goal, whether it's preparing that quarter's budget or developing an airplane's navigation system. No longer constrained by geographic proximity, groups can structure themselves in many ways, sometimes having subsets of members near each other and other times being completely distributed. However, how individuals are distributed has implications for group communication, performance, and perceptions. Walther (1997) looked at various configurations of group members working in international workgroups, and found that not all groups are created equal. Some are configured to create ingroup effects often unanticipated by the organization trying to bring together diverse individuals and skill sets.

The most productive groups were those that were fully distributed—no two group members were in the same geographic location (Walther, 1997). In these groups, individuals were forced to collaborate via CMC, and they not only demonstrated the best results on the task but they also had generally high attributions (e.g., task attraction) toward all other group members. Slightly less well performing were task groups in which just a few members were geographically close: no more than

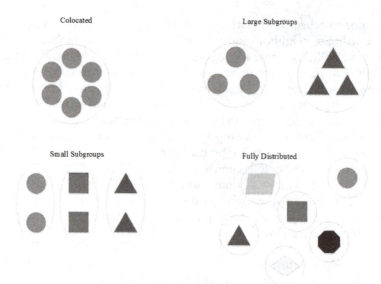

FIGURE 7.8 Physical configuration of various groups.

pairs were colocated in any of the six-member groups, reflecting small subgroups (see Figure 7.8). These groups still had adequate performance and reported generally positive attributions of all group members, though with slightly more positive perceptions of the other person located near them. Surprisingly, the worst-performing groups were those whose six-member groups were split into two large subgroups of three colocated individuals. These task groups had the worst task performance, and evaluations demonstrated that while members thought highly of those with whom they were colocated, members had negative perceptions of group members at other locations. What happened in these groups that led to these effects? The most likely explanation is that these task groups fell victim to intergroup effects within the larger group as a whole.

Unlike completely distributed groups, whose configuration left individuals with no option but to identify with the larger group, groups broken into large subgroups started identifying with their colocated subgroup members over the higher-level task group (Polzer et al., 2006). In other words, members felt part of the "Denver" or "Philadelphia" subgroup more than they felt part of the overarching task group. The challenges of subgroup affiliation are further complicated when organizations are multinational and span multiple cultures (Suzuki, 1998). In the cases where all group members are part of the same organization, but some are located in Tokyo, Japan, and others in Berlin, Germany, the cultural ingroup/outgroup distinctions are even more pronounced and even stronger guides of intragroup communication.

The perception of ingroups and outgroups within an online group or team leads to the SIDE processes discussed in Chapter 5: members affiliate with ingroup members—other individuals located near them or sharing a culture—while disaffiliating from outgroup members—geographically or culturally distant subgroup members. Instead of the members coming together as a single, unified group, intergroup processes were primed and engaged, leading to discord in the task group and to social identity-based perceptions rather than individuals viewing themselves as individuals or part of the superordinate task group. This explanation plays out in the small subgroup conditions. As it may be difficult to divine a social identity or perceive a subgroup based on just yourself and another person, the geographically separated dyads still had *someone* with whom to identify (i.e., "I'm one of the two members from Denver," or "I'm one of the accountants"), but were still forced to primarily identify with the superordinate group, resulting in more equalized perceptions of groups and increased focus on task achievement without the distraction of intergroup perceptions and interactions (Wise, Hamman, & Thorson, 2006).

Task Groups

Sometimes, certain tasks or goals are too big to be achieved by a single person. In those cases, groups of individuals coordinate their knowledge and activities in order to collaboratively accomplish that task. Though tasks groups can be formed by community members, student groups, and groupings of otherwise unassociated individuals, task groups are often formed as subgroups of larger organizational structures. These task groups then can either seek to accomplish (or undermine) organizational goals or fulfill (or frustrate) individual needs not otherwise met by the organization (Gladstein, 1984).

Online task groups are similarly formed to accomplish a specific, often clearly defined task. Members collaborate to accomplish something any given individual may not have the resources or skill set to accomplish alone. In these task groups, which can be either formal or informal, communication serves as a means of structuring and coordinating the group's activities. Either for decision-making or task execution, CMC is useful to facilitate the interaction and coordination that groups may need, without having to colocate or find a unified meeting time. To exemplify the many types and influences of task groups, let's focus on two places task groups are common: organizations and video games.

Task Groups in Organizational Contexts

Task groups are very common in organizations, taking the form of task forces, project teams, quality circles, subcommittees, and other organizational structures. Importantly, task groups within a larger organization are usually not hierarchically structured (i.e., no group member is necessarily "in charge") as would occur in an organization, differentiating task groups from other organizational structures. These task groups can occur in educational, nonprofit, service, commercial, and production organizations; but all share the common trait of bringing together organizational members to achieve a task identified by or for the superordinate organization. When you take part in group work for a class, you're engaging in a task group nested within the larger organization (your class and/or school).

Focused on accomplishing a specific task rather than necessarily in engaging members socially or interpersonally, online task groups can facilitate representatives of a larger system or organization, particularly when that larger organization is geographically and culturally distributed. As CMC tools could theoretically connect every organizational member, an important question then becomes how diverse a task group needs to be to be functional. One response to this question is to consider the principle of **requisite variety**, which suggests that for a group or system to be functional, the diversity within the system needs to match the diversity of the broader system or organization in which the group operates (Ashby, 1958). In other words, the task group should be representative of the larger organization, in whatever ways "representation" matters. For example, a task group at a university exploring student recruitment and retention may benefit by ensuring the task group is comprised of students from rural, suburban, and urban areas, as well as various racial, religious, or educational backgrounds. This diversity of the current student population would reflect the variety of stakeholders within the university. Additionally, the group may benefit from a representative from university marketing, admissions, legal, and student life departments, all of whose perspectives are necessary in the discussion of getting and retaining more students. As CMC tools can make larger groups more manageable, an upper threshold to the size of group membership may no longer be limited by the number of seats at a conference table, and may simply number however many are needed as representative of the relevant social categories in which the organization operates.

Once in the task group, group members may benefit by being aware that online groups are typically more task- than social-focused. In part, this is because of SIPT's proposition that relationships and communicative norms take longer to develop online than in person. As groups may not be able to devote time and effort to building rich, interpersonal ties via CMC, it is perhaps unsurprising that members of online task groups typically report lower group satisfaction than members of online groups (Baltes et al., 2002). Members typically enjoy the group communication process more when offline than when online. The good news, though, is that task groups are often not formed with the outcome of "I was happy being in this group" in mind. Instead, the effectiveness of the group in accomplishing its goal is often the outcome that matters to group members. In that respect, online groups can be just as effective as their offline counterparts, provided they are given sufficient time to exchange messages and information (Baltes et al., 2002), as it takes longer to

Research in Brief

Come Together, Right Now

When groups work together face-to-face, demographic differences among group members—including age, nationality, gender, and work area—can decrease group members' trust in each other and the group, making them less effective and more antagonistic. Though computer-mediated groups can require additional time to write and read messages as their FtF counterparts, being online can also give members the chance to work more effectively by focusing on the collaborative work rather than individual group members. Krebs, Hobman, and Bordia (2006) decided to test whether being online (rather than offline) can therefore increase trust in groups made of dissimilar members.

Testing the effects of being online on trust within workgroups, the researchers had 167 psychology students take part in a creative task. Though drawn from the same Australian university, participants represented five countries of origin, 25 plans of study, and a relatively wide age range. Consisting of three or four members, the groups came into a research lab for three consecutive days to meet either face-to-face (25 groups met in a lab room for 20 minutes each day) or via CMC (25 groups met over a synchronous text chatroom for 60 minutes each day) to develop a radio script for the university. After completing the creative task on the third day, all participants completed measures assessing trust in their group and group members. Findings revealed that all participants' trust with group members increased over the three-day task, in both the FtF and CMC conditions. As expected, age differences increased distrust of group members, but only in the FtF conditions. In the CMC conditions, where participants couldn't visually see other group members, age differences didn't affect the development of trust. Unexpectedly, birthplace dissimilarity—diversity of group members' country of origin—*did* impact trust in the CMC conditions.

What, then, can CMC do to help group communication? Krebs, Hobman, and Bordia's study suggests that CMC can help suppress or hide *some*—but not all—of group members' social cues. Age differences, which impacted group trust in FtF groups, didn't affect online groups. However, nationality did, which Krebs and colleagues were a bit puzzled by. Ultimately, this study tells us that although CMC tools are not a fix-all, CMC tools may help groups overcome certain biases or challenges caused by group composition more so than their offline counterparts.

Reference

Krebs, S. A., Hobman, E. V., & Bordia, P. (2006). Virtual teams and group member dissimilarity: Consequences for the development of trust. *Small Group Research, 37*(6), 721–741. https://doi.org/10.1177/1046496406294886

FIGURE 7.9 Guilds, such as this one in *World of Warcraft*, are groups of players in online multiplayer games who coordinate their individual behaviors to complete difficult game content.

type and read than it does to speak and hear. Particularly as more individuals have experience using CMC, online task groups can be just as effective as their offline counterparts, often at a fraction of the cost.

Task Groups in Video Games

Another common place task groups appear in CMC contexts is within and around online multiplayer games. Games like *World of Warcraft* and *Elder Scrolls*, which require large groups of players (often 10–40 members at a time) to collaborate to complete end-game events, often lead to strong, formal, persistent associations among groups of players, known as **guilds**. These formal associations of players come together to engage in complex, cooperative game play in order to defeat difficult bosses (see Figure 7.9). Communication is used within the game to coordinate play and activities during game play, but it is also used out of the game (and still online, via chat forums and other CMC tools) to schedule play times, coordinate player activities, and generally conduct guild business (Williams et al., 2006).

One notable thing about task groups in online games—particularly guilds—is that because these groups are persistent, their experiences may be substantively different from organizational task groups. Whereas organizational task groups form for a specific purpose and then dissolve when that task is accomplished, task groups in online games have repeating or progressive tasks: once one boss is defeated, there's the next boss to defeat, or that same boss to defeat more efficiently, or a new game to migrate the group to for new challenges. Because guilds are typically more persistent then organizational task groups, guild members often report high levels of satisfaction with their guilds and social attraction to fellow guild members (Carr & Van der Heide, 2009; Williams et al., 2006). Because guild members spend so much time interacting, SIPT processes can engage as the gamers develop communicative norms and mutual understanding among themselves. As the social dynamics and interactions are much more central to the functioning of gaming task groups than to many organizational task groups, group members' social ties and processes reflect that relational development.

Social Support

Online, our groups are not used just to support collaborative efforts toward a particular task. Instead, particularly with common-bond groups, online groups are often means of communicating to achieve social needs. One such need often met in groups—both common-bond and common-identity—is social support. **Social support** refers to communication—both verbal and nonverbal—between recipients and providers that functions to enhance a perception of personal control in one's experience by reducing uncertainty about a situation, the self, the other, or the relationship (Albrecht & Adelman, 1987). Seeking and providing social support is an important part of human communication, as people depend on social support in times of stress and uncertainty (MacGeorge, Feng, & Burleson, 2011). When we end a relationship, start a new job, or face a challenging semester, we seek social support to make that experience more manageable. Social support is often interpersonal: we call our mom, go out for coffee with a friend, or lend a welcoming shoulder to a roommate. But increasingly individuals turn to CMC for social support, in part because online groups allow us to access more individuals or individuals who may be better suited to provide the type of social support we need. Particularly as the perception of social support is often a function of how many people from whom we could seek that support (Eastin & LaRose, 2005), online tools have become useful in seeking out many forms of social support, in large part by giving us access to larger networks.

Types of Social Support

Social support can manifest in many ways, and numerous scholars have tried to categorize the types of social support in equally numerous ways. One of the more well-used typologies of social support is Cutrona and Suhr's (1992) four dimensions of social support: emotional, esteem, informational, and instrumental support. *Emotional support* refers to expressions of interpersonal caring or emotion, including concern, empathy, and sympathy. When we casually refer to "social support," we are most likely referring to the emotional dimension of social support. Being there for a friend who is going through a rough time or sending an email of sympathy following the death of a friend's pet are examples of emotional support. *Esteem support* refers to messages intended to make someone feel better about themself, including promoting the target's skills, abilities, or sense of self-worth. Reminding a friend who failed one exam that they're still passing the class and are still going to graduate, helping a sibling who just experienced a breakup understand it doesn't mean they're doomed to be alone forever and are still worthy of love, and even self-affirmations that you're good enough and smart enough are all forms of esteem support. *Informational support* refers to the provision of facts or advice. Examples of providing informational support could include your friend telling you about a job opening, reminding you of an upcoming test, or passing along the name of someone else looking for a roommate this summer. Finally, *instrumental support* refers to providing material assistance, often in the form of money or physical help. When you lend a friend money or a textbook for class, help move a couch, or drive them to the airport, you are providing instrumental support. As these examples illustrate, these forms of social support can be readily communicated offline. But how do these types of social support occur online—if at all—particularly within the group context?

Communicating Social Support Online

It is perhaps by now unsurprising that early theorizing did not think social support could effectively occur online (Short, Williams, & Christie, 1976). What did the 1970s think CMC was good for?! But it should likewise come as little surprise that we quickly realized support could also be transmitted and received online. Online tools have made social support easier in several ways.

First, online tools allow communication to span geographic and physical distances, significantly reducing the resources needed to seek or offer social support. Rather than driving to meet a friend or even finding a time both parties are free for a telephone call, the Internet readily facilitates synchronous and asynchronous social support. We may send a quick Snap message to a friend we know is taking a difficult test that morning, so that the supportive message is received once the test is complete. An additional benefit of this ability to provide and receive from a distance is that individuals can take advantage of the hyperpersonal processes mediated social support can foster. Being forced to read, consider, and compose a response can lead to individuals evaluating social support more positively as they idealize their support partner and the support provided online (Rains et al., 2019). These opportunities may allow individuals to seek and provide social support in more accessible or impactful ways than they may be able to do face-to-face.

Second, the Internet can facilitate communities of individuals bound by a common identity or support need, enabling support for specific concerns that even those closest to us may not be able to empathize with. For example, Wright and Bell (2003) noted that those with rare medical conditions may find it difficult to talk with others afflicted with the same condition as the nearest patient may be hundreds of miles away. However, support fora and discussion sites online can bring together individuals with similar concerns and needs, providing a sympathetic and empathic ear. Consequently, online social support may be a boon for those with unique support needs and/or those who may not find adequate support from immediate friends and family.

A third and related benefit of online social support can actually be drawn from the CFO approach, as individuals take advantage of the cue-lean environment of the Internet. Individuals interacting online may do so behind a shield of pseudonymity or depersonalization, not having to communicate with others in a way that forces self-disclosures or potential stigmatization. Several studies have looked at depersonalized support communities and the beneficial social support they offer to community members. For example, individuals who have been badly burned or scarred may feel very self-conscious due to physical disfigurement, and shy away from public interaction or FtF social support. However, online, these burn victims need not be identified initially based on their physical traits, and need not even fully or accurately disclose their condition; both affordances allow the individual to interact more comfortably with others and to seek and provide social support (Turkle, 1995). Similarly, "pro-Anna" support groups—online support for anorexic individuals seeking to maintain their lifestyle—can provide members a sense of community for a condition the broad public holds as negative (Tong et al., 2013). Interacting on pro-Anna sites allows individuals to do so without fear of direct reprisal or social sanctions.

For all of these reasons, we see that online tools can be effective for those seeking and providing social support. Such support may be via interpersonal communication, occurring between friends and family, and we'll explore this more in Chapter 10. But social support is particularly germane to group communication. Far from the absence of social support expected in the early days of CMC research, we now recognize online channels can facilitate all forms of social support.

Emotional Support Online Perhaps the most commonly thought of when mentioning "social support," socioemotionally rich messages of caring—emotional support—can be communicated in many ways online. Emotional support may be as simple as an email saying hello or commiserating over a recent sports team loss, or as complex as the multitude of responses and interactions one gets when posting to a support forum. Early on in studies of online groups occurring in text-based environments, researchers identified the exchange of emotional support through members' reciprocal messages (Parks & Floyd, 1996). Off-topic conversations in sites such as The WELL showed that users within these online groups genuinely cared for each other, seeking and providing messages of encouragement and interest in each other's personal lives beyond the online venue. Though emotional support would be expected in common-bond groups, even in common-identity groups, emotional support can be conveyed. It is therefore perhaps unsurprising to us now that emotional support is one of the most commonly offered kinds of support online, accounting for around 40% of the support sought and provided in online groups (Braithwaite, Waldron, & Finn, 1999).

In many social media, comments and replies are means of providing social support. Many individuals may post comments to social media looking for social support, like "I really struggled on that test today." Replies from others can convey compassion, understanding, and caring from various groups within one's social network (Rozzell et al., 2014). Responses like "It's just one test. You'll be fine," "Don't worry—you got this," and, "Yeah, that sucked. Want to grab coffee later to commiserate?" can all provide the initial poster a sense of concern and empathy from those who posted, and these responses all represent a form of social support.

Social support can be provided even via simpler and less verbose cues than comments and written text. One-click cues including Likes, upvotes, and favorites can also be perceived as emotionally supportive, sometimes even in unexpected ways. One of the most common of these one-click tools—the Facebook Like—can be used to communicate social support (Hayes, Carr, & Wohn, 2016). For example, an individual posting "Did great in my internship today" can interpret Likes received as individuals' messages of emotional support, agreeing with and literally "liking" the content and nature of that post. But these one-click cues can also be emotionally supportive in ironic and surprising ways. For example, an individual receiving many Likes to the post "I got fired today" usually does not interpret those Likes as celebrating their unintended unemployment. Instead, the poster likely interprets those Likes as messages of condolences and empathy. To that end, whereas comments from group members may be interpreted very literally and objectively as forms of emotional support, individuals may also get the same emotional support in different and more subjective ways via leaner cues, which are less defined and more open to interpretation.

An important consideration here is that all forms of social support—but particularly emotional support—may take additional time and communication to convey online relative to offline. Recall that social information processing theory (Walther, 1992) proposes it takes time and experience to adapt to online communication as interactants develop unique codes and means of conveying messages to account for the lack of other (e.g., nonverbal) cues constrained by CMC channels. Even in groups, emotional support often comes from the individual ties to group members. And because emotional support relies on interpersonal connections, the support giver and support receiver may find more successful exchanges of support as they experience more online communication. Similar reactions may convey different levels of social support based on the nature of the relationship with the support provider (Carr et al., 2016; Rozzell et al., 2014). However, the extra time to form relationships with group members to foster this support can be a worthwhile investment. Often, individuals turn to online groups for support when they feel their offline networks are unable to provide sufficient support. In the cases where individuals feel the support they receive from nearby individuals is insufficient, they often turn to online connections that may be able to provide deeper and more meaningful emotional support regardless of their geographic location (Turner, Grube, & Meyers, 2001). In your first semester at college, you may not have felt anyone in your dorm could really provide the support you need, and so some emails or texts to high school friends who went to other schools let you access emotional support that better met your needs and made you feel like it was all going to be okay.

Esteem Support Online Sometimes, individuals simply seek to feel valued and worthy: *esteem support*. Esteem support can be communicated in many similar ways to emotional support. Direct messages can validate someone's identity or make them feel as though they belong to the group. Imagine receiving an email that said, "I like you for who you are. You are a great person." Such a message would make you feel supported and affirmed. Similarly, self-statements online can receive feedback in the form of one-click cues to provide esteem support. When your "Feeling cute, but may delete later" photos receive many Likes, you may feel affirmed or valued as the Likes provide an esteem boost. But esteem support can manifest (or not) in other ways as well online.

Esteem support is often sought online in times of transition. As individuals' networks and environments change and they move between groups, they seek out messages of validation and assurance of their value from both existing and new networks, and online tools are well situated to reach out to both old and new groups (Mikal et al., 2013). Computer-mediated communication tools in which others can commend your abilities and successes, such as LinkedIn's endorsements feature (see Figure 7.10), can validate your self-perceptions and confidence in your own skills and abilities. These tools can span multiple networks (e.g., accessing both high school friends and college roommates on Snapchat) as well as focus in on specific groups tailored to particular domains (e.g., an early-career professional support group Reddit forum), and thus target those who need or may be able to provide the kind of esteem support you need at a given time.

Particularly with esteem support online, it may not matter what kind of support or how much support was received, but it may matter very much from whom support was received. Exploring how individuals interpret likes received on Facebook and Instagram, Hayes and colleagues (2018) found that many posts made in social media—particularly those seeking esteem support—are evaluated based on whether the "right" people responded. Those "right" people are often defined as the relevant and salient group members by whom the individual is seeking to be evaluated or legitimized. For example, you may get dressed for work first thing in the morning and then take and post a selfie of your

FIGURE 7.10 If you are ever unsure of or doubting your professional skills, you may turn to your LinkedIn profile as members there can endorse your job skills, providing esteem support and reminding you of your abilities.

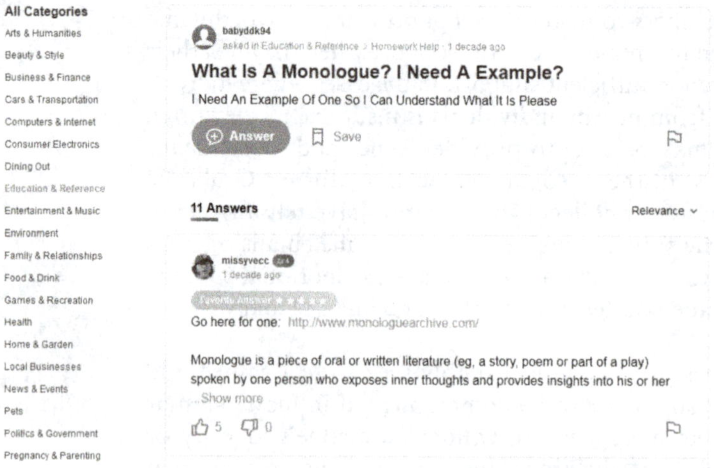

FIGURE 7.11 Individuals can use online groups to seek, obtain, and provide informational support, helping to answer questions and provide advice to group members. In this example from Yahoo! Answers, someone asked and got informational support to learn about monologues.

day's outfit to Instagram. Having made this post, you proceed to eat breakfast, getting your work bag together, and otherwise prepare to leave. Just before you walk out the door, you check on how your post is doing. You look through the feedback you received, including both Likes and comments. Rather than looking strictly at the amount of feedback received, you're looking for approval from the "right" people: specific officemates, coworkers, or friends to whom you look up to. If the right people haven't responded to your outfit by either Liking it or commenting, you assume your attire does not meet the standards or approval of the day's needs, and head back in for a quick clothing change to something you hope that target group will perceive more favorably. Affirming, positive esteem support may not be supportive at all unless it comes from members of the right group whose opinions and messages regarding your character you value. And ultimately, at least in broad, common-identity communities, esteem support may be offered the least frequently relative to the provision of other types of support (Coulson, Buchanan, & Aubeeluck, 2007).

Informational Support Online Providing facts or advice—*informational support*—is another common use of online groups. Some online tools allow groups of individuals connected offline a space to exchange information they may find valuable, such as through the use of RateMyProfessor to provide information and advice about which faculty member offers the best section of a popular course (Edwards et al., 2007). Other sites—including Yahoo! Answers, GlassDoor, and Yelp—crowdsource informational support by allowing users to ask questions that can then be answered by the community of users (see Figure 7.11). When you are in a new area and not sure what's a good restaurant, Yelp members can provide informational support by guiding you to the good Ethiopian place while avoiding the bad shawarma grill. Even if Yelp reviews are not personally directed information, Yelp stars and reviews are still forms of informational support providing data and advice about where to eat from other Yelpers.

Online tools are not always developed to facilitate the exchange of informational support. Yet users can use those same tools to seek and provide informational support anyway. For example, job seeking can be considered a form of informational support, as individuals seek information about new job openings in their field, advice on the organization, or whether to apply. Most often, information about a new job opportunity comes from what Granovetter (1973) calls *weak ties*: individuals with whom we are associated but not tightly coupled in our networks, including casual acquaintances, passing friends, and coworkers. Because weak ties are more likely to be members of different social and professional groups than us, these weak ties are particularly well situated to provide us new information, such as a job opening that hasn't yet been posted. Indeed, even as jobs are posted and searchable online, individuals still report heavy uses of SNSs and both personal and professional connections to learn about and pursue new jobs (Piercy & Lee, 2019). Using LinkedIn, Facebook, Twitter, and other SNSs to let the various social and professional groups of which you are a member know that you are seeking a new job is an effective way to gain new information about recent job postings, provide information to others about your own job skills and experience, and span various networks (i.e., social groups) to do so. Among the many CMC tools, social media are uniquely situated to help us ask questions and find supporting answers or advice, because they can span and tap into so many of our various social and professional groups (Gray et al., 2013; Vitak & Ellison, 2013).

One category of online groups where informational support is some of the most sought is medical support groups. Populated by individuals overcoming personal trauma, battling cancer, or suffering from irritable bowel syndrome, these common-identity groups bring together individuals with shared medical challenges. Interestingly, these groups are often highly suspect of outgroup members—those who do not share the common ailment—viewing the online group as "their" place. For example, in a study of an online cancer support group, Wright (2002) found that cancer patients often preferred to communicate with other cancer patients rather than medical experts (e.g., oncologists, doctors, nurses) or caregivers (e.g., family members, friends, and supporters of other cancer patients), sometimes to the extent of ostracizing doctors or telling people not diagnosed with cancer to leave the group. This shunning of outgroup members was an extreme case, in which the cancer patients using the online group felt stigmatized and treated differently offline, as people treated users as the cancer rather than themselves and offered empty platitudes (e.g., "Imagine what you're going through") for which they had no reference or way to judge. In contrast, members of the online support group were members of the ingroup—those with cancer—and therefore could empathize and sympathize with individuals seeking support, and were more likely to view members as individuals rather than their diagnoses. Because of this use of the support group for access to ingroup members, Wright's finding that informational support was highly exchanged is perhaps unsurprising: users sought functional advice and data on medical information and diagnoses, how to deal with doctors, family members, and their workplaces, and generally exchanged information and advice—rather than sympathy or validation—as they managed their cancer. Similar work into other health support groups found that among the other types of support exchanged, informational support was the primary function of the group as users sought to help each other navigate health care systems and manage their illness (Coulson, 2005).

Finally, CMC can be used not to access informational support directly, but rather to identify who may be able to provide informational support. Some online tools—particularly those developed for use in organizations and groups—are now being developed to make tapping into information support easier, by making it possible to see *who* knows *what* information. Rather than seeking particular information across an entire group, some CMC tools (like enterprise social media; see Chapter 6) provide detailed and searchable user profile elements to help identify those who may hold onto specific knowledge or experience, or even to simply see information or social networks to more efficiently access information within the group structure (Leonardi, 2015). For example, imagine you wanted to find a used copy of a textbook for a class you're taking next year. You could wander aimlessly through your dorm asking if various residents had taken the class and would lend or sell their old textbook, but wouldn't it be more efficient if there was a dorm-wide system or group that archived and let you search based on what courses residents had previously taken? Being able to quickly search or send a message just to those who had previously taken the class may let you more efficiently find someone who may have the book and to get advice on how to be successful in the course. (Note here that *physically getting* the book may be instrumental support, which we'll discuss in a moment. Knowing *who has* the book is what exemplifies informational support.) By helping us identify who has or is positioned to give helpful information and advice, CMC can not only provide informational support directly, but it can also be a tool to identify who may be positioned to provide informational support.

Instrumental Support Online The last kind of support—*instrumental support*—was the dimension that was, perhaps for the longest time, the most difficult to tap into via CMC (Williams, 2006). Though other forms of support can be communicated verbally, instrumental support—sometimes referred to as *tangible support*—often requires the exchange of physical resources, which can be hard to do online. When online, it can be hard to help a friend move a couch or lend someone a cup of sugar. Increasingly, though, online groups are finding ways to provide instrumental support via online channels. One example of providing tangible support is the community members who have food delivered to other community members who are ill or unable to afford a meal, including Imgur's "FreePizzaDude" (https://imgur.com/gallery/7w2c4I8). Another example of instrumental support being provided in groups is the use of CMC tools (ranging from Facebook groups to more niche sites like StudyBlue and Koofers) to provide class notes, flashcards, and even tutoring to other students who may or may not be in your course or university. Though I cannot help you move a couch via TCP/IP, I can hire a TaskRabbit to help you move the couch or send you study resources.

A particularly popular CMC tool for the provision of instrumental support in recent years has been **crowdfunding**: seeking small financial contributions from a larger number of individuals to provide financial support. Popular crowdfunding sites including GoFundMe, Seedrs, and Kickstarter are used to fund new business ventures or inventions, pay medical expenses or help with living expenses, donate to philanthropies, conduct research, or simply obtain loans to be later repaid to campaign supporters (see Figure 7.12). Crowdfunders in North America donated

FIGURE 7.12 Billions of dollars are transferred between individuals each year as they provide food, education, medical expenses, and support for personal and business ventures via crowdfunding sources. https://fundly.com

US$17.2 billion in 2019 alone (Shepherd, 2019), demonstrating tangible support is certainly now occurring online.

Among the many crowdfunding topics, medical expenses is the most common and most donated to, at least in the United States. One-third of GoFundMe's campaigns are seeking tangible support to offset the costs of receiving healthcare (i.e., procedures, medicine, insurance) or supporting healthcare caregivers (i.e., travel and lodging expenses for those taking someone to a clinic, offsetting lost wages while a family member undergoes

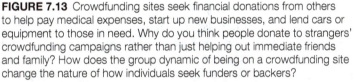

Navajo and Hopi Community Relief Fund

$219,035 raised of $500,000 goal

FIGURE 7.13 Crowdfunding sites seek financial donations from others to help pay medical expenses, start up new businesses, and lend cars or equipment to those in need. Why do you think people donate to strangers' crowdfunding campaigns rather than just helping out immediate friends and family? How does the group dynamic of being on a crowdfunding site change the nature of how individuals seek funders or backers?

treatment). Representing both the largest proportions of campaigns and donations on the crowdfunding site (Zdechlik, 2018), the necessity of seeking tangible support online is certainly a commentary on the state of healthcare in the United States. But the popularity of medical crowdfunding sites is also attributed to the groups that occur online, both within and beyond the crowdfunding site itself. Rather than simply soliciting a donation from immediate friends and family members (i.e., via interpersonal communication), crowdfunding allows messages to be circulated and donations sought in multiple online groups. Your mom shares the campaign with her card circle, your uncle with his car restoration club, your sister with her military unit, and so on. Taking advantage of easy transmission and permeability of groups online, information can be shared much more quickly online. Additionally, tapping into groups whose interests or identities may be shared or of interest to the crowdfunding topic can help spur donations (Gonzales et al., 2018), such as when a family seeking help paying for a child's cancer treatment is introduced to a parents' Facebook group, whose members—identifying as caregivers of children—may be more likely to donate than others whose "parent" social identity is not active (see Figure 7.13).

Concluding Group CMC

In this chapter, we've explored how being online can help you find a group of others who share the same social category as you, how groups form and function for both specific tasks and general social communication, and the types of support or help groups online can provide. Think of the groups with which you associate online—what do these groups say about you, and how do they influence the types of interactions you have online and the types of interactions you do not have online? Groups are powerful influences on how we interact, either alone or when together, and CMC can often exacerbate our social identities by giving us a place to find others similar to us. Whether for a class project, building a new airplane, or seeking support and assistance in times of trouble, a group communication line gives us new opportunities and challenges to interact with others regardless of where they (and we) are.

Key Terms

group communication, 107
social categories, 108
social cues, 110
interactivity, 110

requisite variety, 119
guilds, 121
social support, 122
crowdfunding, 128

Review Questions

1. CMC tools like Instagram and Tik-Tok focus on self-presentation: you send a message to others. How do group processes influence the use of these tools?

2. A common-bond group and a common-identity group may look very similar at first glance to a new user. What evidence would tell an outside observer whether an online group is one or the other?

3. How do groups of video game players—specifically raiding guilds—represent online task groups? In what ways are guilds like work-groups? How are they different?

4. Which type of social support seems easiest to provide or receive online and why? Which seems the most challenging to provide or receive online?

5. How does crowdfunding rely on the power of groups online to solicit donations and contributions to a cause?

CHAPTER 8

Interpersonal Computer-Mediated Communication

On *Sesame Street*, Aloysius "Snuffy" Snuffleupagus and Big Bird play together and talk together, but until 1985 (14 years after he was introduced) only Big Bird had ever seen Snuffy. On screen, Snuffy and Big Bird are best friends. That relationship is now also reflected on Twitter. @MrSnuffleupagus has thousands of followers, but Snuffy follows only one account: @BigBird. For those who grew up with the long-running PBS program, the online relationship between Snuffy and Big Bird reinforces what we've seen on screen. Those of us not living at 123 Sesame Street still may find our relationships reflect Snuffy and Big Bird's: new media tools are influencing how we construct, form, and maintain our ties with others. Those relationships may be friends, family, classmates, passing acquaintances, and our best friends. Digital tools, both old and new, give us additional channels through which to communicate with those we know, sometimes simply moving our relationships online and other times giving us entirely new ways to relate and communicate.

In this chapter, we explore how CMC influences interpersonal communication. Communicating with others dyadically—whether they be family, friends, or relational partners—is an essential part of the human experience. Just today, you've likely already had multiple interactions. Many of them were face-to-face, but even more (at least in the quantity of people with whom you've interacted) were likely digitally mediated in some form. Many people now sleep next to their smartphones and habitually check them—accessing multiple social media sources—even before getting out of bed in the morning. Social media, messaging, and email notifications likely punctuate your days, and a final check of our digital channels is the last thing before bed for many. Clearly, CMC has become an integral part of how we establish, maintain, and even terminate many kinds of relationships.

Friends and Family

The Internet has long been used to establish and maintain interpersonal relationships. Though we often think of contemporary social media tools like Facebook Local and Meetup as used to meet and interact with others, online tools for communication with others go back much farther. One of the earlier tools to achieve broad adoption was The WELL (an acronym of "The Whole Earth 'Lectronic Link"), which at its founding in 1985 was one of the earliest communities connected entirely online, and it retains an active user base today. The WELL and other chat fora, from message boards to Reddit, have served as places where individuals can come together to discuss whatever issues they find interesting and relevant, from the serious to the satirical and from the titanic to the trivial. As a result, numerous works have documented the creation and maintenance of relationships in these communities (e.g., Parks & Floyd, 1996; Rheingold, 1993). Counter to initial suppositions that CMC limits social cues and the formation of meaningful ties (i.e., the CFO paradigm; Short, Williams, & Christie, 1976), today we see numerous relationships formed and/or maintained online. As these relationships are indeed happening, let's dig a bit into the processes by which they occur and with whom.

Online Interpersonal Communication

An obvious way CMC has facilitated interpersonal communication is through its use for interpersonal interaction. Numerous computer-based tools let us communicate with specific others: email, file transfer, and messaging services all allow us immediate one-to-one means of communication with friends, family, acquaintances, and other associates. Email allows for similar forms of written communication, but the nature of computers and computer networks means emails are received almost immediately upon sending, regardless of the distance between sender and receiver. The same time-and-space compression benefits are applicable to other means of CMC, including file transfers (imagine having to exchange a physical medium like a CD every time you wanted to share a new song or funny video with a friend) and messaging (like calling on the phone, but without the need for both parties to be present at the same time). One of the ways CMC has affected the nature of our interpersonal communication is that CMC has speeded up the transmission and receipt of messages. It still takes time for senders to compose and send messages and for receivers to open, read, and respond to them. However, with CMC, the actual time to send and receive messages anywhere in the world is now functionally nil, as computer networks span distance with almost imperceptible speed.

Computer-mediated communication has also affected interpersonal communication in other, less obvious ways. Most notable has been the changes in how we actually communicate—the deliberateness and effort involved in transmitting a message. Interpersonal communication online could be broadly classified based on the effort involved in composing and transmitting a message. On one end of this typology are the direct interactions—similar to FtF conversations or telephone chats—that require substantive effort to compose to convey specific meaning and achieve certain goals. The other end of this typology would be phatic interactions—similar to waving to someone nearby—that still communicate, but with little effort or specific conversational goals attached to them.

Dyadic Interaction "Computer-mediated communication" is a notoriously broad concept. As we discussed in Chapter 1, as more of our communication occurs through digital tools and as new tools continue to emerge that defy prior categorizations, it is increasingly difficult to say what's "in" and what's "out" of CMC. However, some CMC tools have existed long enough and have been such staples of human communication over the past few decades that they are worth a brief mention. Mentioning these three channels is not meant to be exhaustive, but rather to briefly acknowledge the rise and place of several CMC tools that have been heavily used to facilitate direct interpersonal interaction with a relational partner. This one-to-one communication is known more technically as **dyadic interaction**.

Email One of the most prominent means of direct, purposeful interaction between two individuals is email. Email made its debut in 1980, and took almost a decade to reach its first million users (see Figure 8.1), most of whom were members of colleges/universities or the military. Dubbed the "killer app" that brought personal computing into organizations and subsequently our homes (Tassabehji & Vakola, 2005), email is now one of the most-used means of digital communication. You likely have more than one email account yourself: a personal account, another one issued by your school or employer, and perhaps a few more—even some fake or burner accounts. If so, you are like most email users. The 3.9 billion worldwide users of email have 5.59 billion active accounts (Clement, 2020), for an average of 1.75 email accounts per user. And these accounts are used for a lot of communication: an average of 293 billion emails are sent *every day*—though 67% of those emails are likely spam (99 Firms, 2019). Still a "killer app" of CMC in the business world, email remains the dominant means of digital interaction between users (Kooti et al., 2015), even as other digital tools (from Snapchat to virtual reality) have emerged.

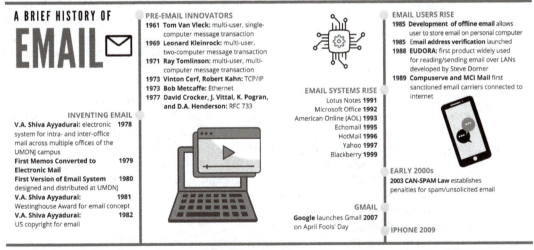

FIGURE 8.1 Email has diffused exponentially since its development in 1980. Now, more than half of the world's population has an account.

Texting/SMSs Another means of dyadic CMC are text messages, originally called the short message service (SMS). Text messaging typically occurs via mobile phones (or other smart devices) between two users, and were originally constrained to 160 characters, including letters, spaces, and punctuation. Texting was developed in the 1980s, but the format was not immediately available on cellular phones due to limitations in the cellular network's infrastructure. The first SMS was ultimately sent in 1992, but was not quickly adopted due to continued infrastructure challenges, concerns over billing, and a difficult interface on most phones (Taylor & Vincent, 2005). Texting was first popularized in Japan in the early 2000s, and from there diffused worldwide (Hornyak, 2019). Now, more than 97% of those with a smartphone use its texting function (Smith, 2015), sending an average of 94 texts each day (Burke, 2018). Texting has become a particularly useful tool for short information exchanges or social interactions between moderate friends and acquaintances—interestingly, for our closest friends, we now often switch back to phone calls or video chats for dyadic interaction (Anderson, 2015). Other tools and services have emerged that function similarly to text messaging, including WhatsApp, Facebook Messenger, and WeChat. These newer messaging tools are tethered to a specific user rather than a telephone number, but they function similarly to SMSs by facilitating short, almost synchronous messages (often text, but increasingly including pictures or videos) between individuals.

Social Network Sites The most recent on this list are **social network sites** (SNSs), which refers to "web-based services that allow individuals to (1) construct a public or semi-public profile within a bounded system, (2) articulate a list of other users with whom they share a connection, and (3) view and traverse their list of connections and those made by others within the system" (Ellison & Boyd, 2013, 152). These three defining characteristics of SNSs distinguish them from social media more broadly: SNSs emphasize individuals by emphasizing the development and exploration of individual profiles, and the interpersonal associations between users—seeing who knows and is connected to whom—is a critical function. An important distinction here is between social *network* sites and social *networking* sites, as network sites typically focus on helping users connect with and maintain connections with those they already know, while networking sites typically focus on creating new connections based on either common interests or attributes (Boyd & Ellison, 2007). Early SNSs include SixDegrees, CyWorld, and Myspace, and popular SNSs including Twitter, Tencent QQ, Pinterest, and YouTube now exemplify these social network sites. SNSs have been rapidly adopted and integrated into our communication. Facebook alone has more than 1.59 billion active daily users (Facebook, 2019), or about one out of every five people on Earth.

Social network sites offer many forms of dyadic interaction, depending on the specific tool. Many allow direct messages that can be sent privately from user to user. Some allow a user to be personally acknowledged in an otherwise public post (i.e., masspersonally; see Chapter 4) to bring the otherwise broad message to the individual's attention (e.g., "Happy birthday, @Willow!"). Others allow private channels for users to communicate synchronously, either via text or audiovisually. Although SNSs may be most visibly known for their more mass or masspersonal interactions, the dyadic communication SNSs facilitate still serves important communicative functions.

Lightweight Interaction *Lightweight interaction* refers to simple and quick messages that do not necessitate significant commitment of time, cognition, or effort to transmit. Offline, such lightweight interactions may be waving to a friend from across the quad, leaving a smiley face on your significant other's door, or even a simple thumbs up to support your roommate during her presentation. Digital tools offer even more ways to provide simple messages or interaction experiences to maintain relationships.

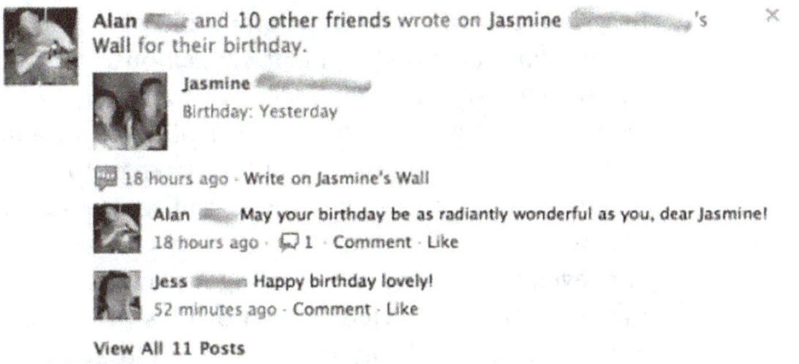

Lightweight interaction is particularly easy online on social network sites like

FIGURE 8.2 A common form of lightweight CMC: "Happy birthday" posts on Facebook.

Instagram, TikTok, and Facebook. Many of these services offer simple and noninteractive means of communicating interpersonally. Two common means of lightweight interaction are posting about someone and commenting on someone's post. Think of all the simple messages you direct toward someone on social media: you may mention someone, tag them in a post, or comment on their page or timeline. But these comments are not meant to elicit lengthy interaction or deep responses. Instead, these quick interactions are simply used to tell the person you're thinking of him or her or to otherwise engage in some relational maintenance. These lightweight interactions are often annually visible as many of a user's Friends/Followers post "Happy birthday" messages on the individual's page (see Figure 8.2). Whereas wishing someone a happy birthday in person may additionally entail a conversation, inquiries into the past or forthcoming year, sharing of sweets, and additional social discourse, by posting "Happy birthday" to the individual's Facebook page, you've acknowledged their special day without necessitating replies or additional communication. Indeed, many recipients simply acknowledge all the well wishes in the aggregate, such as by later posting, "Thanks for the birthday wishes, everyone!"

Lightweight interaction is one of the foremost communicative uses of social network sites, accounting for both a large portion of the messages sent and the value users experience from engaging in the sites (Ellison, Steinfield, & Lampe, 2007). It can be challenging to keep up with too many friends: after about 150 relationships (Dunbar, 1992), it becomes difficult to remember each individual (including birthdays, pets' names, and other individuating details that demonstrate you're maintaining a meaningful relationship), and each additional friend requires more time to engage in relational maintenance activities. At some point, there just aren't enough hours in the week to have coffee or a phone call with so many friends. Tools like SNSs can help us cognitively offload some of that relational information: we rely on a notification to remind us when it's someone's birthday or search someone's profile to remember their favorite restaurant when suggesting dinner later in the month. Additionally, SNSs readily facilitate lightweight communication, which can allow us to maintain those ties without taxing our temporal, cognitive, or social resources. We can quickly go through and wish many people "Happy birthday" on Instagram without having to engage in the more time-consuming discourse required offline. In this way, though SNSs may not necessarily increase the number of strong, close friends we can maintain at a given time (Wellman, 2012), SNSs can make it easier to maintain some form of interpersonal relationship by allowing easy and fast means of communicating with our friends, family, and acquaintances, ultimately increasing the number of weaker relational ties we can simultaneously maintain.

Phatic Interaction Phatic communication refers to messages meant to establish a mood, share acknowledgment, or demonstrate sociability but that do not necessarily convey a specific meaning (Malinkowski, 1972). In FtF interactions, small talk, greetings, and gossip often serve as phatic communication, helping to establish rapport or a sense of familiarity more so than to convey actual information. Consider when you walk through campus and see an acquaintance. That passing exchange, "How are you?" "Fine, you?" is likely phatic, as neither of you really considered or conveyed your true state of being. Instead, that exchange served merely as a social lubricant to acknowledge each other's presence and not walk past each other in awkward silence. From the perspective that CMC filters out most socioemotional cues, phatic interactions would not be expected to occur online. But over time and with experience, users have adapted the channel to find numerous ways to communicate additional meaning, context, and social content beyond the mere words exchanged.

Paralanguage Although visual and other cues can increasingly be transmitted digitally, text remains the dominant channel for CMC. Within our text-based channels—including email, text messages, and instant messaging—some cues are constrained as text alone limits our ability to see, touch, taste (hmmm), and hear others. In doing so, text can constrain our paralinguistics as well. **Paralinguistics** refers to the parts of verbal communication beyond the actual words spoken (Schuller et al., 2013; Trager, 1958). This broad concept can include things like tone and volume of voice, speed of delivery or pauses, and pronunciation and enunciation. If words are *what* you say, paralinguistics is *how* you say it. Offline, we can easily hear and identify the role of paralinguistics. The next time you talk with a friend face-to-face, be mindful of the paralinguistics the two of you use. How do you tell the difference between a brief pause to collect thoughts and continue on, rather than the end of a thought and a signal to reverse speaker-listener roles? Without the person verbalizing the punctuation at the end of a sentence, how can you tell whether a sentence is a statement or a question? (For that matter, as you read these sentences, does the intonation in your head go up at the end of each sentence that's a question?) And paralinguistics makes it so there are clearly two meanings to "Fine" when you ask your significant other how he or she is doing, and which "fine" he or she is depends on the inflection (see Figure 8.4). Offline, we have grown up with these paralinguistics, developing a deep and implicit understanding of how paralinguistics is used to keep conversation flowing, create and interpret meaning, and interact with our language use. Paralinguistics may not be quite as frequent online as offline given that much of CMC is written, which innately strips out a lot of the verbal cues we would otherwise transmit. Remember from Chapter 3 that the early cues filtered out (CFO) paradigm thought online interaction was innately lessened because of the inability to transmit vocalics and other verbal cues that comprise paralinguistics, in addition to a perceived inability to transmit nonverbal cues. More recently, we've seen and developed means of recognizing and even developing paralinguistics to be used via CMC.

One of the most obvious examples of paralinguistics in CMC is the convention of using all capital letters to denote yelling. When someone raises the volume of their voice in person, it is usually easy to hear they are angry, frustrated, or otherwise upset, and they have subsequently spoken louder to convey their words more emphatically. Online, whether in email, a chat box, or a comment replying to a post, when you see something typed IN ALL CAPITAL LETTERS you likely read that part more loudly in your mind, AS IF IT WERE YELLED, rather than assuming the sender's caps lock button was broken. Likewise, ellipses can be interjected into synchronous messages to indicate a purposeful … pause. Unlike in spoken, FtF interaction where the communicator pauses to think of the right word, online you may simply stop typing until you think of the right word. The inclusion of ellipses paralinguistically indicates to the reader a hesitation or hedge in the subsequent clause. The emergence of paralinguistics online has been complex, idiosyncratic, and organic, sometimes a function of what the system allows users to do and other times a function of how users adapt to the system (Carey, 1980). Some of these paralinguistics were adapted from written communication like books and magazine articles; others (including Twitter's hashtag; Burgess & Baym, 2020) emerged naturally in online communities through trial and error, community use, and eventually diffusion beyond the online community (see Figure 8.3). Paralinguistics continues to emerge today as a way of communicating emotion, intensity, purpose, or tone of voice to complement the text entered online.

Social information processing theory helps explain the emergence and use of paralinguistics for interpersonal communication online. As we have more experience interacting with someone via CMC, we start to understand their nuances, particular behaviors, and use of language. As we grow more familiar, we cocreate our own meaning

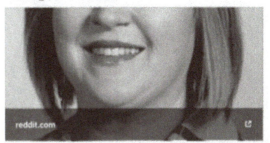

FIGURE 8.3 New paralinguistics sometimes is introduced in online communities (like Reddit or Imgur) and subsequently appropriated for use interpersonally. One example is the textual denotation of sarcasm through the mixed use of upper- and lower-case letters and/or this particular screen capture meme of Spongebob Squarepants.

of language use, including paralinguistics. For example, you and your roommate may begin to use extra letters or typographic symbols alongside text to indicate humor, sarcasm, hesitation, or other emotional states, but this paralinguistics may not be shared and used with a mutual friend. Consider how the same, simple message may have a very different meaning depending on who sent it, and therefore the communicative experience and constructed meaning you have created. Your mom texting you "youre late" could be interpreted as she's hurried or angry, as you know your mother is typically careful and precise with her grammar and would never omit punctuation and capitalization unless she is typing in anger. But your friend texting the same "youre late" may be taken as a simple reminder of the time as your friend tends not to be too concerned with either promptness or punctuation. But these distinctions occur as a function of experience communicating with each other, not from the precise use of paralinguistics as it was identical in each message. In other words, the same phrase can convey different meanings based on how you interpret the unique sender's paralinguistics. Many of the paralinguistics used within our online relationships—even with offline counterparts—may be unique to each particular pairing (see Figure 8.4).

Paralinguistic Digital Affordances (PDAs) One form of phatic communication that is particularly common in social media is the **paralinguistic digital affordance** (PDA), which is a simple (usually one-click or hover-click) cue that is easy to transmit, and

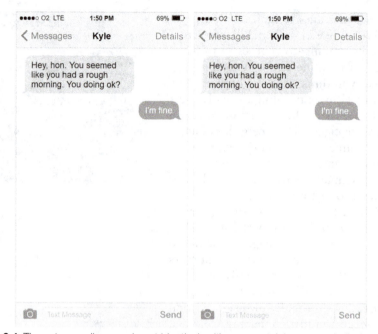

FIGURE 8.4 These two replies are almost identical, with one containing a concluding punctuation mark and the other not. What different interpretations do you have of Kyle's mood between the two responses, based on paralinguistics?

often whose receipt is displayed in the aggregate (Figure 8.5). Some of the most common examples of PDAs include the Like (Facebook and Twitter) and Upvote (Reddit and Imgur). Social media use these PDAs to provide fast, lightweight responses to user-generated content. These PDAs often have verbiage associated with them that can help explain their intended use. For example, Facebook's eponymous "Like" button serves as a PDA that can indicate a user likes or enjoys the content posted. Likewise, Twitter's PDA was originally called the "Favorite," and was used by many to archive or index particularly important tweets (e.g., recipes, life tips, job postings) that could be subsequently searched and retrieved for later use. Whether PDAs are used to convey actual enjoyment, denote content was one of a user's most preferred, or another goal, many users can use them *faithfully*—consistent with their intended design and use—to communicate. You Liking your friend's engagement announcement provides a simple and social acknowledgment of the upcoming nuptials without requiring significant communicative or cognitive effort on your part (as the sender) or their part (as the receiver).

Interestingly PDAs are not always used or interpreted consistent with the verbiage designated by the platform's designers (Hayes, Carr, & Wohn, 2016). This *ironic* use—use not consistent with the associated word or verbiage of the PDA—is perhaps most evident when seeing a negative post on Facebook that receives apparently favorable PDAs. When you post a status, "Grandma died peacefully in her sleep today. She was a good woman, and I'll miss Memaw every day," how do you interpret the

FIGURE 8.5 Liking can be a phatic form of communication showing the poster that you literally "like" the post or merely acknowledge seeing it, and can convey affinity toward the poster, help you index/ save content, or even serve as a public self-display.
Kornburut Woradee / EyeEm

75 Likes that may be provided to that post? You likely do not assume Memaw was a terrible woman and others are happy she is now deceased. Few people likely literally *like* that your grandma is dead. Rather, we often provide and interpret PDAs within the context of the message to which they reply. In the example of a recent family death, Likes are likely interpreted as emotional support from your Facebook network and acknowledgment of your grandma's passing. This is what is meant by *ironic* use:

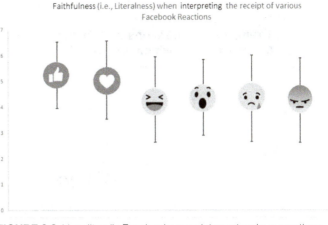

Faithfulness (i.e., Literalness) when interpreting the receipt of various Facebook Reactions

FIGURE 8.6 How literally Facebook users interpret various reactions when received (adapted from Sumner et al., 2020).

individuals often use PDAs phatically to convey sentiment that doesn't exactly match what the term means. After Facebook introduced PDAs beyond the Like (including Love, HaHa, Wow, Sad, and Angry) in 2016, users still did not always interpret a PDA literally, even as they could communicate more emotions and sentiments via these hover-click cues (see Figure 8.6; Sumner et al., 2020).

Emoticons Another way users have adapted CMC to convey nonverbal cues, particularly facial expressions and body gestures, is through the introduction of graphical representations of nonverbal cues. Some of the first surrogates for nonverbal communication used online were **emoticons**: keyboard symbols presented in a way to resemble facial expressions (Rezabek & Cochenour, 1998). Early—and also still some of the more commonly used—examples of emoticons included a smile :), a wink ;), and a frown :((Danet, Ruedenberg-Wright, & Rosenbaum-Tamari, 1997). Over time, users developed emojis that were more complicated both in structure and in the emotions conveyed. A simple :) could be made more anthropomorphic (i.e., more humanlike) by adding a tilde for a nose :~), and one could indicate anger by using additional keyboard characters to denote not just an angry face but someone so angered they flipped over a nearby table (╯ ಠ_ಠ)╯ ︵ ┻━┻. Eventually, emoticons grew beyond simple pictographs of a face to include complex patterns and designs using keyboard characters to depict entire bodies or scenarios (see Table 8.1). These pictures—complex designs composed of standard character sets—are called ASCII art, and they can range from large but simply constructed images to complete and beautiful renderings.

One benefit simple emoticons may offer over more complex symbols or even facial expressions is that emoticons are typically easier to recognize and decode. A study contrasting observer ratings of people's facial expressions (e.g., smiling vs. frowning) and corresponding emoticons (e.g., :) and :() found more consensus between evaluators of emoticons than actual facial expressions (Walther & D'Addario, 2001). Though simple arrangements of keyboard characters may convey less information than the complex glut of information presented by facial expressions and body gestures, it may be that emoticons are subsequently less ambiguous and more objectively understood. However, this clarity of meaning does not seem to be followed by an impact on affect. In other words, though the paralinguistics is clearly interpreted, it does not necessarily influence the receiver's mood as interpreted. In the same study, Walther and D'Addario (2001) found that positively valenced emoticons

TABLE 8.1 **Examples of Emoticons and Textual Nonverbals with Increasingly Complex Composition and Detail.**

Homer Simpson ASCII art horroroso

O_O	(*ﾟ∀ﾟ)ﾉｼ	
Simple facial emoticon displaying excitement	Simple full character emoticon displaying excitement	ASCII art of an excited Homer Simpson from *The Simpsons*®

(i.e., smiles and winks) did not influence how receivers interpreted the corresponding written messages. However, a negative emoticon (i.e., frown) alongside the written message resulted in receivers perceiving the message as more negative than when the message appeared without the emoticon. So although it seems that emoticons may have clearly denoted meaning associated with them, how they influence corresponding written communication is less clear. Were you to type "I like you :)" to a potential romantic partner, though the "smiley" emoticon is clear, whether it's meant to be simply an expression of kindness (i.e., "I like you and you make me happy.") or a more meaningful expression of hope and desire (i.e., "I like you romantically, and hope the feeling is mutual.") may depend more on your relationship and communicative history with the person than the emoticon itself.

Emojis Given communicators' desires for more nuanced and humanlike nonverbals, emojis were developed in 1997 as the next iteration of typed nonverbal communication. **Emojis**, pictographic representations of faces, objects, and other subjects, quickly gained popularity first in Japan and then diffused abroad. Emojis such as ☺ and ☹ allow more graphically intense displays of paralinguistic and nonverbal cues. Rather than simple figures using set keys, emojis are more artistic and nuanced than emoticons. Emojis have begun to replace emoticons (Pavalanathan & Eisenstein, 2016)—so much so that if you now attempt to type an emoticon (e.g., :)) on your word processor, it may automatically change it to the corresponding emoticon (e.g., ☺). Because they are more graphically rich and complex, emojis including faces ☺, flags 🏁, and objects 🔫 have the potential to communicate even more information than emoticons, but their role and influence in communication may not be that different.

Similar to the findings of the influence of emoticons in interpersonal communication (Walther & D'Addario, 2001), research into the communicative effects of emojis (Hu et al., 2017) reveals facial emojis—the direct descendants of and replacements for many emoticons—do not seem to greatly influence the meaning of an accompanying message. Only negative emojis (e.g., ☹) appear to have the intended

impact, causing receivers to perceive negative messages with emojis as more negative than those without. Positive (e.g., ☺) and neutral (e.g., 😐) emojis both make
their corresponding message seem more negative in contrast to the image actually
depicted. This ultimately tells us that emojis may not be a common *lingua franca*,
lacking the same denotative meaning and implications to all users. Even though
emojis are highly (and increasingly) used in CMC (Barbieri et al., 2016), one challenge is that they may be used differently by each user and within specific relationships to communicate different things. As such, emojis are one of the more complex
yet understudied aspects of CMC, and scholars have called for more work to be
done to understand their role(s) in interpersonal communication and online culture
alike (Kaye, Malone, & Wall, 2017).

Summarizing Paralinguistics Ultimately, what all of this means is that paralinguistics
conveys meaning, but interpreting ellipses, emoticons, emojis, Likes, and other forms
of digital paralinguistics depends a lot on contexts and individual relationships, similar to how a smile can be construed as a sign of friendship or smugness. How you use
and interpret nonverbals online with others may be just as much a function of your
prior interactions with each individual friend, partner, or acquaintance as of the innate and denotative meaning of the nonverbals and paralinguistics used. Importantly,
though, whatever form and impact they take, the initial view that nonverbal and paralinguistic cues could not be transmitted online to develop and maintain interpersonal
relationships is clearly not consistent with continued practice. Users have adapted to
the channels and adopted (and created) cues and signals to convey not only information but also emotion, sentiment, and nuance in their written communication.

Extracted Information

One of the most impactful ways CMC tools have changed how we maintain relationships is through being able to passively observe our friends and family without
necessitating direct interaction. Offline, we have long recognized **passive uncertainty
reduction strategies** (Berger & Calabrese, 1975) as means of learning about others
through discreet observation of others' behaviors. Looking over the shoulder of that
cutie in your engineering class to see her/his playlist and learn her/his musical preferences, following them after class to learn where s/he eats lunch, or watching your
interest interact with other students to help plan your own smooth opening line
are all examples of passive strategies. Online, these tactics can help us learn about
others too, though we must go about our passive observation in different ways if we
can't physically colocate to observe their behaviors and other interactions.

Obviously, we can't (ethically) see all of someone's Internet use behaviors. What
we *can* do is engage in **extractive uncertainty reduction strategies** (Ramirez et al.,
2002) as a means of learning about others, observing the traces and records of their
online self. Sometimes that information is self-generated, such as the contents of
an individual's social media profile, the posts made to a class discussion board, or
drawings uploaded to a DeviantArt account. Other times information extracted from
digital sources is about the target but authored by others, such as when her/his hometown newspaper publishes (and digitally indexes) the hometown students on their
college's dean's list, stats from a league's database on its season's athletic performance,
or others' pictures of the individual posted to and tagged in Facebook. Such data can
provide information about an individual you are getting to know, present new information about an established relational partner, or just maintain a connection.

Importantly, extractive information was put forward as an active process.
Individuals engaging in extractive information seeking purposefully search the

Internet and Web for information about a target individual. This may be as simple as a Google search or as complex as cross-referencing small bits of information from multiple sources to develop a more complete profile of the individual. Though the individual seeking to extract the information may not necessarily know what data they are looking for, they have some motivation and intentionality in their search behavior. After all, imagine how hard it would be to run across Taylor Swift's entry from just randomly following Wikipedia links—though easily searched, her entry is quite difficult to find simply playing Wikipedia roulette. Ultimately, the two distinguishing properties of extractive information seeking are that information is sought online rather than offline, and such seeking is done deliberately.

Extractive information seeking can have strong effects on our relationships, particularly in their early stages. As a means of getting to know others, taking advantage of the many bits of information about others found online can help us quickly and efficiently learn about others and guide our initial impressions and interactions. For example, when you first learned about your new roommate, you likely did a little bit of online searching to see what you could learn about the person you'd soon be sharing a room with, even before your first call, text, or email with them. Getting an early sense of their taste in music, fashion, politics, and other parts of their lives often readily discernable from text and pictures online gave you an early sense of who you were dealing with: if they were serious or goofy, prim or laid back, and whether you would likely get along. Such extractive information seeking can guide initial impressions and early dialogue between two people (Walther et al., 2011).

Extractive information seeking can also be used in established relationships. Whether it's someone we haven't interacted with in a long time (e.g., looking up an old high school friend to see what s/he has been up to) or someone we interact with daily (e.g., seeing what your roommate has posted today, or learning his/her location by geotagged posts), extractive information seeking can be integrated to influence ongoing relationships (Wise, Alhabash, & Park, 2010). In both cases, extractive strategies can help us learn about and be more confident in both new and existing relationships as we purposefully seek out and obtain information about our relational partners.

Relational Development

Relational development refers to the life cycle of interpersonal connections, addressing the entire process from beginning to end of associations among individuals. Several means and models can be used to think about the relational development process, but one of the most popular is Knapp's (1978) relationship escalation model (see Figure 8.7). The relationship escalation model can be used to explain how relationships form, develop, are maintained, and even deteriorate and end. The model is comprised of 10 stages, 5 of which deal with relational formation and 5 of which deal with relational termination. Briefly, *relational formation* involves creating an initial impression (initiating), getting to know each other (experimenting), becoming less formal and with increasingly deep disclosures (intensifying), increasing the level of intimacy and interdependence in each other's lives (integrating), and publicly recognizing the relationship (bonding). *Relational termination* involves beginning to disentangle the partners' lives and thinking more individually (differentiation), establishing boundaries to formally and informally increase the distance between them (circumscribing), a lack of development or novel communication (stagnation), beginning to avoid physical, emotional, and even communicative contact (avoiding), and finally ending communication and the relationship (terminating).

Knapp makes several important notes about the model and its application that are worth noting with regard to our application to interpersonal CMC. First, and

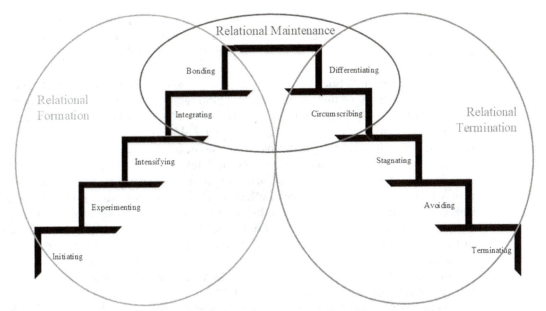

FIGURE 8.7 Knapp's (1978) Relational Escalation Model of Relational Development and Dissolution

most important to our use and application of the model, is that Knapp's model applies to all forms of relationships, including familial, friendship, romantic, and occupational. Much of the application that follows focuses on romantic relational development as it's one of the most common and readily distinguished, but one of the model's strengths is that it's applicable to other types of relationships. A second important note is that most relationships are not a direct linear progression from step 1 to step 10. Instead, many relationships—even established ones—move back and forth among stages, such as when married couples may suddenly discover a previously unknown trait, habit, or interest that causes relational turbulence and may cause them to move to a lower level (Planalp, Rutherford, & Honeycutt, 1988). However, movement is usually sequential between stages, meaning we do not typically jump directly from initiating to integrating, but we may move through the intervening stages quickly. (This is especially true for Disney princesses, who seem prone to love at first sight. See *Frozen*'s Anna.) Next, what we consider *relational maintenance*—or the communicative and social effort to stabilize and normalize interpersonal functions with another person—is typically at the middle few stages. Bonding and destabilizing may be common stages for relational maintenance, as individuals establish and reassess their relationships but ultimately work to keep relationships in the bonding or integrating stage, pending the relationship is viewed as valuable and worthwhile. Finally, these relational stages are descriptive of relationships regardless of the medium used to develop, maintain, and/or end a relationship. To that end, they can help us consider and explain the role of CMC in our interpersonal relationships.

Relational Formation

Initiating romantic relationships requires a few events and processes: mutual awareness of existence, shared interest and/or attraction, and relational development. As Knapp and Vangelisti (2005) aptly put the formation of romantic relationships: "Holding hands will generally precede holding genitals" (p. 155). Offline, individuals historically have found relational partners and initiated romantic relationships with those they are frequently near: classmates, neighbors, coworkers, fellow congregants, teammates, and so on. Computer-mediated communication has given us access to new networks and new means of finding and beginning romantic relationships.

Finding Each Other While the relational process may not have radically altered, the means by which we meet others and develop relationships has altered as more relationships are initiated online. One of the most evident—though not necessarily new—ways romantic relationships have been altered is through online dating sites. Early dating websites (including Match.com, Jdate, and eHarmony) initially served as online personal ads. These digital updates to the newspaper classified ads or video dating service analogs provided users a means of creating personal dating profiles and searching the profiles of others to find potential suitors. Online dating sites continue to provide a way for individuals to self-identify as seeking a romantic partner and to actively seek out potential matches without the constraints of space or time of a classified ad or short recorded video message, and sometimes regardless of geographic distance. The use of online dating tools has increased (and been somewhat destigmatized), with about half of single Americans now reporting they have used some form of online dating (Vogels, 2020).

As the market (including total users, sites, and revenue) for online dating developed, sites tailored themselves toward specific market segments: FarmersOnly sought to connect amorous agriculturalists, Ashley Madison targeted married individuals to help with mutual infidelity, ChristianMingle helped daters date within specific religious affiliations, and Grindr provided a dating platform specifically for LGBTQ users. In this way, the proliferation of dating sites has created niches—specialized markets and sites—that can reduce users' search costs and preestablish certain criteria. A Jewish individual seeking dating partners who are also Jewish can use Jdate and be reasonably confident that any user found has already met an important criterion by sharing a relevant social category. Even more general dating sites like eHarmony or OkCupid often have search features that allow users to seek within potential paramours' profiles to identify important social categories (e.g., gender, religion, educational level) to help refine dating efforts. Though offloading some of these initial selection criteria to the system may mean missing an unexpected and wonderful dating partner, it also can help daters quickly prescreen potential partners or match up with a more desirable partner, more so than the more resource-intensive strategies FtF dating requires. Rather than finding out 10 minutes into coffee that your date is a smoker—a deal breaker for you—and then having to either awkwardly leave or wait out the natural end of a date you know won't go anywhere, online dating may let you quickly screen in or out among potential and important criteria.

While many CMC dating tools help you find or connect with someone regardless of geographic distance (just in case your true love resides in Ulaanbaatar, Mongolia), the rise of smart devices has allowed online dating services to be closely associated with physical location. This has allowed opportunities for meeting potential suitors based on their geographic distance even more so than personal characteristics. These services, including Proxidating and Bumble, use either global positioning (i.e., GPS) or open Bluetooth connections to identify other users geographically near your present location. This can be helpful to identify who around you may be available and interested without emblazoning scarlet letters on their clothing to indicate interest, further reducing search costs for proximal relational partners. Ultimately, these **geolocated services**—tools that connect your information and communication to your physical location—transcend the online/offline boundary, with much of the search offloaded to the computer tool.

Finally, CMC tools can provide online equivalents of offline processes to find relational partners, most commonly through simple association. Historically, relational partners tend to come from our extended social networks: Your significant other is typically a friend of a friend (Felmlee, 2001). Offline, we may encounter these individuals at functions that span multiple social groups and networks like social events, weddings, or other parties comprised of usually separated groups of friends. It is even easier to span these social groups and meet individuals who are not already part of your extant social networks online. Several SNSs, from Facebook to Last.fm, allow you to seek out new connections based on shared interests, from movies to music to hobbies. In this way, CMC tools may make it easier to meet people, but this ease comes with a caveat. As easy as it may be to encounter new people online, converting those new connections to friendships can still be challenging, requiring more than just an introduction. As Baym and Ledbetter (2009) found studying Last.fm users, simply sharing a common musical taste is not sufficient to foster sustained relational contact. For Last.fm users to truly develop friendships on the SNS, they have to go beyond interacting solely on Last.fm and engage in communication across multiple channels. Consistent with media multiplexity theory (Haythornthwaite, 2005), which we'll explore in a moment, users were more likely to become friends with those they met on the SNS when they additionally communicated via text, alternate social media, and even phone calls, all of which help form and foster a meaningful and sustained friendship.

Getting to Know Each Other Computers now facilitate new ways to learn about others with whom we begin a relationship, regardless of whether the relationship begins online or offline. Even when the relationship is initiated offline, individuals may use online tools as abundant sources of information to learn about potential dating partners. Indeed, many individuals use SNSs like Facebook and Instagram to learn about a prospective partner's interest before pursuing a relationship. Taking time to look at another person's profile and online activity is a strong predictor of an intent to pursue a relationship offline (Backstrom & Kleinberg, 2014), so if your crush is doing some snooping on you, it can be a good sign.

When a relationship begins online, CMC can play an even larger role in learning about potential dating partners. Online databases such as SNSs and personal websites can provide information seekers with **extractive information**—information about a target individual obtained from online tools. Many daters, upon finding a potential partner, exert considerable effort exploring the partner's online profile and the information contained therein before deciding to initiate contact (LeFebvre, 2017). Extracting information from web pages, social media profiles, and online search results about a potential relational partner can help identify cautionary personal details as well as potential tidbits of information from which to begin the relationship. For example, you may glean from a guy's online presence that his previous three relationships did not end amicably, you share radically different political views, or he likes music you find repulsive (e.g., Nickelback, Justin Bieber, polka, more Nickelback). From these data points, you proactively decide the two of you are incompatible and do not pursue even a first date. Alternately, perhaps you find pictures of his adorable puppy, enjoy his sense of adventure (as noted through his tracked and publicized Fitbit runs and Instagramed kayak outings), and appreciate the same types of movies (anything with Scarlett Johansson), and you subsequently decide to ask him out—perhaps to the newest ScarJo movie. Extractive information seeking can therefore influence relationship initiation by helping individuals learn about each other, as well as providing a tool to both prescreen and identify mutual interests.

Although CMC may alter much of the process of getting to know new relational partners, it is important to remember that many things—particularly the more fundamental aspects of getting to know relational partners—remain the same. Just as in offline relationship development, individuals online are often seeking dating partners similar to themselves (Tidwell, Eastwick, & Finkle, 2013), and these perceptions of similarity are fostered through direct interaction and observation of the other person. So although some things—such as the control over information presented to someone new to guide initial impression formation (Ellison, Heino, & Gibbs, 2006)—can change by being online, the general process of creating relationships through human communication remain generally stable and resilient, even as more communication occurs via digital technology.

Making the Switch

Modality Switching One challenge with online dating is the inevitable **modality switch**, moving interactions from one channel to another. When should you move from emailing and online chatting with a romantic partner to actually having a physical and colocated first date? A driving concern for the timing of this communicative change is the hyperpersonal model. Recall from Chapter 4 that when individuals initially and exclusively interact online, they can overidealize each other, leading to impressions that are ultimately more positive than would be possible interacting face-to-face. In support of the hyperpersonalization that can come from CMC prior to meeting face-to-face for a date, daters typically report modest reductions in the perceived physical and social attractiveness of partners after a first date relative to their pre-date online-only impressions (Sharabi & Caughlin, 2017a), partially because the real date likely can't fully live up to the idealized partner they envisioned. Consequently, being mindful of when to move an online relationship offline is important for romantic partners seeking not to have false impressions of each other.

Though the exact timing of this switch surely depends on the nature of the specific relationship and its participants, it is important that the switch from online to offline occurs early enough that partners do not develop hyperpersonal, idealized perceptions of each other. Waiting "too long" (whatever that means to the specific relationship) can create perceptions and expectations to which relational partners are ultimately unable to live up once they meet face-to-face, ultimately reducing their attraction and harming their relationship (Antheunis, Schouten, & Walther, 2020; Ramirez & Zhang, 2007). Switching modalities earlier on—naturally taking into consideration personal safety and privacy—before hyperpersonal processes are activated may thus be important to romantic relationships that are expected to eventually move offline, preventing dating partners from idealizing each other before meeting face-to-face. Better to know they chew food like a cow—and that you're comfortable with that—before you start getting fitted for rings.

Media Multiplexity Beyond simply moving from online to offline, the initiation and development of romantic relationships leads to other changes in media use among romantic partners. **Media multiplexity** refers to the tendency of individuals who are tightly connected relationally to use more available channels for communication (Haythornthwaite, 2005). Alongside the proliferation of wireless telephony and smart devices, media multiplexity has become increasingly present in our society and relationships. We see multiplexity in our platonic relationships, as you may simply group chat with members of your class project group, but you email, text, Snap, call, hang out offline, and more with your close friends. During relational development, similar media multiplexity effects occur as romantic partners begin to utilize more channels as their relationships develop, including increased use of cell phones and text messaging to foster direct, interactive, and interpersonal interaction (Coyne et al., 2011). Similarly, as relationships develop, romantic partners use asynchronous tools less often, and instead use more synchronous tools for relational maintenance, as the more cue-rich technologies can foster the intimate and reciprocal disclosures that are the hallmark of more developed relationships (Ruppel, 2015). In a broad study of partners across five relational stages, Sharabi and Caughlin (2017b) found that as individuals moved deeper into their relationships, impersonal and masspersonal tools like email and social media tended to subside, while interpersonal and dyadic communication tools like texting, telephone, and FtF interaction increased. Ultimately, these findings suggest romantic partners' uses of CMC tools change as they move from the development stage to the maintenance of the relational escalation model.

Relational Maintenance

Computer-mediated communication tools certainly can be used to meet new people and form new relationships (Parks & Floyd, 1996). However, Tong and Walther (2011) note that "users are more interested in maintaining ties with existing offline contacts rather than forging new ones" (p. 106). Indeed, one of the most common reasons for using popular SNSs, including Facebook, Instagram, and Tumblr, is not to discover new people with whom to form a relationship, but to keep in contact with existing relational partners, including current and former classmates, childhood friends, and past coworkers (Ellison et al., 2007). Thus, particularly over the past decade, CMC tools have moved to the fore as means of maintaining relationships, both new and old. Whether the relationship began via CMC or face-to-face, in our digital age, established relationships are almost innately maintained, at least in part, online. This is not to say you do not or should not go out for coffee and chat with your significant other. Rather, it simply acknowledges that *relational maintenance*—the act of keeping a relationship in satisfactory condition (Dindia & Canary, 1993)—now also occurs on Instagram, over Snaps, and even via Likes. Computer-mediated tools provide many channels and means by which we can keep our existing relationships in satisfactory condition, using communication to stay connected and engaged in our relationships and with our relational partners.

Before getting into some of the processes and effects of CMC in relational maintenance processes, let's do a coda on our frequent refrain thus far: media are neither innately good nor innately bad for individuals. Instead, it's all about how individuals use those media and the communication those media facilitate. I bring this up both because it's a good frequent reminder to have and because of the framework we'll use to discuss relational maintenance via CMC. Though CMC is not inherently "good" or "bad," the way digital tools are used to maintain our established relationships can have both positive and negative outcomes. The good, the bad, and the ugly of online relational maintenance are discussed next in respect to romantic relationships. It will also be helpful to remember that the communicative processes used to maintain romantic relationships are similar to those used to maintain non-romantic relationships. A romantic relationship may not be the same as a close friendship, and may be different still from a loose acquaintance, but the relational processes are all similar and the concepts transferable. Consequently, it may be helpful as you read the following subsections to consider how similar processes and effects manifest within your various interpersonal relationships with your parent, roommate, significant other, coworker, or teammate.

The Good: Keeping Up One thing CMC has done is allowed us to utilize communication to maintain relationships—also known as social ties—across time and distance. Analog media and offline interactions all have some shortcoming: written letters are slow and asynchronous, phone calls take coordination to set up and execute with both parties committing significant time for a meaningful interaction, and going out for coffee or a shared event takes even more time along with geographic proximity. Digitally, we can now cross distances and time zones in a functionally immediate way: small lags in text messages, social media posts, or videoconferences are present but often unnoticeable. Computer-mediated communication, including email, social network sites, and even video games, can facilitate the communication with our ties required to maintain relationships in "good working order," and do so without being constrained by what time zone partners are in, when they're free

Research in Brief

Love Is a Battlefield

Social network sites have become integral components of our relational maintenance: we use tools like Facebook and QQ to communicate with everyone from passing acquaintances to close friends and distant family members to our live-in significant others. But many of these SNSs have multiple channels within them through which we can communicate in different ways. For example, in Facebook, users can communicate masspersonally through publicly viewable timeline posts and comments, or interpersonally through private messages. Which tools, within the social medium, do we then choose to use?

Tong and Westerman (2016) answered this question by surveying Facebook users about their relationships and use of Facebook to communicate with romantic partners. Results of 309 surveys found that several relational outcomes (social presence, relational satisfaction, and relational certainty) were predicted by public (i.e., masspersonal) communication with romantic partners. In other words, when conversations were aboveboard and publicly accessible, the quality of the relationship tended to be higher. However, private messages did not have the same impact on relational outcomes.

To understand why these beneficial relational outcomes were found for public—but not private—communication, Tong and Westerman (2016) asked a different group of 99 Facebook users to provide the last positive and negative communication they

had via Facebook with their partner, and how that message was communicated. Coding the types and channels of messages, the researchers found that masspersonal messages (i.e., timeline posts and comments) tended to be positive and display affection. Alternately, private messages were more often negative, involving coordinating activities, relational conflict, or discussion of the relationship itself.

Together, these findings reveal that we may use the same medium for the same relationship but different channels within that medium for different parts of the relationship. We typically don't air our "dirty laundry" or discuss the intimate nature of our relationship publicly, instead using more private channels for raising or resolving problems, being negative, or complaining about our day. Alternately, positive messages like professions of love, positivity, and mutual admiration are more commonly addressed masspersonally: directed to our relational partner but in a more performative way so that others can see. Given all this how do we use SNSs in our romantic relationships? It seems the channels we use are a reflection of the type of messages being communicated.

Reference

Tong, S. T., & Westerman, D. K. (2016, January 5–8). Relational and masspersonal maintenance: Romantic partners' use of social network websites [Paper presentation]. 49th Hawaii International Conference on System Sciences (HICSS'16), Koloa, HI.

to chat, or even whether both parties need to be aware they are communicating.

Direct Communication Digital technologies are particularly useful to maintain long-distance relationships (see Figure 8.8), and though CMC is used with both nearby and distant ties, CMC is used more as we get more geographically distant from our friends, family, and loved ones (Billedo, Kerkhof, & Finkenhauer, 2015). The increased use of CMC to maintain relationships with those not physically near us may be

FIGURE 8.8 CMC tools can be used to maintain relationships and are especially useful in long-distance relationships.
tyle_r (CC BY-NC 2.0)

a simple matter of channel displacement: when we can meet a nearby friend for coffee, run into them while running errands, or frequently see them at work or school, we may not feel the need (or benefit) to use digital channels to communicate. But as we are less able to use other channels, we rely on CMC tools as the primary means of interaction. We can use simple and complex tools to maintain our relationships across distances, as an email costs the same and takes the same amount of time to transmit whether the sender is next door or on the other side of the globe. Important to both of these considerations is that, although CMC can let us maintain relationships across time and space, it takes time to do so, consistent with what we know about SIPT (Walther, 1992) and mediated interactions. As Wright (2004) notes, the frequency of mediated interaction is one of the best predictors of the quality of and satisfaction with ongoing online relationships.

One context in which the communicative value of CMC for maintaining direct interpersonal interaction is immediately evident is communication with military members deployed abroad or domestically (Faber et al., 2008). Synchronous channels like Skype or GroupMe chats can be used regardless of where the servicemembers and family members are located and help maintain a sense of immediacy and interactivity through rich socioemotional exchanges, even across vast distances. This effect is similar to how more frequent cell phone conversations between college students and their parents can improve relational satisfaction and family closeness (Miller-Ott, Kelly, & Duran, 2014). Even asynchronous and leaner channels, from email to social network site messages, provide means of interaction among deployed friends. Particularly given the vast differences in time zones (a servicemember calling from Kabul, Afghanistan at 5 p.m. would be calling home in Santa Barbara, CA, USA at 5:30 a.m. due to the different time zones) and daily routines (the servicemember may not have many 45-minute breaks with which to have a phone call in her day while actively deployed), messages sent when convenient to be received and replied to when the other party is able can be even more efficient ways to communicate.

Both deployed servicemembers and their family at home identify asynchronous tools—especially social media—as critical means of "staying in touch" and "staying connected" with each other during deployment (Rea et al., 2015, 332), which can help maintain the relationship. The communication these asynchronous tools facilitate can also reduce uncertainty and anxiety for all parties by increasing the amount of information exchanged (Faber et al., 2008). Particularly when active deployment means instability in the lives of both the servicemember and her/his family and friends at home, frequent interactions can help maintain relational stability during the deployment and even lead to smoother returns home and reintegration of the servicemember (Houston et al., 2013). One final way CMC can help deployed servicemembers is by allowing them to see what friends and family back home may be doing, and in doing so remaining aware of and feeling connected to even minor events back home (e.g., a niece's Little League game, the new dog on the block, weather) while stationed abroad (Rea et al., 2015), but that sense of presence and inclusion requires a different type of communication and awareness.

Ambient Awareness Another unique benefit of CMC to relational maintenance has been **ambient awareness,** or the awareness of knowledge of social others as a result of frequent and repeated exposure to fragmented personal information, particularly via social media (Levordashka & Utz, 2016, 147). Ambient awareness can occur offline (Figure 8.9), such as when a mutual friend updates us on someone we've not seen in a while. But online—particularly within SNSs—we are constantly exposed to snippets of information about others, even without specifically seeking out that information or target (Figure 8.10). Rather than actively seeking out information, every time we log on to sites like Instagram and Twitter, we are presented with recent posts from our network members. Their recent posts are *pushed* to us, meaning that we do not actively seek them out and instead the systems and algorithms expose users to the content of others, so that we are incidentally exposed to various updates on jobs, relationships, pets, and other interpersonal ephemera sporadically. It does not take any time or effort to be exposed to this information, but through receiving that information we stay informed about what our relational partners are doing and stay updated on their recent life events.

This ambient awareness has several benefits. First, staying updated on what our various ties are doing helps us feel socially connected and socially fulfilled, even when we simply passively receive updates without acting on them, and the more you use social media the more updated and connected with others you feel (Krämer et al., 2017). A second benefit is that ambient awareness can provide you knowledge or access to activate later (Levordashka & Utz, 2016). Imagine you're interested in a job with a big firm, but don't quite know what they really are looking for in an applicant. However, you remember from your LinkedIn feed that an old high school teammate got a job there last year. The ambient awareness of your old friend's employment status helped you know *who* to ask for insight as you construct your cover letter. This example also illustrates

FIGURE 8.9 Ambient awareness then.
Israel_photo_gallery (CC BY-ND 2.0)

the third benefit to ambient awareness: reactivating latent ties.

Latent ties refers to those social connections that are not currently active (Haythornthwaite, 2002). This may include relationships we once had that have fallen dormant or to potential relationships (e.g., mutual friend or someone on a work email list) we have not yet explored. Particularly for the former—old relationships we have let lapse and with whom we no longer directly interact—ambient awareness allows us to maintain that latent tie and awareness of the target and quickly activate that tie if needed. Before SNSs, calling an old high school friend with whom you had not interacted in a decade to ask about a job opening may have been considered impolite or awkward: after all, you hadn't talked in more than 10 years. Now, as you can stay connected without significant effort or direct communication, that call is not a "cold call," but rather the activation of a relationship that has been passively maintained, and with someone who has at least some familiarity with you now and what you have been up to. Staying passively aware of someone can make them seem more approachable (Levordashka & Utz, 2016), and if they have been ambiently aware of you, it can make that approach less awkward. Ultimately, much of what we do while scrolling through our SNS feeds is keeping up with others through ambient awareness, feeling like a part of our social networks and quickly keeping tabs on a larger number of social connections than we could offline or via other channels.

FIGURE 8.10 Ambient awareness now.

The Bad: Keeping Tabs Though direct communication and ambient awareness can have positive effects on communicators and for relational maintenance, CMC can also have detrimental effects on the same when used in other ways or contexts. For example, use of CMC can create uncertainty among relational partners, particularly when novel information is discovered. Consider if you were carefully following a friend's online behavior only to find s/he frequently contributed to a discussion board of a political group with whose ideology you strongly disagree. The discovery that your friend may not share your political views may destabilize the relationship (Cover, 2012), resulting in moving the relationship into a differentiating stage. It can be difficult to keep a relationship—whether romantic or friendly—in good standing after learning new things about others, especially facets of their lives they may have otherwise not revealed to us (Planalp et al., 1988), which can lead to perceptions of greater differences and relational distance between us and our friends. Computer-mediated communication tools can sometimes let us see our relational partners in different or uncommon contexts, and in doing so affect relational processes.

One way in particular we may experience negative consequences to relational maintenance due to CMC use is when we move beyond passive ambient awareness to actively and problematically keep tabs on and surveil our relational partners. **Online surveillance** involves "conscious, furtive and deliberate [online] surveillance" to increase awareness of another's offline and online behaviors (Bevan, 2017, 168). Online surveillance may include benign things like checking Facebook to see if your friend is back yet from spring break travels, but often online surveillance has an antisocial or problematic motivation that makes the behavior challenging to relational maintenance. Unlike ambient awareness, which infers passive observations often fostered by system properties, online surveillance (often called *creeping* or *Facebook stalking*) is a very intentional, patterned behavior of making yourself aware of another through online tools. Online surveillance seems to be equally pervasive among many demographics: individuals are not more (or less) likely to engage in online surveillance based on gender, amount or frequency of Internet use, geographic distance, or even experience with infidelity (Tokunaga, 2011). What does seem to affect online surveillance are age and integration of the online tool into daily behaviors: younger users are more likely to engage in more online surveillance activities, as are those who feel more involved in and committed to a particular channel (e.g., Instagram *or* TikTok *or* Twitter).

Though many CMC tools—especially social media—are primarily used to keep in touch with existing friends, going further and deliberately seeking out information about others may go beyond the established bounds of the relationships. By checking up on what our friends—especially relational partners—are doing habitually and/or without context, someone engaged in online surveillance may assume or impose negative relational outcomes due to what is surveilled. For example, seeing a romantic partner talking to someone else while walking through the quad may be dismissed as polite chitchat, but continued monitoring online in which repeated interactions are observed may be misconstrued as cheating, errantly introducing thoughts of infidelity into the relationship (Fox & Warber, 2014). Even in non-romantic relationships, online surveillance may lead to unfair social comparisons. For example, as users engage in more social media use to learn about peers and see what they are doing, they may experience a fear of missing out (FOMO), and wonder why a peer was able to get a certain job, go somewhere awesome for spring break, or meet someone famous. This social comparison, particularly from continued online surveillance, can lead to depression or reduced sense of self-worth (Oberst et al., 2017).

Online surveillance is often less problematic or relationally turbulent than other relational monitoring behaviors, as access has already been granted to the online information (e.g., by becoming Friends/Followers) and therefore is not seen as an invasion of privacy or snooping. Also, there is less chance of being caught conducting online surveillance than other types of relational monitoring (Bevan, 2013). Importantly, though, online surveillance *is* strongly related to perceptions of jealousy (Elphinston & Noller, 2011; Tokunaga, 2011), so even if it is more socially or relationally tolerable, it can still have undesirable impacts on relationships.

The Ugly: Jealousy One of the bigger challenges to successfully maintaining relationships in the contemporary media landscape is that our interactions—both ours as well as our partners'—are often made visible and public, raising opportunities and awareness in ways not possible in offline interactions. In any given day, we interact with dozens of people: friends, family, classmates, and strangers. These banal interactions aren't often brought into the awareness of our relational partners by nature of their ephemerality—the interactions only exist in the moment, with no record left once they're done. Online, not only is there often a record (e.g., comments, Likes, system-recognized associations like Friends or Followers), that record can be seen or imputed by many. This is why jealousy is so problematic when it comes to social media. It is not that social media makes us jealous, but rather the communicative properties of social media and their use create an environment that can enable jealousy, even in stable romantic relationships.

Though such jealousy can come from online surveillance, findings have identified jealousy emerging even from common use or ambient awareness online. The time spent using an SNS appears to generally increase feelings of jealousy (Hoffman & DeGroot, 2014; Orosz et al., 2015; Utz & Beukeboom, 2011). However, if you've paid attention in research methods, you know that correlation does not equal causation. Indeed, the time spent using online tools likely also parallels other relational factors, which ultimately results in perceptions of jealousy, which may destabilize a relationship. For example, the time one spends using Facebook can expose the individual to more interactions between the relational partner and potential relational alternatives (i.e., Beckys with the good hair). As you use Instagram or Twitter, knowing that you know your partner, the algorithms may increase the visibility (i.e., show more often) of your partner's interactions with others. Even if these are simple friendly chats and platonic interactions, that the partner is (a) not attending to you and (b) interacting with other people who *could* be relational alternatives can spark feelings of jealousy. The other mechanism that can increase jealousy is when your partner's mediated interactions with others are not visible, in which case you don't even have the buffer of seeing the banal and platonic interactions (Utz, Muscanell, & Khalid, 2015). Instead of a visible Facebook message exchange or series of Likes, you may simply see that your partner has added someone to a Snapchat thread, but are unable to see the exchanges between your partner and potential rival. Knowing these hidden interactions exist can further enhance jealousy and introduce turbulence to relational maintenance processes.

Relational Termination

Ending a romantic relationship is even more common than maintaining one (Sprecher, Zimmerman, & Fehr, 2014), as more relationships fail than are successful. Although this realization is a bit of a bummer, it means the process of moving from a couple to being single is a common human experience. Relational dissolution has always been difficult, particularly among close romantic partners. Breaking up can involve simply withdrawing a relational status (e.g., reverting to just friends or ceasing interaction altogether), but ending a romantic relationship becomes more complex as partners' lives are entangled: shared friends, belongings, residence, pets, children, finances, and Netflix accounts may need to be reconsidered in the wake of the relationship's end. Even for entirely offline relationships, the disentanglement of relational partners online and the CMC with or around relational partners can be complex and challenging.

Digital technologies play an increasingly paradoxical role in the initiation of breakups. Broadly, individuals consider mediated channels as unacceptable means of initiating a breakup with a relational partner, but nonetheless use the same tools to break up (Weisskirch & Delevi, 2012). This paradox—that people do not see CMC as an appropriate means to break up even though almost 25% of individuals have experienced a breakup initiated through some digital channel—may speak to the ubiquity of CMC or the lean nature of CMC channels. For either motivation, the increase in the use of digital tools to initiate romantic relationships is mirrored by their use to end them as well.

Breaking up in the digital age does not have to mean sending a "Thank U, next" text to let the other person know you are breaking off the relationship. Instead, relational dissolution can take the form of a common communicative axiom: you can *not* communicate. Individuals can withdraw from a relationship indirectly to focus on their own needs (Regan, 2017) by simply breaking off contact, ultimately ghosting their partner. **Ghosting** refers to "unilaterally ceasing communication (temporarily or permanently) in an effort to withdraw access to individual(s) prompting relationship dissolution (suddenly or gradually) commonly enacted via one or multiple technological medium(s)" (LeFebvre et al., 2019, 135). In other words, we *ghost* when we end our relationship simply by not responding to texts or calls, interacting on social media, emailing, or otherwise interacting with our partner, and simply fade the relationship away. Ghosting can serve as a means by which individuals indirectly cue relational partners to their termination of a relationship. Unlike simply avoiding an ex on the quad, this fading away can be more notable in the always accessible CMC environment. For the partner initiating the breakup, ghosting can be an effective way to end a relationship—particularly a toxic or harmful one—provided she or he is not concerned with the former partner's subsequent feelings or otherwise maintaining the non-romantic parts of the relationship. For the dumpee, ghosting can be especially challenging, given the sudden and high degree of uncertainty about the causes for the breakup and the change in relational status, as well as the more permanent digital reminders of the relationship, including online photographs, events, and shared network connections (LeFebvre et al, 2019). For the one being dumped, ghosting can also increase the use of surveillance behaviors following the breakup (Brody, LeFebvre, & Blackburn, 2017; LeFebvre, 2017) as a means of reducing the uncertainty that comes with the breakup strategy of simply fading away.

Computer-mediated communication can be further used to identify the process of a breakup, not simply to communicate the relationship's end. Changes in users' communication patterns and use of CMC tools can now be used to identify when a relational termination is about to occur and how the dyad goes about communicating the breakup. For example, users' behaviors on social media can now predict the beginnings and endings of romantic relationships. When beginning relationships, the frequency with which someone posts to social media first spikes (talking to or about the suitor and potentially showing off) and then plummets as media displacement occurs: you're out on dates and spending time together rather than posting (see Figure 8.11). The reverse happens right before breakups: social media posts tend to spike as interactions between partners diminish preceding a breakup and the partners individually increase their communication with others, only to drop off at the breakup point as the former partners process the relational termination (Garimella, Weber, & Dal Cin, 2014). However, shortly after the breakup, interaction with other network members increases as the former partners seek out or receive social support from their relational network (Friggeri, 2014). Importantly, this happens from the time of the actual breakup, not from the time of Facebook statuses switched from "In a relationship." In other words, if you examine your SNS interaction behavior over the past few weeks, you may actually have a good predictor of beginning or ending a relationship, just based on the patterns of how often you post, with whom you interact, and how frequently. No longer do our breakups play out and occur only offline. Instead, online traces in our behaviors and communicative patterns may be indicative of a breakup—even before we ourselves come to realize the relationship is ending.

Computer-mediated communication can be used both actively and passively with an ex, even after the breakup. One of the most common means of post-breakup CMC is simply staying connected online. Almost 80% of college students remain connected (i.e., stay Snapchat or Facebook Friends, Twitter followers, etc.) with their ex once they have broken up (Fox & Tokunaga, 2015; Tong, 2013). Sometimes that connection is used to maintain non-romantic relationships among former romantic partners—to just stay friends. This would include active

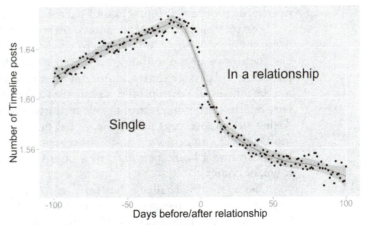

FIGURE 8.11 When you are about to enter into a romantic relationship, your online posting behavior spikes and then suddenly drops off as you formalize your relationship. When you break up, the same process happens, but in reverse.
Facebook Data Science

communication between individuals through interaction: messaging, commenting, sharing content, and so on. Another communicative strategy is to remain connected online but not directly interact. This maintenance of a latent tie can then be used to passively observe the former partner. Such ambient awareness patterns are common, but partner monitoring is more common for those who did not initiate the breakup, who may consider the relationship as possible to recapture, or who are ruminating over the breakup (Fox & Tokunaga, 2015; Fox & Warber, 2014).

Finally, ending a relationship can have consequences for one's online networks. Often following a romantic breakup, individuals experience a small decline in their connections on social media, usually around 15–20 people (Garimella et al., 2014). In an online equivalent to "picking sides" following a breakup, individuals will either defriend or find themselves defriended by mutual friends who have remained closer with the other partner, while retaining unique network ties themselves. In other words, breaking up can lead to reconfiguration of the people with whom an individual can interact in online spaces, particularly social network sites where social groups may span former relationship partners.

Casual Encounters

A final form of relational formation (and termination) online is those seeking casual romantic encounters without the corresponding long-term relationship and requisite relational maintenance. Casual hookups certainly occur offline, ranging from strip clubs (potentially parasocial relationships?) to phone sex to one-time consensual encounters. But the Internet has made the ease and diversity of accessing potential partners more immediate. Early online hookups were arranged in online chat rooms and virtual bulletin board posts (e.g., Craigslist), but numerous CMC tools now utilize personal information—including both relational and aerobic interests, demographic considerations, and geographic location—to help partners find each other.

The recent proliferation of dating apps, including Tinder, Bumble, Happn, and Grindr, have made evident the growing interest in individuals seeking a fleeting or casual encounter. Similar to traditional dating websites in that they offer varying degrees of profile development and narrowcasting (e.g., Grindr is used primarily by those seeking same-sex encounters, thereby proactively filtering potential partners), these tools allow individuals to develop profiles, browse potential partners, and then swipe or tap to indicate an interest, which is then relayed to the potential partner. If the partner shares an interest after receiving the notification and observing the individual's own profile, a mutual match then enables direct communication between the parties to arrange details for further interactions. Unique to Web-based dating sites, these apps utilize the global positioning system (GPS) within users' mobile devices to geolocate and limit search results to a designated radius, so that users can find a hookup, a date, or a board game partner within the same block, town, or county.

One interesting challenge has been the "outing" of people through their dating app profiles. From teachers to servicemembers to clergy, many have found themselves under an undesired spotlight after a community found their profile, potentially impugning the reputation of their organization, group, or even selves (Fowler, 2012; Tandoc & Jenkins, 2018). In one notably problematic occurrence, attendees of the 2016 Republication National Convention in Cleveland, Ohio, found themselves under scrutiny after many members of the conservative group (often not strong advocates for advancing the rights of individuals in the LGBTQ community) had their Grindr profiles identified by locals and reporters in the Cleveland area during the convention.

Scintillating as such outings may be to the mass media, one thing they hopefully reinforce is the masspersonal nature of seeking interpersonal relationships online. When one posts online they are DTF (whether that means "Down to family" or "Down to fornicate"), that search is accessible to anyone using the site, not simply those legitimately seeking a similar dating experience. We'll discuss privacy more in Chapter 13, but for now, it's enough to simply realize that when you interact via CMC in a public sphere, such interactions can be accessed by multiple individuals and different parts of your social network. This may include audiences and individuals you didn't intend to receive that message when you posted it.

Concluding Interpersonal CMC

On someone's birthday, it's common to see lots of "Happy birthday" posts on social media, to send an email or digital card, or even text a quick "HB!" message or cake emoji. This past year, I celebrated my friend Chad's birthday by texting him "Happy birthday." However, I communicated this one letter an hour over the course of 14 hours (I even texted the space). About 10 a.m., once he realized what I was doing, Chad asked why I couldn't just send a single message like any of his other friends. I replied that anyone could send a single message (and many had—he's a nice guy), but it took some special commitment to remember to text 14 times on schedule. At the end of the day he followed up to let me know how much he'd appreciated my stupid antic. Though I'd sent him the same message many others had, I did so in a unique way that left an impression far beyond the message itself.

Computer-mediated communication tools give us new and interesting ways like this to communicate and to build, maintain, and even dissolve our relationships. Though this doesn't necessarily allow us to maintain a larger number of meaningful relationships—we can still only maintain about 150 close ties at any given time (Dunbar, 2016)—digital communication does let us maintain larger and broader networks of connections with others. Whether it's a dear friend (either Snuffy or Chad), family member, or (ex) relational partner, our relationships now often play out to some degree online.

Key Terms

dyadic interaction, 133

social network sites, 134

phatic communication, 136

paralinguistics [digital affordances], 136

emojis, 140

geolocated services, 145

passive and extractive uncertainty reduction strategies, 141

extractive information, 145

modality switch, 146

media multiplexity, 147

ambient awareness, 150

latent tie, 151

online surveillance, 152

ghosting, 154

Review Questions

1. How can you tell if an interaction online was dyadic, masspersonal, or mass communication? What characteristics distinguish dyadic interactions from other forms?

2. How are phatic cues affected by the sender and the context of the communication? When you receive a "Like" to a post about a negative life event, what influences how you interpret that cue?

3. If you meet a potential partner online, what is the communicative or relational concern for waiting too long to meet face-to-face? What online relationships might you never expect to meet offline?

4. What is the benefit of ambient awareness for our relationships? When and how may that benefit become a challenge to our relationships?

5. Trends in your use of CMC can reflect offline events, such as the start or end of a relationship. Why do your offline behaviors influence your online behaviors such as frequency of use or posting?

CHAPTER 9

Intrapersonal Computer-Mediated Communication

Before his fame as *How I Met Your Mother*'s womanizing Barney Stinson, actor Neil Patrick Harris played the title character in the 1990s series *Doogie Howser, M.D.* about a child prodigy who became a doctor. At the end of each week spent navigating the challenges of the emergency room, first loves, and teenage angst, Doogie turned on his computer to compose a diary entry about his experiences and the lessons he learned (see Figure 9.1). As computers became networked, people used online tools to keep similar journals, using blogs, digital calendars, and social media posts to record and reflect on daily events. Though available to anyone with an Internet connection, almost three-quarters of all blogs are written primarily with the sender as the intended audience (Papacharissi, 2012).

We so often think of CMC and online tools as social tools allowing us to communicate with others in broad social networks. But take a moment and think about all the times you've used an online communication tool today when you communicated with *yourself*. Perhaps you've set or received a calendar appointment, written a blog entry for a class project, emailed yourself a reminder or memo to read later, or even posted a status message somewhere more for yourself than for anyone else. All of these demonstrate the capability of CMC to be used for a social tool through which we can communicate with ourselves. Indeed, for all of the social and *inter*personal uses we often consider CMC to offer—email, social networking, file transfers, and more—we often forget all of the *intra-personal* uses of CMC. From bookmarking files to posts to see later or online diaries to how we create ourselves online, the intrapersonal uses of CMC are almost as prolific as the interpersonal ones.

Defining Intrapersonal

Intrapersonal communication remains one of the most understudied domains of communication (Honeycutt, Zagacki, & Edwards, 1990), even though it purportedly serves as the foundation for all other forms of communication (Honeycutt & Hatcher, 2016). **Intrapersonal communication** refers to the communication we have with and within ourselves (Wenberg & Wilmot, 1973). Intraper-

FIGURE 9.1 Each episode of the 1990s series *Doogie Howser* ended with Dr. Howser (played by Neil Patrick Harris) using a computer to journal his experiences and lessons learned—a form of intrapersonal communication.
ABC Photo Archives / Contributor

sonal communication may be as explicit as self-talk, in which you verbalize your own ideas aloud and to yourself. But intrapersonal communication can take other forms, occurring whenever you serve as both the sender and receiver, encoding and decoding your own messages. You are engaging in intrapersonal communication when you write in a diary or a journal, as you seek to capture and structure your own ideas on the page and with no intent of sharing them with others. Likewise, individuals often engage in *imagined interactions* before an interpersonal event, a form of intrapersonal communication where an individual rehearses and prepares for an interaction by internally developing mental scripts of how the interaction is expected to play out (Honeycutt & Hatcher, 2016). Even communicative acts that seem to have a social component, such as public self-statements, may really be (or have strong elements of) intrapersonal communication. We have long made statements about ourselves publicly, but also intending ourselves as the audience as we publicly commit to an identity. When you stand in the quad and yell, "I am smart, and I *will* study for my tests better this semester," you are engaging in intrapersonal communication, as you are likely not directing this affirmation at any random passersby, but rather at yourself. The same is true when you wake up on Friday afternoon and post to social media, "Ugh. I'm never drinking that much again." Both of these communications exemplify broad messaging without any intended audience, and psychologists have long acknowledge intrapersonal communication's ability to increase an individual's commitment to a publicly made position (Bem, 1972).

One hotly debated question about intrapersonal communication is its differentiation from mere psychology. How is self-talk different than simply thinking? Go ahead—ask some of your faculty their opinion of "intrapersonal communication," and see whose pulses spike. I'll wait. See what I mean? It's a debate. But let's assume for a moment that intrapersonal communication is actually a communicative concept. If we do so, *psychology* is identifiable and distinct—it focuses on unspoken or unconscious thoughts, such as those ideas you just can't put into words. *Intrapersonal communication* therefore is identifiable as the use of language to communicate—to actually interact—with or about one's self, including self-talk, visualization, and imagination (Honeycutt et al., 1990; McLean, 2005). Some have even considered intrapersonal communication simply as interpersonal communication where the individual is concurrently the sender and receiver of messages (Ruesch & Bateson, 1968). But ultimately, let's consider intrapersonal communication to

FIGURE 9.2 How you can be linked to your statements can influence both how you see yourself and how you interact with others.
Igorq Stevanovic / EyeEm

be that which involves some form of message transmitted from the individual back to herself/himself, whose act of sending, transmitting, and receiving subsequently affects the individual, others, and/or both.

Identifiability

One of the most important constructs underlying intrapersonal communication is the matter of identifiability: do you and other communicators know who you and each other are (Figure 9.2)? One way to think about how known and individuated we are online is to follow the advice of Anonymous (1998) and consider identifiability as a continuum ranging from identifiable to anonymous, with pseudonyms falling in the middle of that continuum. To understand interpersonal communication online, it is first beneficial to understand how we and others understand ourselves online.

Identifiability

When we are **identifiable** online, the source of a message is distinguishable and persistent. Said another way, when an identifiable communicator sends a message, receivers know (1) who the sender was, (2) to connect the message with the sender, and (3) that both that message and previous/subsequent messages come from that same sender (Anonymous, 1998).

One common set of CMC tools today are SNSs, defined in part by the ability they provide users to create individualized profiles (Ellison & Boyd, 2013). Many users take advantage of both forced-choice fields like "What is your gender?" or "Employment Status" as well as open-text fields where they can describe themselves however they want by freely entering whatever content they choose (Lampe, Ellison, & Steinfield, 2007). And sometimes, the system even enters information on the individual's behalf, such as generating the number of friends or followers, the date the individual generated an account, or other information (like Facebook's default educational institution affiliation when it was initially open only to students of select colleges and universities). Ultimately, these profiles serve as significant identifiers, providing cues by which users can distinguish themselves from other users and interact in the medium. In many popular SNSs, including RenRen and Instagram, the system indicates a message's sender using her or his profile image and name. Increasingly, other sites—like discussion boards and news comments—connect to and rely on SNSs to make posters identifiable.

Implications of Identifiability in CMC Identifiability online is important for many of our most frequent interactive processes. For example, emails are frequently a form of identifiable communication. Emails come from a defined, persistent email address, so that you assume every email to or from Buffy@ UCSunnydale.edu is a correspondence with the same person. In this case, the email address also identifies the sender's organizational affiliation: University of California–Sunnydale. In addition, most emails are sent with a signature block indicating the name of the sender and other personal data, especially in initial emails. Social network sites—including Twitter, Instagram, and Reddit—likewise

rely on identifiable pro-
files to help find the people
we intend, not only help-
ing us find just *some* "Katy
Pearce," but *the* specific
"Katy Pearce" we meant to
find among others with the
same name. Interacting via
email and SNSs would both
be very different experiences
were users not identifiable,
as identifiability lets us
know with whom we're con-
necting, helping us intrapersonally communicate about
the interaction by crafting

FIGURE 9.3 Social media profiles make us highly identifiable. They often contain rich personal details that allow us to distinguish ourselves from other users, including pictures, names, locations, interests, and more.

our messages and expectations guided by the specific communication partner.

Another implication of identifiability online is much more intrapersonal in nature. Identifiable CMC can give us an opportunity to present ourselves as we want to be seen. We have much greater control over how we present ourselves online compared to offline. Computer-mediated communication gives us a chance to present only those features about ourselves that we want to display, putting our proverbial best foot forward. In other words, it's not just that we are identifiable online, but rather by what we are identified. For example, Turkle (1995) noted those with very visible physical stigmas (e.g., those with physical abnormalities, members of suppressed racial or ethnic minorities) may likely feel or be treated differently when interacting face-to-face, as that physical trait can overshadow other conversation. Online, though, individuals do not need to identify themselves by characteristics they do not believe are relevant, and as such they can often choose how they are identified. In an obvious example, many social media pro-vide users profile fields to which they can provide information about themselves. Sometimes those fields are closed-ended, such as when you enter your age in years or a predetermined age range, or when your organization affiliation is determined based on the email address used to register. Other times, these fields are open-ended, meaning users can write whatever text they want to identify themselves. On popular SNSs, many of these means of identification have been readily used to help individuals distinguish themselves and construct their identity online. For example, on Facebook, users must enter a name, but they frequently additionally provide their hometown, relational status, and favorite movies/music/books (see Figure 9.4; Lampe et al., 2007). Users also use open-ended fields to further dis-tinguish themselves from others. Perhaps nowhere is this desire to identify and individuate one's self as clear as in dating profiles (see Figure 9.3), where per-sonal information, quirky profiles, or mentions of relational attitudes or goals can be the difference between a left swipe and a right swipe. While these have inter-personal implications, what these profile elements—what's completed and what's left blank—mean intrapersonally is that individuals can more strategically choose how they want to see themselves and curate that identity to guide communication.

A final implication of identifiability online is that knowing who communicators are can lead to communication patterns more similar to their offline counterparts. Because our messages are linked with our persistent identities, we may be more mindful of how we communicate and with whom. One is not likely to call a boss's

Field	Description	%
Sex	Gender of the user.	93.80%
Status	Type of institutional member	100.00%
Member Since	Auto field listing account creation date	100.00%
Last Updated	Auto field listing last profile update time	100.00%
Hometown	Town of residence before joining MSU	83.30%
High School	School attended before college	87.10%
Residence	Specifically addresses on-campus housing	45.10%
Concentration	Major field of study	89.50%
About Me	Open-ended field allowing users to describe themselves	59.80%
Interests	Items separated by commas become linked, enabling search for other users listing same.	77.70%
Favorite Music	Same as above	78.20%
Favorite Movies	Same as above	80.10%
Favorite TV Shows	Same as above	46.50%
Favorite Books	Same as above	66.90%
Favorite Quotes	Content not linked.	73.80%
Political Views	Drop down list of political spectrum options	60.90%

FIGURE 9.4 Facebook profile elements completed by percentage of users. From Lampe, Ellison, & Steinfield, 2007.

idea "stupid," because that may result in personal and professional repercussions. Online and identifiable, individuals may likewise be more mindful of their communication because that communication (and the consequences thereof) can be attributed back to them. Consequently, identifiability online tends to reinforce offline communicative patterns, roles, and processes (Baltes et al., 2002).

Pseudonymity

When we are **pseudonymous** online, the source of the message is distinguishable and persistent, but not faithful to the offline sender. A **pseudonym** refers to a "fictitious alternative identity" (Anonymous, 1998, 384). It is a name, an image, a persona, or an identity that is not necessarily our own but is how others see and know us. Importantly, a pseudonym is persistent, so that although receivers may not know the true identity behind a pseudonymous message, receivers can attribute messages over time to the same source. Pseudonyms give us an opportunity to be someone else online.

Pseudonyms are not unique to CMC, having occurred in numerous other media. Radio communication has long used pseudonyms, including radio DJs' on-air names and the handles Citizen band or ham radio operators use to identify themselves. Authors, including "Samuel Clemens" (better known as Mark Twain) and Stephen King (whose first novels were published under the pseudonym "Richard Bachman"), have long used pen names to mask their identities. In television, movies, and theater, actors have long used the names of the characters they portray, in addition to using a pseudonymous stage name in their credits. For example, Norma Jeane Mortenson was better known by her pseudonym, "Marilyn Monroe," just as most people know Nicholas Coppola as "Nicolas Cage." Pen-and-paper or live-action role-playing games have likewise used pseudonyms so that players can act as their characters (Vorgar, Destroyer of Worlds and Wreaker of Havoc!) rather than themselves (Tim from Accounting!). Pseudonyms give the individuals behind them an opportunity to try new things, to disassociate themselves from their work, and sometimes to protect their privacy. Although pseudonyms are not unique to CMC, pseudonymity is much more readily executed online, as offline others can still see the speaker, and thus there is some degree of identifiability. Whether the speaker is known as "Chad Stefaniak" or "Aaron Hergenreder," the audience can still clearly identify and link the speaker to his message. Online, where individuals are less likely to be linked with a physical appearance, pseudonyms have been commonly used since the early CMC tools.

Some of the earliest CMC tools were discussion fora. When first logging on to these online discussions, users got to create their online personas. Their pseudonyms could be as simple as a username, but in places like The WELL and other online communities, pseudonyms often additionally included the development of rich, complex biographies further used to create the character. Using the pseudonyms allowed users a sense of privacy, as they were not linked to their real names or home addresses unless they voluntarily disclosed that information. However, using pseudonyms also provided users an ability to know to whom they were talking and begin to foster a sense of community. Though you may not know the name or physical location of the user known as "Persephone," you can be confident that messages from or to "Persephone" have all been sent from or to the same person. This persistence of identity, even if that identity is not fully faithful to the individual offline, allows individuals to develop identities through which to guide online interactions over time.

Implications of Pseudonymity in CMC Turkle (1995) has long asserted that pseudonymity can provide individuals with unique opportunities online that they may not have offline. One of the foremost of these opportunities is what Turkle called **identity trial** or identity exploration. Because it is difficult to connect the statements and actions of pseudonymous individuals online back to their offline selves, individuals can try out facets of their identities online without fear of those messages or displays being connected back to them (Figure 9.5). In this way, in-

dividuals can use chat rooms, discussion boards, and social media as relatively safe places to try out parts of themselves they may not be fully comfortable with or feel safe disclosing to known others. For example, consider a young teen living in a small, conservative town and wrestling with issues of sexuality. The teen may worry about discussing sexual orientation with family and friends, concerned they may view or treat the teen differently after learning it is a topic being wrestled with. And what happens if the teen experiments with a potential orientation, only to realize it doesn't actually fit with her/his identity, and wants to maintain the original sexual orientation? In such a circumstance, publicly trying out a different identity in a small town may have lasting and undesirable impacts for the teen. Alternatively, the same teen can go online and create a new social media profile or interact with others in a discussion board under a pen name to try out the different facets of her/his identity. Testing out the different identity or sense of self pseudonymously in an online space may give that teen a chance to experiment with a different sexual orientation without concern of family and friends judging or treating her/him differently. Also, if the identity trial is unsuccessful, the teen can drop the pseudonymous identity without much effort. Turkle asserts that online spaces give individuals opportunities to try out or explore parts of themselves they may not be comfortable with or want to portray offline, ultimately being able to adopt those identity traits the users deem successful while rejecting those that were not.

FIGURE 9.5 The book (and subsequent movie) *Ready Player One* illustrates how avatars give individuals the chance to look and act differently online. Players can communicate as different genders, races, or even species, all without other players knowing the user behind the avatar. The virtual world also gives players a chance to try out identities that may not faithfully portray themselves as they are offline. ventonero2002 (CC BY-NC-SA 2.0)

Another opportunity pseudonymity presents, and perhaps linked to identity trials, is the potential for individuals to act out usually hidden parts of themselves that may not be socially or personally acceptable or valued in their offline interactions. Similar to the deindividuation processes discussed by SIDE (see Chapter 5), pseudonymous CMC can lead to a **disinhibition effect** whereby individuals are less concerned about the outcomes of their actions and can thus act more freely (Suler, 2004). In such cases, pseudonyms may provide opportunities for individuals to act out parts of themselves they usually keep hidden. A person who is typically a curmudgeon offline may enter an online forum for gambling addicts and display tremendous empathy and patience. Likewise, an individual everyone always thinks of as kind and considerate may use a pseudonymous online profile to rant and rave about politics or social movements in a way they feel they could not with friends and family (Leavitt, 2015). The disinhibition effect thus explains some of the more negative experiences of online interaction, such as trolling or flaming. However, the disinhibition effect also creates opportunities for cathartic communication, as individuals "blow off steam" or vent pent-up negativity or attitudes online in a way that may not be appropriate offline or consistent with how they want to be seen—by themselves or by others. Whether pseudonyms allow us to try out the best of ourselves or show off the worst of ourselves, pseudonymous CMC gives users opportunities to be different versions of themselves or someone else entirely when online.

Anonymity

When we are anonymous online, the source of a message is unknown and unspecified. One way to think of this may be: "Someone spoke, but we do not know whom." In anonymous communication, nothing connects the message sender to the message itself. Consequently, not only do we not know who made an anonymous statement but we may also not know if subsequent messages are made by the same or a different sender. Unlike identifiable communication, where the sender is known, or pseudonymous communication, where the sender is persistent but obfuscated, in anonymous communication, the sender is simply no one and/or anyone. Offline, notes from secret admirers are a form of anonymity: you know someone sent you flowers but not who, or even if multiple bouquets are from the same individual or multiple paramours. Particularly online, with the ability to separate the sender from the message, anonymity can be easier.

Types of Anonymity Anonymity is not an absolute, objective property of a message. Messages may be identifiable to one party in an interaction but anonymous to another. In conditions of **self-anonymity**, the sender perceives herself/himself as anonymous to the receiver. In conditions of **other-anonymity**, receivers cannot distinguish the sender. Returning to the example of a note from a secret admirer, you may think you are anonymous as you send the unsigned note (self-anonymity), but if the receiver recognizes your handwriting, you may not be other-anonymous. Particularly online, subtle cues over which the sender has no control or awareness may give away clues to their identity or pseudoidentity. For example, someone who consistently contributes to a discussion online at exactly the same time each day may believe herself/himself anonymous, but others viewing the posts may quickly realize all the otherwise-anonymous posts occurring at the same time are by the same sender, resulting in self-anonymity rather than complete anonymity. Other metadata—information about the nature of an online message, but not the contents of the message itself—such as internet protocol (IP) addresses that correspond with the physical location of a computer can be used to identify a sender, thereby reducing other-anonymity. We'll talk more about de-anonymizing users later in this chapter, but for now it is enough to recognize that "anonymity" can be subjective, informed by to whom a source is anonymous: the sender, the receiver, or both.

Another way to think about anonymity is to consider the mechanism by which a sender is anonymous, regardless of whether it results in self-anonymity, other-anonymity, or both. One typology was offered (Anonymous, 1998) based on how the sender's identity is obfuscated: is the sender *physically* or *discursively* anonymous? Particularly in text-based CMC, discursive anonymity may be more critical to consider, but as more CMC becomes multimodal, physical anonymity becomes equally important to communicators.

Physical Anonymity Physical anonymity refers to the state in which one cannot determine or sense the presence of the physical message of a source. In other words, though you know what's been said, and perhaps even by whom, you cannot tell who or what that source is. Offline, physical anonymity may occur when talking on the phone or over an online voice chat: you may (or may not) know to whom you're speaking, but you cannot see the sender. Another way physical anonymity has been used offline is by putting a mask or bag over the communicator's face so as to not identify them. We see this use of physical anonymity to obscure a sender in masquerade balls or via Halloween costumes, as well as by placing communicators in different rooms or bags over their heads. When we cannot see with whom we communicate, we may act differently toward them as we cannot necessarily tell who they are (Zimbardo, 1969). You may have sensed physical anonymity when making a phone call, as even though you know to whom you're speaking by voice, the nature of a disembodied phone call can lead to a sense of distance and separation with the person on the phone.

Online, physical anonymity is quite common, as most systems by default limit the physical cues to individual identity. Whether through email, social media, messaging service, or even video chat, we rarely have a physical cue to ourselves we do not put (or allow) there ourselves. When you create and update your email account, you choose first whether to have a picture associated with your email messages and then choose *which* picture. To that end, you choose whether to be physically anonymous or identifiable, and if the latter, you choose *how* you are identifiable. In many SNSs, individuals use profile photos—typically posed, contextually appropriate, and isolating the user (Hum et al., 2011)—to physically identify themselves and link their messages to their photograph and physical self. Alternately, an online dater may choose to not post a photograph to her/his profile, omitting a profile photo for concerns of personal safety or to let potential suitors consider the content of the profile before being exposed to physical attributes. More commonly, CMC tools that provide only user names—even tools as simple and commonplace as email—offer physical anonymity. Such visual anonymity may be disconcerting for those used to having a physical cue to communication partners' identities, such as in a classroom setting, but it can ultimately help communicators increase their disclosures and perceived closeness. Activation of hyperpersonal processes by visual anonymity can lead to more positive interactions and frequent disclosures, as well as idealization and frankness among communicators (Chester & Gwynne, 1998). Consider a Zoom audiovisual teleconference conducted either *with* (i.e., physically identifiable) or *without* (i.e., physically anonymous) the video feature on (See Figure 9.6). How does having the video feature on change the nature of communicators, including their appearance (e.g., dress, posture, personal hygiene), their

FIGURE 9.6 How does the way you interact with someone change based on whether the communicator is physically identifiable or physically anonymous? What do you think is different in this conversation when the person on the right has her/his video turned off? Rawpixel

nonverbal communication behaviors (e.g., eye contact, gestures), and their engagement in the conversation? In other words, what's the communicative difference between a visually anonymous Zoom session and a phone call? They should function very similarly.

Discursive Anonymity Discursive anonymity refers to the state in which specific comments or statements cannot be linked to a particular sender. In other words, though you know what's been said, you don't know who said it. Offline, discursive anonymity may be difficult to achieve, as usually we can see who's speaking a message. However, FtF discursive anonymity is possible, such as by yelling in a crowd when nobody can discern the identity of who shouted by sight. Another example of this may be the "stranger on the train" phenomenon, wherein you speak with someone you've never met and never will again. Although you can clearly see each other, you have no idea who each other are, and as such you may communicate more freely and without concern for what is said being linked back to you—because the communication partners do not know each other's identities (Derlega & Chaikin, 1977). But ultimately, the examples of discursive anonymity in FtF and most forms of dyadic communication are few and challenging to contrive. Disassociating a message from a sender is much easier online.

Discursive anonymity has been long possible—and much more common—via CMC. Early text-only bulletin board systems (BBSs) allowed users to either create pseudonyms to link their statements to some username (thereby offering discursive pseudonymity) or simply have the name of the poster left blank, the latter of which gives users a form of discursive anonymity. Discursively anonymous posts may not have any sender name attributed to the message. This distancing of a specific user and her/his message can even further liberate thought and contribution by disinhibiting users: if they cannot be linked to their own statements, there is little fear of procedural or social repercussions as they can't be fired or judged if others dislike what they've said (Misoch, 2015). However, because messages aren't linked to specific senders, a challenge to discursive anonymity is conversational coherence. In other words, because users can never quite be sure who shared one message, it is impossible to tell if a subsequent message is offered by the same person. Consider the following exchange (Figure 9.7), and read through it twice: once with discursive anonymity (i.e., cover the usernames with your hand) and then again a second time with discursive pseudonymity (i.e., alongside the usernames). How does the meaning and flow of the conversation change when you can't track who said what and to whom?

Implications of Anonymity in CMC Whether one is physically or discursively anonymous, the inability to connect senders to messages can lead to a disinhibition effect, just like pseudonymity. Anonymity may increase the disinhibition effect over pseudonymity, making a communicator even more likely to transmit messages they perceive receivers may find unwelcome or negative. Unlike pseudonymity in which an identity is masked but persistent, anonymous communication has no connection to the sender or even consistency in identity. Consequently, individuals may be much less concerned about repercussions of negative behavior. One place such disinhibition has been discussed of late has been in online discussions, particularly around politics and other news items. Exploring readers' discussions on the websites of 900

User24601:	I just heard the new Lizzo album.
Javert:	It's good!
User24601:	I know, right!?
User24601:	But is it better than her last one?
Javert:	I'm not sure. Her albums have a way of taking a while to really land on audiences, and perhaps this one needs more time, too.

FIGURE 9.7 Chat log with pseudonymous discussion. How does the flow of discussion (including who is asking/answering questions) change if you read the chat with the user names covered?

newspapers (half of which were anonymous and half of which were identifiable), Santana (2014) found that discussions were much more civil and measured when posters were identifiable, and that conversely discussions were much more uncivil when users could post anonymously.

Identity Online

So far in this chapter, we have focused on the nature of identifiability as it relates to CMC. But that discussion has considered the nature of identity and the degree to which your identity is detectable to yourself and others. We still need to address your identity and the identity you communicate online. It would be simple enough to think of your identity as monolithic: singular, unchanging, and definitive. That attitude would follow the axiomatic approaches of pop culture to assume "I am what I am, and that's all what I am" (Fleischer, 1933), or the more contemporary "We are who we are" (Ke$ha, 2010). But, as we now address, who *you* are (or at least that part of you that you use for interaction with others) may change based on to whom you are talking, the context of the interaction, or even how you see yourself.

Dimensions of Yourself

In *Alice in Wonderland* (Carroll, 1865), the Caterpillar asks Alice, who is having a very curious day, "*Who* are you?" (p. 61; see Figure 9.8). How would you answer the Caterpillar's question? It may seem simple and straightforward on the surface, but as you start to ponder an answer you quickly realize that the question and its response are much more complicated than they appear. One reason for this is that our identity, specifically how we think and communicate about it, is not a singular thing. Higgins's (1987) self-discrepancy theory gives us one way to begin to think about and answer the Caterpillar's very fundamental question: Who are we?

Higgins (1987) argues there are three domains of the self, and self-discrepancy subsequently argues that when these three domains of the self are not aligned or consistent, individuals experience agitation, feelings of dejection or rejection, negative emotions, and other undesirable outcomes. For our purposes, though, let's focus on the three domains of the self: actual, ideal, and ought. The **actual self** is the representation of you (including your physical, personal, social, and psychological attributes) someone believes you *actually* possess: the person you truly think you currently are. That "someone" may be either yourself or someone else, and so these domains of self may either be us as we perceive ourselves or us as others perceive us. The **ideal self** is the representation of the attributes someone would like you to possess: the self you strive to be and become. Finally, the **ought self** is the representation of you that someone believes you should possess: the self perceived as needed in a given context. These selves are often distinct: we may not spend a lot of time seeking to become our ideal self (i.e., self-actualizing), instead focusing on being comfortable with who we are in that moment (Maslow, 1965). However,

3

THE CATERPILLAR.

FIGURE 9.8 In *Alice in Wonderland*, the Caterpillar wants to understand who Alice is. But Alice realizes she isn't even the same person she was at breakfast because of her day's adventures down the rabbit hole. Like Alice, our sense of selves—our identities—are somewhat malleable. Whiteway

online, these domains can be strategically presented and used to achieve relational outcomes.

One example of where these domains of the self are strategically used is in online dating. In online dating profiles, individuals often face a tension: presenting themselves in such a way as to be perceived as favorably as possible by potential dates, but making sure not to actually be deceitful as doing so could ruin the subsequent relationship. (Recall the previous chapter's discussion of online dating and modality switching.) Users of online dating sites report constructing profiles that are not completely honest: they underestimate (or underreport) weight, use the most flattering (sometimes old) pictures, and exaggerate about the frequency with which they work out. However, these users do not perceive these representations as lies. Instead, these profiles are seen as promises or commitments—more to the user than to potential suitors—about who the individual wants to become. Interviewing online daters, Ellison, Heino, and Gibbs (2006) found that this discrepancy was not an attempt to be deceitful, but rather a discrepancy between the actual and ideal self. Online daters are pretty aware of who they currently are: their actual selves. But recognizing they have some time between when they find someone online and when they ultimately decide to meet that dating partner, users often present their ideal self rather than their actual self. They put forward the person who may be five pounds lighter or have read recent bestsellers, so that when someone shows interest on the dating site, they will commit to going to the gym, eating better, and reading some new books (see Figure 9.9). By the time they meet, they have become this better version of themselves. In other words, daters present their ideal selves strategically to motivate themselves. Ultimately, daters online do not see themselves as "lying" because the selves they put forward *will* be accurate by the time they eventually meet anyone in person. Importantly, all of this is done under the auspice of the ought self: daters put forward information in dating profiles and interactions they perceive as needed or valuable in the dating context. This is why daters use their most flattering/attractive photos as profile photos (Hancock & Toma, 2009) and often address relational goals and desire for shared experiences (Griffin & Fingerman, 2018). Daters believe these attributes are normal and expected by other daters.

Another place where the tension between the actual, ideal, and ought dimensions of the self is manifest is as individuals engage in identity trials online. When individuals use the opportunity of pseudonymity online to test out identities or present facets of themselves they usually do not, they are potentially presenting an ideal self. Whether it is someone wrestling with gender identity, trying out a political affiliation her/his family does not ascribe to, or just testing out being more or less social than they usually are, pseudonymity gives individuals a chance to present an ideal self, seeking to identify who they or others think the individual should be. As Turkle (1995) notes, these identity trials—which could be considered presentations of an ideal self—can help us determine whether our ideal selves are really ideal, both to us and to others. If the identity trial is deemed successful, we incorporate that identity into our ideal self and strive to make that ideal self our actual self. If the identity trial is not considered successful, we consider our actual self sufficient and maintain that dimension of our identity. Ultimately, the pseudonymity of online interaction can give us a space to see if our ideal self is really ideal.

Another way we see our identity dimensions play out is when we are identifiable. Popular SNSs like Facebook and Twitter are notable because users are typically highly identifiable (both visually and discursively), and they also have to present a singular self across multiple social contexts. We often have audiences that are kept separate offline: a tweet is accessible to our friends, family, religious, sports, and hobby groups all at the same time. We often act slightly differently in each group, perhaps being more conservative while at Friday prayers and more ani-

FIGURE 9.9 Profiles on dating sites are often tensions between selves. People often present the idealized selves they hope to become once a potential partner is found, which may not represent their current actual self. AndreyPopov

mated while playing on our Wednesday night softball league. But on Facebook, where both our imam and shortstop can see us, individuals often present what Hogan (2010) refers to as our *lowest common denominator self*. This is the self that is likely the least objectionable and most consistent across all of our social contexts. Another way to think about this lowest common denominator self is as our actual self: it is the core self that is consistent regardless of social context and who we think we should be in a given scenario (i.e., our ought self). Perhaps this is why on social media, blog postings, and other online identity performances, we communicate a more simplified version of our selves, but that simplified version is typically identified as an accurate depiction of our actual characteristics by both ourselves and those who know us (Bargh, McKenna, & Fitzsimons, 2002). In other words, when we are identifiable online and our communication crosses multiple social contexts, our actual self is likely to guide our interactions. For all the concerns of selective self-presentation, SNSs including Instagram and QQ may actually be some of the truest and most faithful representations of our actual selves. When we know that how we present ourselves will be seen by many people whom we know, we present that which we find to be the most faithful to communicate, and in doing so engage in intrapersonal communication as we reflect on and then present that actual self.

Identity Shift

A particularly noticeable way CMC can affect our interpersonal communication is that CMC serves as a popular and persistent means of making and seeing our own statements. From personal websites to blogs to social media posts (see Figure 9.10), large amounts of the content we post online are actually messages about ourselves (Carr, Schrock, & Dauterman, 2012; Papacharissi, 2002). But unlike the comments we think or speak aloud about ourselves, the things we post online are things we can later see and reflect on. "My hair looks great today" is fleeting and easily forgotten when said aloud, but is a much more long-term statement when we take time to compose and write down the message, and can then later scroll back and see that statement. This ability to be both the sender and receiver of statements can affect how an individual sees herself/himself and ultimately lead to changes in attitudes and behavior about the self, all via a mechanism known as identity shift.

Identity shift refers to the process of self-transformation occurring when an individual makes mediated claims about herself/himself (Gonzales & Hancock, 2008). In the common identity shift study, an individual is asked to type in a description of herself or himself as either an extroverted, outgoing person or an introverted, shy person. Afterward, individuals are asked to complete a standard scale of extroversion. Multiple studies have shown that those asked to remember and describe a time they were outgoing and social rated themselves as more extroverted, and those asked to reflect on and write about times they were quiet and not outwardly social subsequently reported greater introversion. In the studies, the simple act of making statements about oneself ultimately changed how individuals perceived themselves. Though introversion/extroversion has been the most common personal attribute studied with identity shift, similar findings have also been found considering changes on participants' appreciation for art (Johnson & Van der Heide, 2015), identification with a corporate brand (Carr & Hayes, 2019), and even gender identities (Fritz & Gonzales, 2018). Consequently, identity shift occurs when individuals make statements about themselves and subsequently experience a change in their self in line with the selectively presented attributes.

The process of identity shift is more than just self-affirmation or the power of positive thinking, as it possesses specific psychological mechanisms to make mediated self-statements affect self-perceptions. When writing statements about yourself, you typically have time to compose and consider them, resulting in more thought and internal cognition put toward your own understanding of yourself and the attributes you are communicating. Additionally, once the statement is put forward—whether it's on paper, online, or recorded audiovisually—you can actually see/hear the words you write about yourself, which can cause additional reflection. In other words, identity shift is more than just thinking about yourself as you would like to be. Rather, identity shift is self-transformation consistent with specific, deliberate self-presentation. Built on the mechanism of self-presentation, and given its grounding in the hyperpersonal model, identity shift can also be significantly influenced by feedback to that self-presentation.

"Mrs. Carstairs will read her blog of the last meeting."

FIGURE 9.10 Typing statements about yourself, such as through blogging or social media posts, may result in identity shift, changing the way you see yourself. Statements about performing well in a meeting may make you more confident in future ones.
Cartoonist: Caldwell, John

Self-Presentation

Identity shift is grounded in the hyperpersonal model (Walther, 1996), and specifically its selective self-presentation component. According to the hyperpersonal model, mediated interactions allow individuals to selectively self-present. Whether you are typing an email or writing a letter, you get to choose what you say and how, and you can revise it until it is exactly how you want it. This selective self-presentation means that individuals online spend some time reflecting on and internalizing their self-presentation even before it's made. These choices in how an individual presents herself/himself are the most central component to identity shift, as that mindful and deliberate consideration of what and how facets of one's self are portrayed is an important form of intrapersonal communication (Bem, 1972). This transformation results from not only considering the self-presentation but then also subsequently seeing that facet of yourself you have presented, whether it's on a page or screen. Mediated self-presentation lets you actually see and consider that self-presentation, making you both the sender and receiver of the self-presentation even if that message is never shared or made public.

Merely thinking about and taking time to compose a statement demonstrating a particular identity trait can result in slight adoption and incorporation of that trait into your sense of self. In a blog post or social media update, thinking about how to describe a time you were energized by inward focus rather than social events can lead you to perceive yourself as more introverted (Gonzales & Hancock, 2008). Tweeting about the times you engaged in good study habits may make you see yourself as slightly more scholarly. These effects also seem to have some duration, as making even brief comments about a piece of art can result in greater art appreciation at a later date (Johnson & Van der Heide, 2015). In all, selective self-presentation can therefore substantively influence how you see yourself and behave. Consequently, identity shift is an inherently intrapersonal effect, as even if self-presentations are not shared (e.g., a journal entry or private online post), identity shift processes still occur (Carr, in press).

Feedback

Identity shift can be a purely intrapersonal effect, but our statements about our selves are increasingly made online in channels that invite some form of interactivity, whether replies, comments, or upvotes. Because of the ability for others to provide feedback to these self-presentations, and with consideration of the hyperpersonal model's feedback loop, there are also interpersonal effects that can intensify identity shift processes. Walther and colleagues (2011) found that while statements about our selves can cause identity shift, feedback agreeing with the stated trait intensifies identity shift's transformative effect. In other words, if you tweet, "I'm an introvert," you may perceive yourself as slightly more introverted, but if someone replies, "You sure are," your self-perception would shift even more so that you see yourself as even more introverted. Confirmatory feedback intensifies identity shift, as the sender is both selectively self-presenting the trait and receiving feedback to reinforce that identity claim. But what happens when feedback is not consistent with the trait presented?

People may not always agree with you or affirm you online, and as your audience gets bigger, the likelihood of a dissenting opinion or rebuttal increases. Take a look at any celebrity or politician's social media presence: any post about themselves ("Looking good today!") is met with a mix of support ("Yeah you are! Rock it!") and criticism ("I've seen better-dressed hobos."). Though the intrapersonal effects of identity shift are measurable, the effects of feedback are typically stronger. In an instance where an individual makes a claim like "I'm a Pepsi person" and receives disconfirmatory feedback like "No you're not—you always drink Coke when we're out together," may experience a stronger self-transformation from the feedback rather than the statement (Carr & Hayes, 2019). In this example, the disconfirmatory feedback would suppress the change toward seeing yourself as a Pepsi fan, and may even push you slightly toward seeing yourself as a Coca-Cola drinker.

Limitations and Boundaries of Identity Shift

Though work into identity shift has been promising, some important limitations to the process and our understanding of it are worth mentioning. First, identity shift effects tend to be small, so identity shift is not a magic bullet for self-transformation. When you tweet, "Because I'm a good student, I'm going to do great on my tests this week," identity shift predicts that you will see yourself as a *slightly* better student, but the transformative shifts are so small following a single post that they may be undetectable. What may help more than a single post is repeated public commitments to your identity as a deliberate and focused student. Social media seem to give us channels where weekly or even daily statements about our selves may be acceptable, and so it is much more likely that the composition, public posting, and (hopefully) agreement from others with public posts about your commitment to your education would actually result in your seeing yourself as a good student—and hopefully even improve your grades as a result. But such transformation takes time.

Another important limitation to identity shift effects is that you must identify with your self-presentation as your own. Hiding behind anonymity may prevent identity shift effects, as even though you make the statement, you don't strongly connect the message to your sense of self. In other words, you don't actually engage in the intrapersonal communication required for identity shift, failing to connect the presented self to your offline self. Social media make self-identifiability almost unavoidable: any statements users post typically appear connected to their profile, including their name and profile photograph. As such, identity shift may be almost constant in social media. However, in channels like chat rooms or email where identifiability is not as strong, identity shift effects may occur. During a 2016 conference panel, a group of communication scholars discussed their identity shift experiments that had not supported identity shift effects. One of the most common "failures" of identity shift appeared to be that participants were not asked to identify themselves in their blog posts, emails, diary entries, etc. When these studies were conducted again and participants were asked to sign off all posts with their names and hometowns, identity shifts reappeared. Therefore, it seems that a strong connection between the message and the self is required for the self-transformation predicted by identity shift.

A final boundary to identity shift is that the individual has to also identify with the statements being made. In other words, s/he has to actually see the self being presented as a viable and faithful articulation of the self to experience transformative effects (see Schlenker & Trudeau, 1990). In studies where participants were asked to simply make up attributes about themselves, no identity shift effects were found. Only when individuals drew on their own actual life events and experiences did they experience the expected identity shift. This limitation manifests in two ways. First, people can't magically transform if they don't believe in or associate with the self they're presenting. Returning to the earlier "I'm a good student" example, a failing student can't just keep posting "I'm a good student" and expect to change. Instead, s/he needs to make specific statements about personal experiences in which s/he was *actually* a good student to exemplify that identity. The second manifestation of identity shift's requirement that self-presentations are connected to the individual's sense of self is that the self-transformation must be possible. One cannot simply continuously post, "I'm a great pianist" and expect to become a virtuoso—actual piano practice would be required.

Summarizing Identity Shift

Identity shift research is still relatively new, and current research continues to unpack the transformative effects of intrapersonal communication. Fifty years ago, it could likely have been said that identity shift effects stemming from periodic diary or journal entries were so small and so particular to activate that they may not have even been worth mentioning. However, as so much of our self-presentation is now mediated, and as social media represent environments in which we constantly make statements about ourselves, identity shift is a process that deserves continued attention. As children grow up alongside CMC, linking their selves with the selves they present across the cornucopia of media at their disposal is likely an impactful question. After all, spending 18 years making self-claims likely impacts how we see ourselves. As such, CMC—especially social media—do not represent a new means of intrapersonal communication, but the processes and effects of identity shift can be much more pronounced and prolonged online.

Digital Realities

Identity shift is one-way intrapersonal communication occurring online, particularly through more traditional text-based tools like blogs, web pages, and status posts. But newer tools are emerging that make the online—and even offline—experience very different today than it was a decade ago. New tools are giving us the ability to go beyond text, audio, and pictures in our interaction. New devices allow the encoding and transmission of smell, taste, and touch online, and advances in older devices can make immersive experiences more accessible and lifelike. With that in mind, let's explore how digital technologies are shaping and changing our realities, both online and offline, and how we may communicate with and understand ourselves.

Virtual Reality

Virtual reality refers to "a digitally created space that humans could access by donning sophisticated computer equipment" (Fox, Arena, & Bailenson, 2009, 95). Users of virtual reality (VR) systems typically wear some form of headgear that provides visual and audio input to let them "see," "hear," and even "feel" the virtual environment in which they are placed. Users' movements are then tracked and the surroundings rendered—or created—in the virtual environment according to the user's movements. Virtual reality systems began in the mid-1980s, and were often bulky, cumbersome headsets. Advances in display and motion-tracking technologies now make VR systems more lightweight and portable (see Figure 9.11). The advanced computing power and motion-tracking technology in many console gaming systems (e.g., Nintendo Wii, Xbox Kinect) now enable high-quality VR experiences. Additionally, inexpensive headsets (e.g., Google Cardboard) as well as more expensive systems (e.g., Oculus Rift) can be used to make many smartphones into VR systems, letting the user tour and interact with the dinosaurs of the fictious *Jurassic Park* or experience a murder mystery as if they were on the famed Orient Express train.

Virtual reality can be as simple as putting on a headset to see a virtual environment. But as tools and inputs become more technologically advanced and less costly, the ability to provide multisensory stimuli and responses makes VR more **immersive**. Immersion refers to the use of digital devices to simulate multiple senses so that users can experience sight, sound, touch, taste, and/or smells in the virtual

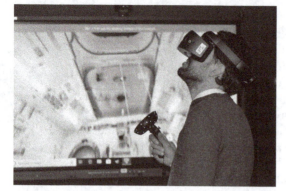

FIGURE 9.11 Virtual reality systems can place the user into a digital environment and let them interact with elements of that environment, from walking around the International Space Station (ISS) to practicing bomb defusal.

Laurens De

environment around them (Blascovich & Bailenson, 2011). As more sensory tools are used, the VR tool is said to be more immersive. Merely being able to look at a rendering of the Bayeux Tapestry in VR may not be very immersive, but being able to manipulate it to see the embroidery stitching on both the front and back, to smell the fabric and the stale air of the museum gallery, and to hear the murmurs of other museum goers and docents would be very immersive. Though a high degree of immersion is not necessary for a VR experience (again, consider the many VR experiences and apps that can be downloaded for free to a smartphone), as the VR system provides more sensory stimuli, users can feel more involved in and connected to the experience.

Virtual reality can be an effective and engaging way to interact, not only with others but also with yourself. For example, VR may be used to help individuals address and overcome phobias in safe and controlled ways (Ahn & Fox, 2017), such as by allowing individuals with speaking anxiety to practice delivering a presentation to an audience in VR prior to delivering it in person. Beyond simply rendering an environment and objects digitally, VR also gives users the ability to manipulate time (accelerating it or slowing it down), feedback (both in channels and intensity), and the objects themselves in ways that would not be feasible outside of VR (Parks, Cruz, & Ahn, 2014). For example, an individual who is afraid of spiders may use VR to experience positive interactions with virtual spiders, which can then reduce their arachnid apprehension. Similarly, a pilot or car driver who has had a bad experience may use a VR simulation program to take a test flight or drive to overcome their apprehension of being behind the stick/wheel once again, or to practice what to do in the event the problem reoccurs. And Parks and colleagues (2014) point out that photorealistic avatars that look like the user can be adjusted to be either heavier or more physically fit to promote more healthy eating and fitness habits for the users offline. Consequently, VR can be a unique way to interact with others, but also to see and reflect on your own self intrapersonally, sometimes even creating change in your self-concept.

Proteus Effect

An intrapersonal CMC phenomenon particularly relevant to immersive virtual environments is the Proteus effect. Named after the Greek god of ever-changing rivers and streams, the **Proteus effect** refers to changes in self-behavior and self-perception based on changes in self-representation (Yee & Bailenson, 2007). In other words, the Proteus effect occurs when you change your thoughts or actions online based on how you physically *see* yourself. Unlike identity shift, which relies on textual or verbal self-presentation, the Proteus effect is about how an individual visualizes herself/himself and the self-transformation that occurs within the virtual space. Because the Proteus effect occurs due to visual stimuli, it has become of particular interest to the study and application of immersive VR, where individuals can go in and see themselves and their world. Critical to the Proteus effect, an individual must see her/himself embodied in the avatar and connect with the avatar (i.e., be able to move or control the avatar) to demonstrate that the avatar's actions or appearance are really the user's. **Embodiment** refers to actions taken and experienced through digital representations of a physical self, allowing the user to take on the perception of the embodied avatar (Ahn, Le, & Bailenson, 2013). Though immersive VR can lead to embodiment, they are conceptually distinct, as emersion deals with the number of sensory inputs while embodiment emphasizes perspective. When an individual feels embodied in a virtual avatar, transformation can occur when the individual feels the avatar they control is acting as them, and thus that the avatar's actions reflect their own (see Figure 9.12).

Research has demonstrated some interesting Proteus effects. For example, individuals seeing themselves as physically attractive avatars are more sociable (e.g., providing more intimate disclosures, standing closer to their interactant) in interaction tasks in the virtual space than individuals represented by unattractive avatars (Yee & Bailenson, 2007; Yee, Bailenson, & Ducheneaut, 2009). Likewise, individuals assigned to taller avatars were more confident (e.g., negotiating more aggressively) in their virtual interactions than individuals assigned to shorter avatars (Yee & Bailenson, 2007). These behaviors parallel what we know of FtF interactions, where attractive individuals are typically more social and taller individuals are often more confident.

But the Proteus effect can also influence self-perceptions in ways that do not reflect their offline counterparts. One notable example of this discrepancy is the effect of a "power pose." Popularized in 2010, the idea of a power pose is that body posture can influence individuals' behavior, most commonly exemplified through the process of standing with fists on the hips like a superhero (see Figure 9.13), which subse-

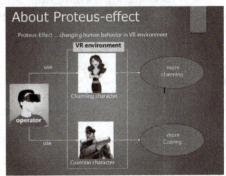

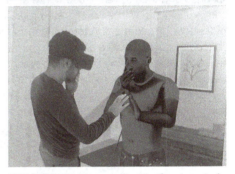

FIGURE 9.12 Named after the Greek god of changing rivers, the Proteus effect notes that changing how an individual physically sees themselves can influence their self-views and subsequent behavior. The Proteus effect is especially applicable in VR, where the user's embodiment can be manipulated.
Kengo Obana
Business Wire

quently increases the individual's confidence and even hormone production (Carney, Cuddy, & Yapp, 2010). The power pose was quickly picked up by the popular press and in leadership and empowerment seminars; however, subsequent efforts failed to replicate it (e.g., Garrison, Tang, & Schmeichel, 2016). It seems the power pose may not have the effects people originally thought. And yet one place striking the power pose does seem to have a stable and measurable effect is in VR. In a series of studies, Peña and Chen (2017a, 2017b) demonstrated that individuals placed into a VR environment who performed power poses via their avatar did demonstrate the expected effects once put into subsequent interactions out of VR. Though research is still emerging and it is not yet understood *why* VR effects manifest when offline effects of the power pose do not, it does seem the Proteus effect is exhibiting a measurable impact on how people see themselves and act offline.

Though interpersonal effects like socialness or communicative dominance can occur due to the Proteus effect, the Proteus effect is an innately intrapersonal

I'm Growing Older but Not Up

It can be hard to save money, especially when you're younger. How can we convince someone to put away some money in investments or savings for retirement, when that avocado toast looks so good right now? One way is to change how individuals see themselves, which we can do online.

One challenge with saving for retirement is that individuals often discount their future selves (e.g., saving for retirement) to focus on their current selves (e.g., a new car, travel). Hershfield and colleagues (2011) wondered if this discounting would still occur if people could visualize themselves as older. Fifty participants' faces were digitally scanned and then superimposed onto an avatar's body, either as they look now or age-progressed to look about 70 years old. After entering the virtual space, participants were asked to look at themselves in a mirror, seeing themselves either as they were now or much older. After seeing themselves, participants were allocated USD $1,000 to spend among four options, including "Invest it in a retirement fund." Participants in the age-progressed condition invested more than twice as much (an average of $172) than participants who saw themselves as they were ($80).

What does this study tell us, and how can we harness it to affect our own behaviors? Physically seeing ourselves as older seems to cause us to be more cautious and forward-thinking in retirement investments. If you're starting your career—or simply revisiting your retirement planning—taking time to visualize yourself as older before making your investment decisions may help you communicate with yourself about your goals, not just in the immediate but in the future as well. You can still buy that jet ski, but perhaps wait until you have maxed your 401K contributions for that year, or at least visualize your 70-year-old self riding that jet ski.

Reference

Hershfield, H. E., Goldstein, D. G., Sharpe, W. F., Fox, J., Yeykelis, L., Carstensen, L. L., & Bailenson, J. N. (2011). Increasing saving behavior through age-progressed renderings of the future self. *Journal of Marketing Research, 48*(SPL), S23–S37. https://doi.org/10.1509/jmkr.48.SPL.S23

process, just like identity shift. The way we present and see ourselves—either by our choices or the systems'—in virtual environments can affect our own self-perceptions and behaviors, even without interacting with others. For example, when participants were brought into a VR environment that aged them (i.e., rendered their avatars to look like their offline self, but age-progressed to be 50 years older than they were) and looked in a virtual mirror to see themselves as older, they tended to be more fiscally conservative and more likely to invest more money for their retirement (Hershfield et al., 2011). In other words, actually *seeing* and interacting with older versions of themselves, individuals became more focused on their future self. As more of our interaction online is embodied, the Proteus effect will continue to influence how we see ourselves, both literally and figuratively.

Gaming

Beyond immersive, multisensory VR experiences, video games can be a means of embodiment through which individuals can enact intrapersonal communication. **Avatars** refer to virtual representations controlled by humans, and are common means of playing and interacting in many video games, particularly multiplayer games. Avatars are the means by which players interact with the game and with other players' avatars, and avatar selection is often the first thing a player does in a game.

Many games let players go beyond simply picking among a preset gallery of avatars, allowing players to customize an in-game avatar through which they experience the game. *World of Warcraft*, *SecondLife*, *The Sims*, and *Fallout 4* all allow players to alter the physical appearance of the avatar they will use to interact with the game world. Customization can vary by game but often includes both broad demographic characteristics (e.g., race and gender) and small details (e.g., eyebrow shape, haircut style). This allows players to create and control an avatar that can vary broadly in the degree to which it physically represents their offline self. Players often take advantage of avatar creation to influence their own self-construct and intrapersonal communication, which

FIGURE 9.13 "Power posing," or taking on expansive and assertive postures, does not seem to work offline. But standing like a favorite superhero in VR does seem to engage the Proteus effect and make people more confident and assertive in the offline interactions that follow. Choreograph

may subsequently influence gameplay. For example, interviewing *World of Warcraft* players, DiGiuseppe and Nardi (2007) reported that several female players created a more masculine avatar (either actually selecting a "male" avatar, or altering physical characteristics of a female avatar to present more aggressive and masculine traits) to avoid identifying as more feminine or "squishy" as the female players often felt societally obligated to do offline (see also Eklund, 2011). Other players customize avatars as means of identity performance, using avatars to see their ideal self rather than their actual self (Bessière, Seay, & Kiesler, 2007). As players spend all of their time in the game *as* this avatar—seeing the avatar, controlling the avatar, and interacting with the game through the avatar—players can often strongly identify with their avatars (Banks & Carr, 2019; Tronstad, 2008).

The ability to customize avatars and the amount of time players spend embodied in that avatar suggest video games may be an additional venue in which identity shift and Proteus effects may occur (Figure 9.14). The identity shift and Proteus effect processes do not change simply by being in a video game rather than social medium or VR environment, but the amount of time individuals spend gaming suggests extension and consideration of these concepts within gaming environments is worthwhile. Who we play as may affect not only how we play, but even more fundamentally how we understand and communicate with ourselves.

Augmented Reality

A final way digital tools are altering our realities is through changing how we actually see and experience our offline world. **Augmented reality** refers to the use of digital tools to influence or affect how we see and interact with the corporeal world around us with computer-generated information that appears to coexist with the "real world" (Azuma et al., 2001). Unlike VR, in which our environments and selves are entirely digital, augmented reality often uses smart devices like smartphones, tablets, or glasses to complement and alter our perception of the physical world around us. If you have ever watched a sporting match, you have likely seen augmented reality at work, as the television broadcast overlays virtual first-down

FIGURE 9.14 How does how you choose to physically present yourself online influence (a) how you see yourself and (b) how you interact with others both in the virtual environment and later offline?

and scrimmage lines on top of the actual American football field, tracks and superimposes pitch data over batters in real time to detail where baseballs are being thrown, or displays drivers' speeds atop their cars in Formula-1 races. These sporting examples illustrate how mass media have used augmented reality for a long time, but personal media have increasingly made augmented reality accessible and usable on an individual basis.

One of the more evident examples of augmented reality at the personal level are many of the programs and apps on smartphones. Snapchat and Instagram filters represent augmented reality, overlaying a selfie with dog ears or accentuated eyes in real time. Games like *Pokémon Go* and *Harry Potter: Wizards Unite* allow users to hold up a smartphone while playing a game, and the camera's display shows the world with virtual pocket monsters or magical creatures as if they were in and interacting with the world in front of them (see Figure 9.15a). Tourism is trying to further integrate augmented reality, both to increase engagement with sites and to personalize travel experiences. Apps will now allow visitors to learn detailed information about particular sites, paintings, and neighborhoods without a personal tour guide (see Figure 9.15b). Augmented reality can even overlay renderings and/or movies, to let a visitor to Rome's Colosseum actually watch a chariot race in the restored arena or explore Machu Picchu as it was when the Incas inhabited it. As these examples illustrate, the applications of augmented reality are limitless and range from the trivial (e.g., pop-up characters on cereal boxes) to the critical (e.g., aiding in medical procedures). From the perspective of human-computer interaction, much more can be said about augmented reality. But we are going to keep our focus on the intrapersonal implications of augmented reality.

Like VR, augmented reality can give individuals a chance to personally experience things they may not be able to do otherwise due to safety or

FIGURE 9.15 Augmented reality tools overlay information onto images of the world around us, and can create new ways to see and interact with our environment.
Wachiwit, Georgijevic

practical concerns. For example, driving while intoxicated is not a safe behavior, but having that experience may help individuals be more aware of their impairment and decreased response times after consuming alcohol, which may reduce future instances of drunk driving. Augmented reality can be used to provide firsthand experience as if one were driving a car while inebriated, all in a safe and controlled environment (Höpli & Cuervo-Alvarez, 2016). This type of personal experience can cause an individual to reflect on himself/herself, potentially changing future attitudes and behaviors. No longer would an individual be able to say, "It was only three beers—I'm fine to drive." Instead, s/he can recall the augmented reality experience and the problematic real-world interactions that occurred, and wisely choose instead to get a ride to the next destination. In this way, augmented reality can help us learn about ourselves with low cost and low risk.

Similarly, though hopefully less dire, cosmetic brands L'Oréal and Sephora have developed augmented reality apps to allow customers to superimpose makeup and deliver tutorials over selfies taken with a camera phone (see Figure 9.16). These apps serve as a chance for customers to try out new products or styles without the more significant time or financial burden associated with an entire styling session (Pearl, 2019). Offering benefits to the consumer, these apps can also increase customer engagement and identification with a brand. Taking time to "use" that particular brand of cosmetics may influence how an individual sees herself/himself in relation to the brand and influence future use. In other words, an individual may use the Sephora app to "try on" makeup and subsequently identify more with the brand. Additionally, such applications of augmented reality are a chance to alter how you see yourself. Usually someone who eschews makeup for the more natural look? Using the makeup app to see how you look wearing more products may change your attitudes on makeup, your own physical appearance and aesthetic, and even your behaviors. Someone who thought your face was unattractive or awkward? Having the app point out that your cheekbones are high and clear and that your face is nicely pear-shaped and symmetrical may cause you to reevaluate how you perceive yourself. As a performative act, trying different makeup, hairstyles, and clothing choices using augmented reality serves as an episode of interpersonal communication, where you the user choose the appearance to take, and may subsequently provide yourself feedback about your appearance, your esteem, or other facets of yourself based on what you see on the screen looking back at you. And as an informational tool, these augmented reality apps could help you learn new techniques firsthand, overlaying what you *should* be doing on top of your actual photograph in real time, and subsequently increasing your skill and self-efficacy with your makeup technique.

FIGURE 9.16 Sephora's app now uses augmented reality to overlay makeup trials onto your selfie, letting you see how makeup may look on you. How could such augmented reality influence how you see yourself?

Summary

Whether it takes the form of augmented reality blending the real and the digital, immersive VR, or simply text typed onto a screen, digital tools are providing us new and readily available means to communicate with and about ourselves. Importantly, these channels can have effects on how we see ourselves, how we talk about ourselves, and how we relate to ourselves. All of this is in addition to how *others* may see, talk about, or relate to us. No longer is CMC simply other-focused channels; rich and complex tools can blend and blur the lines between offline and online and virtual and reality. Those devices give us new channels to use in old ways (e.g., using a blog to journal rather than a physical leather-bound book) as well as new ways to communicate with ourselves.

Concluding Intrapersonal CMC

The recipient to whom I most often send emails is myself—I send files I will need later, reminders of tasks to do, or gift ideas for family and friends. Even a tool that is so helpful for interpersonal communication is one I use frequently to communicate with myself. Email and most other tools for online communication can be used to communicate with ourselves, either directly (as in emails to yourself) or depicting and reflecting ourselves (as in online avatars and profiles). Though surely other users will see intrapersonal messages online, they may influence them differently than they influence you. Meeting another avatar in virtual reality who is standing in a power pose may not have nearly the effect on the perceiver as it does the person standing with fists on hips. Though others may read your blog capturing your college experience, the writing of that blog—whether it's for commemoration or catharsis—is likely for you. And intrapersonal communication may even emerge as we prepare to interact with others online. Before messaging someone, you may browse their social media profile to learn about them and rehearse an imagined interaction to prepare for the real thing (Walther, Van der Heide, et al., 2010). All of these represent intrapersonal communication occurring online. For all of the communicative processes that follow, they start with within ourself. Even in the dynamic social spaces of the Web, intrapersonal communication still plays an important—yet understudied—role.

Key Terms

Review Questions

1. What types of anonymity can occur online? Who (or what) has control over each form of anonymity?

2. What do pseudonyms indicate about the user? What function do pseudonyms serve? When answering, consider both the functions the pseudonym serves for the individual (i.e., the person behind the pseudonym) and for other CMC users.

3. How we present ourselves online can affect how we see ourselves literally and figuratively. What are some of the ways in which you frequently present yourself online? Which of those self-presentations are text-based and which are visual? Are any of them immersive? How do these different forms of presentation potentially affect your self-view and attitudes?

4. What mechanisms or characteristics of CMC allow individuals to engage in identity trials that they may not be comfortable trying offline?

PART IV

Applications

Social Media

Many questions surround the nature of ownership of your Instagram profile, its information, and the interaction within. Do *you* have control over it? What about your Instagram friends? Other users? Does Instagram itself control your profile? The restaurant chain Applebee's had to wrestle with similar questions during a public relations debacle in February 2013. A server at a St. Louis franchise posted a receipt to social media on which a customer had crossed off the automatic 18% tip for a large party and instead wrote a comment into the tip line of the charge receipt (Figure 10.1). The franchise fired the server for the incident, citing a violation of the customer's privacy. However, the damage to the chain's image was already done. Within hours, the story spread across the Internet, and users flocked to Applebee's Facebook page to express their dissatisfaction with the firing of the waitress. Publicly visible comments railed against Applebee's. Applebee's tried various strategies to stem the tide of negative sentiment on its Facebook account, including deleting users' comments and reviews, responding to individual users' criticisms, and making blanket statements about the incident and the company's policies. However, the users kept commenting. Deleted comments were screen captured before deletion and re-linked or reposted, new comments streamed in, and the Facebook comments turned from focusing on the fired waitress to Applebee's broader online presence and the policies of customer engagement within Facebook. Quickly, the ire of Facebook users turned from the specific situation to broader concerns about the ability to post comments to Applebee's Facebook page and the ethicality of Applebee's curating them.

How did Applebee's social media presence get so out of control? While this situation certainly could be (and has been) explored as a public relations nightmare, it also is of interest to us as it exemplifies one of the benefits and banes of social media: they are social. *Social* in this chapter does not simply mean "to be used for interpersonal social actions," like going out for coffee. Instead, we'll treat *social* as

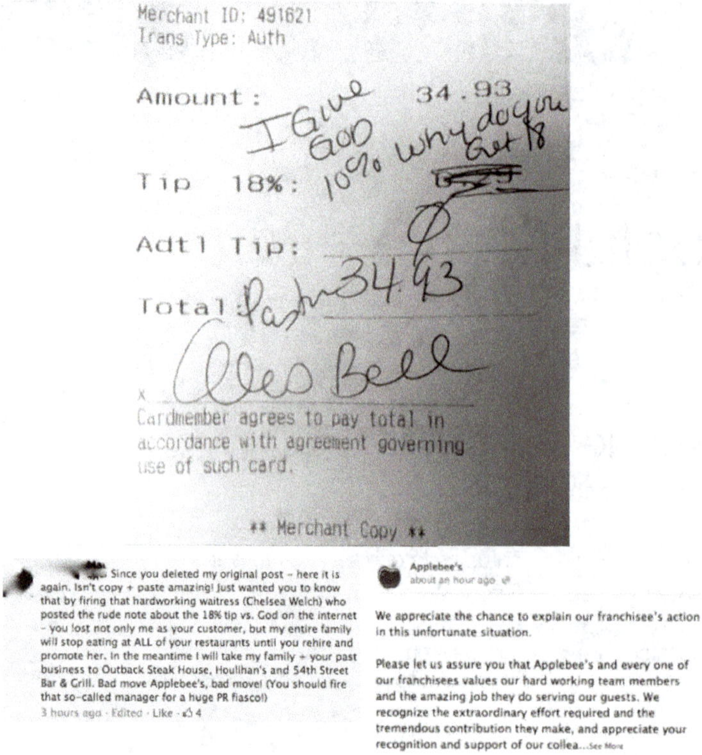

FIGURE 10.1 The Applebee's restaurant chain found itself under fire on social media in 2013 after sacking an employee who had posted a customer's receipt and poor tipping behaviors. The public disagreed with Applebee's action, and let it know via Applebee's social media accounts. Applebee's further inflamed the situation by trying to delete or poorly respond to critiques of its handling of the sitaution, which the public viewed as inconsistent and disingenuine.

it refers to complex, interdependent interactions among large sets of actors—both people and organizations. The very nature of social media means the public can interact at a very personal level that may be accessible to many more. Consequently, one issue important to our study of CMC is the understanding that social media belong to the society using them. The community can have the most influence and ownership of social media and their interactions, which affects the uses and users of social media.

Defining Social Media

After Greek philosopher Plato defined man as a "featherless biped," sassy Cynic philosopher Diogenes found a chicken, plucked its feathers, and returned to the Academy to toss the defeathered fowl at Plato's feet saying, "Behold! I have brought you a man" (Laërtius, 1925). Such is the problem with defining something based on a set of physical characteristics: the definition quickly becomes problematic when the characteristics change or are found as properties of additional things. This problem is perhaps most clear in understanding the nature of social media. We can likely list many social media platforms quickly from memory (Figure 10.2), yet we struggle to explain *why* those channels are social media, and concurrently why other digital channels are not included in our list of social media. Said another way, while most media can be used for social purposes, not all media are *social media*, so what makes Snapchat distinct from email or text messaging? One challenge in discussing, researching, and exploring social media is that there is no clear definition of what constitutes a social medium and what does not. Although there is likely a general understanding of what social media are, scholars in diverse fields have differing qualifications for what constitutes a "social medium," and thus have had problems defining its central elements (Aichner et al., in press; Kaplan & Haenlein, 2010).

Attempting to look beyond specific technologies and emphasize the communicative, interactive properties they share, Carr and Hayes (2015) offer a definition that both encapsulates present-day social media and is robust enough to account for the changing landscape of social media as well as platforms yet to come. Per Carr and Hayes, **social media** are:

Internet-based, persistent and disentrained channels of masspersonal communication facilitating perceptions of interactions among users, deriving value primarily from user-generated content (p. 49).

This definition is specific enough to encompass the current social media landscape, but inclusive enough to account for the next generation of social media platforms. Notably, several features and functions are absent from the definition as it does not address the modality of communication (e.g., text- or image-based), the style or purpose of the medium, or the particular forms of access or interface devices. What this definition *does* offer are several elements worth highlighting and delving into individually.

FIGURE 10.2 Social media come in many forms. By focusing on the attributes and affordances they share, we can more productively understand how they facilitate communication among users, and also consider social media that are still emerging.
EThamPhoto

Internet-Based, Persistent, and Disentrained

This definition of social media begins by addressing some of the technical features necessary for facilitating communication processes. Carr and Hayes (2015) specified that social media are Internet-based (rather than Web-based), noting emerging social media tools often run independent of the WWW. (Thank heavens we covered the Internet-Web distinction in Chapter 2!) Whereas early social media tools like Friendster and Bebo were only accessible via Web browsers, both some of the earliest (e.g., The WELL) and the most recent (e.g., TikTok) are not housed in web pages, but rather utilize the backbone of the Internet to transmit messages. Particularly as more individuals communicate through apps on tablets and mobile smart devices, social media need not be tethered to a "www" address, and many operate independent of the Web.

Social media channels are also *persistent*, in that they allow communication even when individual users are not online and allow users to create stable identities and means of connection. Some digital communicative tools are only meaningfully used when both communication partners are online. For example, video chat services Chatroulette and Omegle do not allow users to go back and reconnect with someone they've met. Because there is no persistent identity or means of reestablishing communication, these media, while social in nature, do not constitute "social media," because they cannot facilitate continued interaction among individuals. However, channels like TikTok and Instagram persist online even once you close the app or log off and therefore meet the "persistence" criterion to be considered social media.

Finally, social media are **disentrained** because they do not require synchronous communication and therefore allow users an opportunity for some degree of selective self-presentation (see Figure 10.3). Although synchronous messages can be exchanged via social media, synchronous interactions are not required for use. For example, although you can message a friend via Snapchat in real time, Snapchat does not lose value as a communicative tool when your friend isn't actively using their phone or the channel at the same time as you. Asynchronous communication allows social media users increased time to carefully construct and maintain their identities online. Though social media are often used to interact with people we know offline or in other relational contexts (Ellison, Steinfield, & Lampe, 2007; Subrahmanyam

kenna!! @makennabirddd · Apr 18, 2017
PSA, torts doesn't like dog filters. 🐶 @BlueJacketsNHL @TheCBJArtillery

79.2K views   0:00 / 0:05

💬 20   🔁 739   ♡ 1.9K   ⬆

FIGURE 10.3 Channel disentrainment does not necessarily mean the channel is asynchronous, just that it gives you an opportunity for self-presentation by disconnecting your self from your self-presentation in some way. Tools like Snapchat filters or Zoom's "touch up my appearance" can facilltiate disentrainment even in synchronous interactions by altering your appearance or making your favorite hockey player look like a puppy. (Incidentally, we use similar technology to do robotic surgery. But yeah, the doggie filter.)

et al., 2008), they still afford a chance of putting our ideal self forward. Consequently, users of these disentrained channels can create profiles, messages, and responses in a careful, purposeful manner to best meet their communicative needs. Significant work has applied the hyperpersonal model to understand self-presentation and interactions in social media (e.g., Fox & Vendemia, 2016; Lyu, 2016; Scott & Fullwood, 2020). When we're not physically tethered to our self-descriptions and self-statements, we can strategically present ourselves on social media.

Masspersonal Communication

Chapter 4 dealt with masspersonal communication directly, but here in Chapter 10, we further unpack it as an integral element of what constitutes a social medium. Earlier media typically fell cleanly into tools for either interpersonal communication (e.g., email, text message) or mass communication (e.g., group chat, discussion board), but social media emphasize facilitating masspersonal communication. Newer social media tools—even those approaching their third decade in service, like Facebook and YouTube—allow individuals to post messages to other individuals as readily as broad user groups and likewise for broad user groups to post a message to a single individual. This is one reason social media have been so notable for acts like bullying.

Bullying has always been a problem, and the trope of the schoolyard bully is well understood. However, the phenomenon of **cyberbullying** has been of great concern as it allows individuals and groups alike to continuously take hostile actions toward a target individual (Whittaker & Kowalski, 2015; Wingate, Minney, & Guadagno, 2013), even after they've left the same physical space or the victim leaves the online space (i.e., logs off). Snapchat, Reddit, and TikTok thus provide a means for both kids and adults to continuously pick on, tease, and name-call classmates, coworkers, and random targets. (Remember, social media are persistent channels. Unlike the playground or the watercooler, the messages continue even when someone—either the tormenter or the target—leaves.) Cyberbullying can have substantive negative psychological and physiological impacts, particularly among children and young adults, for both the perpetrators and victims (Marciano, Schulz, & Camerini, 2020).

These same masspersonal channels can have positive outcomes. For example, when an individual posts a Yelp review of a restaurant to the mass audience of Yelpers, their message can inform and guide others' dining choices. In the workplace, the masspersonal nature of enterprise social media can foster a more positive and engaging work environment by allowing workers to develop social as well as professional connections, fulfilling our need to assimilate and belong (Song et al., 2019). In all, the masspersonal communication of social media is neither good nor bad: masspersonal communication is just a bit more common and accessible on social media as compared to FtF or older CMC channels.

Perceived Interaction

An interesting element of our definition of social media is that interaction need not be present or manifest—only *perceived*. Social media can allow us to feel like we are interacting with another person, even if that interaction is not real, and we may be interacting with a computer or not engaging in an interaction at all. This element is perhaps most readily evident when thinking about whom you Like on Facebook or follow on Instagram. You may follow Selena Gomez on Instagram, receiving daily updates about her thoughts, projects, travels, and breakfast. Moreover, you can mention @SelenaGomez in your Insta to send a message toward her, though whether she'll ever read that message is a bit of an unknown. However, by reading messages Selena Gomez posts that seem to be directed to you (although they're really directed to millions of followers, and thus masspersonal), you start to feel close to her. We call these relationships **parasocial** as they seem like a social interaction, yet are really one-way. Early explorations of parasocial relationships looked at how viewers relate with characters on soap operas and news anchors (Horton & Wahl, 1956): figures with whom viewers became intimately familiar yet could not interact. Although we can interact with celebrities (see Figure 10.4), organizations, and others now via social media, this interaction merely must be perceived—the celebrities, organizations, and others need not alter their communication based on our messages or responses, though they may.

On social media channels, we do not need to actually interact with everyone we follow. Indeed, interacting with all of our social media connections in frequent and meaningful dialogs would be impractical and implausible. Consequently, what is critical to social media is simply the perception that you could interact. Even if a friend doesn't like your Snap or Snap back, they *could*. And that potential for interaction—knowing that others can see and engage with our content, even if they don't— is sometimes just as communicatively enticing as direct

FIGURE 10.4 As you follow Beyoncé on Instagram or Twitter, does being aware of her travels, family, experiences, and insights make you feel closer or more connected with Queen Bey herself? Do you feel like Beyoncé is talking directly to you, even when her post is visible to her 154 million followers? If so, you are likely in a parasocial relationship with Beyoncé.

communication itself. Offline, there's a low chance that Mark Hamill will ever interact with me directly; but online, any time I tweet and mention @HamillHimself there's a chance that he *could* respond. (And he once did, liking my November 8, 2018, tweet, thereby fulfilling one of my top three life goals.) This potential for both known and unknown others to interact with us and our content makes social media more unique than typical interpersonal or broadcast CMC channels.

Value from User-Generated Content

Finally, the value of a social medium for its users should be derived from the content generated by the users themselves. It is important to specify here that the primary content of the medium need not be user-generated. Sites like *Huffington Post*, *Gawker*, and popular blogs are not user-generated and therefore have the same gatekeepers and barriers to entry as traditional media like television and radio. However, when the users of a medium perceive the value or the gratification of the medium as derived from user-generated content, then the site meets this criterion of a social medium. Indeed, you may visit several sites not for the primary content, but for the content posted by other users. For example, TripAdvisor constantly crawls the Internet looking for hotels, restaurants, and attractions to index. TripAdvisor's list of points of interest and hotels, as well as their phone numbers, addresses, and other contact information, is all generated by the service itself. However, the value of TripAdvisor is not from having the names and contact information of accommodations, as YellowPages.com and hotel websites do the same. Rather the value of TripAdvisor is derived from the reviews, suggestions, and travel tips left by users. Though TripAdvisor may let you know there is a quaint bed-and-breakfast in the town you are visiting, its value is provided through the person who stayed there last week letting you know that there is no en suite bathroom and everything smells like feet.

The gratifications and value may depend on the site or service. For example, the value of Instagram may be derived from social connection, LinkedIn's value is generated through job networking, and the value of Yelp is in its food reviews and restaurant recommendations. Regardless of what the particular value of a given tool is, all social media derive their primary value from user-generated content. Your favorite social medium would be much less fun and engaging if all the content you and your friends posted were stripped away.

Going Viral Increasingly, one form of gratification users may derive from interacting online is their content either going viral or being part of a viral trend. Much as in its epidemiological roots, **virality** refers to content that yields a large number of views or iterative instances within a short period of time due to sharing (Tellis et al., 2019). Early viral social media content included memes (see next chapter), YouTube videos like "Chocolate Rain" and "Evolution of Dance" (I'll wait a moment because I know you're desperate to check them out. …Yep. Those were viral videos.), and, more recently, PSY's *Gangnam Style* music video. The 2014 Ice Bucket Challenge, begun by the ALS Association to raise awareness of amyotrophic lateral sclerosis (ALS), also known as Lou Gehrig's disease, quickly went viral as individuals challenged their friends to pour a bucket of ice water on themselves to simulate symptoms of ALS and encourage others to donate (see Figure 10.5). Scholars (e.g., Kwon, 2019) have noted the Ice Bucket Challenge went viral as it capitalized on several unique elements of social media, specifically the early and visible incorporation of celebrities into the challenge and the use of explicitly calling on one's network ties to perpetuate the challenge. Similar viral challenges followed to various effects, including prosocial (e.g., the Trashtag challenge to clean up litter), entertainment (e.g., Harlem Shake, flipping a water bottle to land upright), and distasteful (e.g., the Tide Pod Challenge, licking and returning ice cream to stores' shelves) viral acts. Perhaps most recently the importance of virality on social media has been demonstrated on the Tik-Tok platform as younger users share dance routines and other content in brief videos, partially with the intent of inspiring others to emulate and perpetuate that routine. Already a popular online medium for the sharing of short videos, TikTok gained exponential popularity during the COVID-19 pandemic, as users turned to social challenges and videos as a means of social engagement while physically isolating (Kennedy, 2020).

FIGURE 10.5 The ALS Ice Bucket Challenge went viral on social media, as users poured buckets of ice water over themselves and then challenged members of their social networks to do the same, all to raise awareness of and donations for ALS research. How did elements of the Ice Bucket Challenge maximize its opportunity to go viral?
Office of Governor Baker

The virality of content can be hard to predict. Not all challenges garner the publicity and social involvement of the Ice Bucket Challenge, and not every TikTok video gets millions of shares. What's the secret to going viral? Frankly, there's not a magic formula, but viral content has commonalities that can increase the chance of going viral. One of the dominant factors that seems to predict if social media content will go viral is the arousal of emotions the content engages (Berger & Milkman, 2013; Guadagno et al., 2013). Tapping into human emotions, either within the content itself (e.g., showing something happy occuring) or through the evocation of viewers' emotions (e.g., showing something that makes users happy), can increase the chances of that content going viral. Moreover, positive content is more likely to go viral than negative, so uplifting or prosocial content tends to get shared more. Another element that can increase the virality of social media content is novelty, as content that captures users' attention is more likely to be consumed more diligently and subsequently shared (Brady, Gantman, & Van Bavel, 2020). Briefly, if it is interesting enough to grab your attention amid the cacophony of social media content, it's likely you also think it may also be of interest to your social network, and so you share it. Things perceived as creative or informative are also more likely to go viral, though not necessarily because of the content. Rather, users share content they consider creative (e.g., a unique advertisement) or informative (e.g., a news article) with their networks when they think their network members will subsequently think more highly of them because of that content (Moldovan, Steinhart, & Lehmann 2019). Users may perceive themselves as gatekeepers or providers of quality content and share to elevate their social status among their peers.

Finally, in addition to characteristics of the content being shared, virality is partially determined by the communicative network structures that transmit and retransmit the content. Actors central in their relational networks sharing content can speed up the retransmission or awareness of the content. Actors too far removed from their social network or with networks that are too homogenous (i.e., similar) may be challenged to diffuse and spread social media content, either by having too few individual connections or too few connections with slightly different interests who have not already been exposed to the content. Viral content is more effectively spread by having several key users at the centers of their respective and heterophilous networks diffuse the content to their many disparate social groups, who can in turn spread it to their unique network ties (Liu-Thompkins, 2012). In this way, social media's ability to collapse our social network into a single environment helps information and messages go viral more readily than they will through offline networks.

Again, though, there's no magic formula for what makes content viral. Particularly given the continual deluge of content on social media and the changing societal standards that govern what is emotive or interesting, the fact that a kid on

Research in Brief

You Gotta' Fight, for Your Right!

This chapter focuses mostly on how social media are used in interpersonal or micro-level communication. But how do social media tools facilitate communication at larger scales or for social purposes? One way to examine this question is to explore political and protest movements, such as the 2010 Arab Spring, in which citizens of multiple Arab countries in the Middle East and North Africa (including Tunisia, Egypt, Yemen, and Syria) demonstrated public opposition to dictatorial governments and poor standards of living. Although governments can exert substantive influence over traditional media (e.g., censoring or preventing television broadcasts, newspapers, or radio coverage), the Arab Spring demonstrated that user-generated, distributed social media were more difficult to control and more effective at informing and coordinating the citizenry.

Noting that social media provides citizens new ways to document and share information, particularly to coordinate large groups into political participation, Tufekci and Wilson (2012)—with the help of local nongovernmental organizations—surveyed 1,050 Egyptians who had protested in Cairo's Tahrir Square as part of the Arab Spring about their demographics and media use. Their findings help explain how social media were used to organize and execute the demonstration. About half of the respondents (48.4%) indicated they'd first heard about the demonstration via FtF communication, but interpersonal media—including Facebook (28.3) and the telephone (13.1%)—accounted for most of the remaining respondents' first exposure to the Tahrir Square demonstration. Consistent with the decreased salience of traditional channels for anti-government news, few reported initially

hearing about the protest via radio, television, or newspapers.

Almost all respondents (92%) had phones, which were used to communicate about the demonstration. Eighty-two percent of respondents used mobile phones to communicate about the protest, and all respondents who had a smartphone and a Facebook account (52%) reported using Facebook to access information about the protest. Twitter, still in its early adoption stage, was used by 16% of respondents—but almost all of them used Twitter to communicate about the protests. Ultimately, most Cairene protesters used personal social media accounts to share and receive information about the protest, relying on interpersonal and masspersonal channels more than traditional news outlets.

Social media continue to be used to facilitate communication for civic and political engagement: Black Lives Matter, the Me Too movement, and resistance and reform efforts in response to continued school shootings in the United States have all used social media to establish connections among those who share the same views, to disseminate information (including documents and videos), and to coordinate social movements and protests. By circumventing traditional news media (and gatekeepers), utilizing interpersonal networks, and capitalizing on the viral dissemination of their messages, social media have created effective new ways of organizing and information sharing to effect social change.

Reference

Tufekci, Z., & Wilson, C. (2012). Social media and the decision to participate in political protest: Observations from Tahrir Square. *Journal of Communication, 62*(2), 363–379. https://doi.org/10.1111/j.1460–2466.2012.01629.x

Novocain is viral this month does not guarantee a similar video will go viral next month. In fact, because the second kid may be considered derivative, subsequent attempts to make similar content go viral may fail. But whatever processes or elements drive the virality of content, the many forms of social media tools today have perpetuated viral content more so than offline or Web 1.0 tools.

Social Network Sites

One of the most common—or at least most top-of-mind—subset of social media tools in use today are social network sites. Boyd and Ellison (2007) define **social network sites** (SNSs) as "web-based services that allow individuals to (1) construct a public or semi-public profile within a bounded system, (2) articulate a list of other users with whom they share a connection, and (3) view and traverse their list of connections and those made by others within the system" (p. 2). Based on these guidelines, many of the Web services we use today constitute SNSs, including Instagram, Twitter, and YouTube. These tools remain SNSs even if we access them via an app rather than a Web browser. All three of these sites allow users to create a profile, make connections with others (either through Friending, Following, or Subscribing, respectively), and view others' connections. It is worth noting that although Instagram, Twitter, and YouTube may be the more top-of-mind SNSs, hundreds of other SNSs exist. And though tools like Facebook may seem ubiquitous and like they dominate the globe, many SNSs are popular within specific regions or cultures. As you can see in Figure 10.6, if you live in China, most of the people you know will use the SNS QZone (also known as QQ). Russian speakers log on to the difficult-to-pronounce Odnoklassniki and easier-to-pronounce VK SNSs to connect with friends and family. And in South Korea, though Facebook is now the most-used SNS, the Cyworld SNS has remained popular since its release in the early 2000s. Although Facebook may have the largest global user base and brand recognition, it is important to note that many other SNS services also have large, diverse, and active user bases. As you think about what social media your family, business associates, customers, and users may utilize, it is therefore critical to remember that the social media *they* use may not be the same as those *you* use. You likely need to go the SNSs *they* frequent to communicate with them effectively.

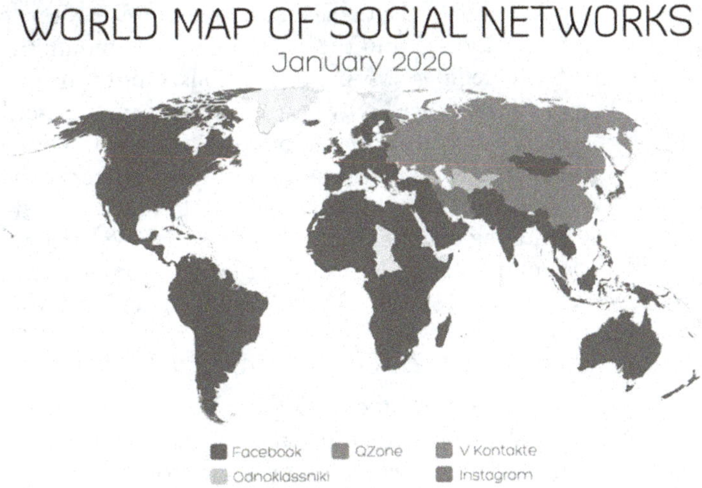

WORLD MAP OF SOCIAL NETWORKS
January 2020

■ Facebook   ■ QZone   ■ V Kontakte
■ Odnoklassniki   ■ Instagram

FIGURE 10.6 Though we may think of Facebook as one of our most common "social networks," Facebook is not the dominant social medium in all regions of the world. Different countries and regions of the globe use different SNSs.
Vincenzo Cosenza, vicos.it

Network and Networking Sites

A small but important distinction to make here is the difference between a "social network site" and a "social networking site." Social *network* sites typically supplement, reinforce, and maintain our existing offline relationships (discussed more in the next section), whereas **social networking sites** are specifically used to create new connections and meet new people (Boyd & Ellison, 2007).

Just as their name suggests, social networking sites are often used by individuals seeking to expand their social networks. These sites can be used to expand one's social (e.g., Meetup.com), professional (e.g., LinkedIn), and romantic (e.g., Match.com, Chemistry.com) relationships.

Although most social networking sites are SNSs, not all SNSs are social networking sites. For example, most dating sites such as eHarmony or Jdate allow—and in fact encourage—users to create profiles to introduce themselves to potential suitors. This feature—profile creation—meets the first criterion of an SNS. However, dating sites do not often allow users to view and follow others' network of contacts. It would be a little odd (But perhaps informative!) to be able to view your date's previous dating history and friend earlier paramours. Consequently, dating sites often do not meet the second and third criteria of an SNS. This bit of sophistry may seem trivial, and most of the time whether a service is an SNS or a social networking site is not an essential distinction. However, sometimes the difference between the two can be critical to their designers and users. Designers should pay careful attention to how the network system is set up and what features are included, as what the designers let the users do and the connections they allow users to form and display will influence how the system is used. Recalling the dating site example, exes being able to view and follow the network connections of subsequent dates may make users' dating lives pretty complicated. Alternately, users should carefully consider the purpose of the sites they choose to use, thinking not only of who they are trying to connect with but also who may try to connect with them. Many people may present themselves one way when seeking to meet new people, and present themselves another way to people they already know. But more on that later. For now, it is sufficient to know that the differences between SNSs and social networking sites have implications for what the services are used for, how they are designed, and who uses them. In the next section, we explore one common use of SNSs and social media more broadly: creating and maintaining social connections.

Connecting with Social Media

One—perhaps the most significant—reason individuals use social media is to take advantage of their social nature and utilize these tools to communicate and connect with others. From the earliest online communities, online tools have been used to facilitate interpersonal connections. One of the first virtual communities, The WELL (well.com), used only text to allow users to describe (and in doing so create) themselves and their interactions and to present ideas and hold discussions with others, but it fostered the connections of its members (Rheingold, 1993). Social media tools have subsequently intensified the Internet's role in bringing people together and facilitating communities. Though social media can aid organizational and workplace communication (see Chapter 6), two common uses of social media are creating and maintaining social ties and providing and receiving social support.

Finding and Keeping in Touch with Others

Take a moment and think about your Instagram or Snapchat use. But instead of thinking about *how* you use that social medium, think about *with whom* you communicate through it. Do you use it to keep in touch with family and close friends? Friends from high school who went away to another college? Do you have friends you don't talk to often but you will sometimes look at their feed? If you are like most SNS users, you likely answered "Yes" to all of these questions (Alhabash & Ma, 2017). What about another type of online community you are part of, perhaps a Discord server, an online multiplayer game, or a group chat for your high school class? Do the people you communicate with through those tools seem like friends to you? Do you feel close to some of the people you interact with online, even if you have never met them face-to-face?

Given some of these common experiences, it is likely unsurprising to you that social media are used to create and maintain social relationships. As online communities and interactions grew and developed, we began to understand that CMC can and does in fact facilitate interpersonal relationships and communities (Parks & Floyd, 1996; Rheingold, 1993). Now, you can meet, interact with, and develop friendships with individuals in SNSs (Burke & Kraut, 2014), massively multiplayer online games like *World of Warcraft* and *Elder Scrolls* (Carr & Van der Heide, 2009; Williams, 2006), and virtual worlds like *Animal Crossing: New Horizons* and *SecondLife* (Kim, 2014; Zhang, Marksbury, & Heim, 2010). We have seen that online communities can foster relationships that eventually continue or move offline (Parks & Floyd, 1996) or that continue exclusively online (Rheingold, 2003; Turkle, 2005). It is perhaps because of individuals' experiences with rich online relationships that CFO paradigms have lost dominance in favor of SIPT and hyperpersonal perspectives (Walther, 2011). Although online relationships can occur in many social media tools, it is perhaps in SNSs that relational maintenance is most obvious and has received the most attention.

Though MySpace, ClubPenguin, Facebook, and other SNSs were originally developed to allow users to meet new people based on common interests or friends, they have repeatedly been found to facilitate connections with people we already know, rather than to create or foster new relationships. In a seminal study of Facebook users—specifically college students, Facebook's dominant early user base—Ellison and colleagues (2007) found college students' Facebook friends were comprised primarily of friends they already had offline: a mix of new friends attending the same college and old friends from high school. Given these findings, the researchers concluded (and subsequent research has supported them) that SNSs are heavily used to support specific types of ties.

Relational Ties Broadly, **ties** are relationships between two individuals, and have been the subject of significant sociology research since the mid-1970s. Granovetter (1973) noted four types of ties: strong, weak, latent, and absent. **Strong ties** are close interpersonal relationships between two individuals, signified by significant time spent together, mutual emotional and social support, reciprocal communication, and relational closeness. Our strong ties are often comprised of family and close friends. Alternately, **weak ties** can be thought of as acquaintances or more general friends: those individuals with whom you may not have as much in common, but still maintain the relationship. These weak ties are often sources for new information, resources, and experiences, and help us broaden our horizons. It is interesting to note that weak ties have been found to be the most helpful in finding a job, as they allow access to the broadest set of networks that may know about job opportunities (Putnam, 2001). A third type of ties—latent ties—have recently begun to receive attention, given the ability of social media to facilitate and maintain them in ways offline channels often

FIGURE 10.7 SNSs like Facebook, Flickr, and Instagram can help us stay connected with friends. Even passively checking friends' social media feeds can facilitate relational maintenance.
scyther5

struggled to do. **Latent ties** are those relationships that we used to have, but may have let slip and no longer actively maintain. They are latent, though, as the ties are lying dormant and can be reactivated. A friend from middle school with whom you may not have spoken in several years may represent a latent tie, and it's certainly easier to accidentally encounter a latent tie in SNSs (Ramirez, Sumner, & Spinda, 2017), either by seeing a mutual friend interact with that latent tie or by the SNS's algorithm recommending them or their content. Finally, **absent ties** are relationships that do not exist, such as people in your class whose names you do not yet know and with whom you haven't even talked yet.

In looking at relational networks of Facebook users, Ellison and colleagues (2007) found users primarily communicated with weak ties, but also used Facebook heavily to maintain latent ties, particularly with high school classmates who may have attended a different college or university. These findings indicate SNSs have allowed new ways for us to readily maintain relationships with minimal effort. Though Ellison and colleagues' data were specific to Facebook, their findings and the same general human communication processes continue to apply and play out in more contemporary social media (Ramirez et al., 2017), from Instagram to Weibo. The simple and quick acts of reading content someone has posted, liking someone's picture, or providing a quick comment may be enough to maintain latent ties in a way not possible 20 years ago (Makki et al., 2018; Pennington, 2020). Before social media, individuals would need to engage in the very time-consuming (and potentially costly) acts of sending written letters, finding someone's email address and maintaining correspondence, or visiting distant friends. With social media tools, we can now readily and discreetly observe how a latent tie is doing and know that they may be doing the same to us, all the while being a wall post or private message away from reactivating that latent tie. Consequently, social media—especially the subgroup of SNSs—have proven to be excellent tools to maintain relationships that we do not want to go away, but do not necessarily have the resources to maintain (McEwan, 2020).

In addition to maintaining latent ties, social media can effectively help us keep up our relationships with weak ties. Weak ties make up the largest part of our social network, and include the friends and acquaintances we have and interact with that may not be in our core social network. Weak ties often give us a sense of belonging and community. Social media provide us an efficient and effective way to maintain and interact with weak ties. For example, while you may feel obligated to throw a party for a friend's birthday (after all, she threw you that great breakfast cereal-themed party last year), for an acquaintance, you feel your friendly duties have been met simply by sending them a quick message or Snap on their special day. As evidenced in this example, weak ties can be maintained with relatively low effort via social media, allowing us to quickly and cheaply maintain these relationships (Makki et al., 2018). Consequently, the majority of people we communicate with on SNSs are weak ties (Johnson et al., 2013; Utz & Breuer, 2019); though whether this is because SNSs are best for weak ties or we simply have more weak than strong ties is yet unclear.

Finally, social media can support strong ties and provide social support (Figure 10.8). By definition, we should spend considerable time interacting with our strong ties. We go out for coffee with close friends, vacation with our significant others, and spend holidays with our families. However, this time spent offline does not mean we do not interact with strong ties online. Rather, we often use social media to complement our offline relationships with strong ties (Kujath, 2011; Reich et al., 2012; Yu, 2020), utilizing social media as another means of maintaining close relationships. As media multiplexity (Chapter 8) tells us, we use more channels to communicate with others as our relationship becomes closer, and social media are yet another channel in our proverbial relationship toolbox. The next section will delve more into one of the primary outcomes of communication with strong ties via social media: social support.

Social Support

Although social media are important for providing access to latent ties and keeping us connected with weak ties, they also can help us communicate and interact with strong ties. These strong ties, whether we know them offline or not, can be a source of social support. **Social support** refers to the verbal and nonverbal behaviors taken to help someone in need of aid (MacGeorge, Feng, & Burleson, 2011), and pro-

FIGURE 10.8 Social media tools, from text-based asynchronous messages to synchronous audiovisual chats, can help us both access and provide social support.
Jasmin Merdan

vides us with a means of fulfilling our emotional and psychological need for support. We can see numerous examples of social support offline, such as when we meet a friend for coffee to figure out if we are in a major that is good for our career, when family members come together to celebrate our graduation, or when all of Hogwarts turns out to comfort Molly Weasley after Fred's death (that last one was a totally real and well-documented event). Although the ability to laugh and cry together, get and give a hug, or simply sit near someone all represent powerful forms of (mostly nonverbal) social support, they also take time and resources, as we have to physically meet up with someone, which can take time and money. We explored social support online more broadly and as a group phenomenon and resource in Chapter 7. In what follows, let's take a moment to consider how social support may specifically and uniquely occur via social media.

Social Support in Social Media Though the Internet has provided means of effective social support, social media specifically have enabled unique access to social support and novel forms of providing it. Though they may not be the only tools to communicate with some individuals, social media can allow unique access to social ties. For example, couples in long-distance relationships can feel isolated from their significant others, disconnected from the social support their loved one provides, even if that's as simple as knowing your partner is near and can provide support. Even without being colocated, knowing your partner is emotionally close can provide feelings of support and benefit your relationship.

As you recall from Chapter 4, *electronic propinquity* refers to the perception of psychological closeness with another, which can be critical to feel needed, supported, and as if you belong. Perceptions of electronic propinquity can be facilitated even via narrow, simple channels, creating a sense of presence and support (Kaye et al., 2005). Just knowing a close relational partner is on the computer at the same time facilitates the feeling that social support is just a quick message away, even if no support is actually needed or sought. Alternately, personal disclosures made on social media, such as through status messages or posts, can create perceptions of intimacy among members of one's social network, enhancing relationships even through indirect communication and ambient awareness (Lin & Utz, 2017).

Social support can also be provided directly via social media from strong ties—those most likely to provide impactful social support. Although strong ties are often those geographically close to us, as society and our personal networks increasingly globalize and mobilize, we find the individuals emotionally closest to us often are not physically near us. Those going away to college often find themselves separated from strong ties like family and close high school friends, and they use social media to communicate with those distant ties. Rather than needing to drive or fly home for a weekend, when faced with a situation where you need a sympathetic ear or to share a recent achievement, you may simply tag someone in an Instagram post, knowing that a distant loved one can read the message and reply, thereby giving you the needed social support.

Social support is often sought and obtained from strong ties via social media. Internet communities have been used since the mid-1980s as a means of providing social support, often facilitating close, deep, and personal strong ties amongst community members, even if the individuals have never met face-to-face (Rheingold, 1993). Though it may not seem possible to have a strong tie with someone you've never met, according to Granovetter's (1973) original explication of strong ties, the hyperpersonal model provides a rationale and theoretical support for the phenomenon. Individuals who interact over time primarily online and around a common interest (such as a shared medical condition or life circumstance) may begin to utilize selective self-presentation and feedback to develop idealized perceptions of each other, fostering close relationships—perhaps closer than could be maintained offline. Consequently, in many online support groups, members continue to interact even after their support need is gone (e.g., after beating the cancer that initially brought them to the cancer support group), both to provide support to new members and to continue relationships and supportive interactions with old friends. Within SNSs more directly, strong ties are the most frequent targets of individuals seeking social support from their established network, as well as the providers of the most substantive support (Kammrath et al., 2020). When you have your entire social network available to you on a social medium, you can target the person who will give you the best social support, regardless of where they are geographically.

Though strong ties are important for social support, social support can also be sought and received from weak ties. In a pair of studies, researchers at the University of Oklahoma looked at how college students sought social support and from who it was received, via Facebook. Unsurprisingly, strong ties were most often sought out for social support, even when using the relatively mass communication channel of a Facebook status message (Johnson et al., 2013). In other words, when students far away from home and family posted a message to Facebook seeking validation or a caring word concerning turmoil in their lives, they posted a status message visible to their entire Facebook network, but with a specific, limited audience in mind: their close friends and family. However, the same study found that social support was *received* from strong and weak ties alike (Rozzell et al., 2014). Both close friends and mere acquaintances commented on status updates with supportive messages, and the researchers found comments from strong ties were perceived as slightly more supportive than comments from weak ties. However, so many more weak ties than strong ties provided social support that in total the weak ties arguably provided more social support (Figure 10.9). Consequently, while strong ties may be sought out more and provide *better* support in social media, because the same media allow access to and with more diverse parts of our social networks, weak ties may provide *more* support in social media.

So far in this chapter, we have looked at some of the beneficial, prosocial outcomes of social media as means of communication. We have looked at how social media allow us to connect with others and maintain relationships. We have also explored the social support we can seek and receive via social media tools (as well as the Internet more broadly) and how online tools can help us develop interpersonal connections that would be challenging offline. Yet social media also introduce an increased need for awareness and concern over our privacy and security, as the very attributes that make social media a powerful interpersonal and relational tool also make them potential liabilities, both for individuals and for organizations.

FIGURE 10.9 Social support in social media can take many forms. Paralinguistic digital affordances including Likes and upvotes can allow network members to quickly and easily provide social support, and commenting and replies allow more substantive efforts to provide social support to an individual. Whether from weak or strong ties, all social support can help somewhat.
anyaberkut

Security

One topic often discussed in tandem with social media is that of privacy and security. As individuals put more information online, and that information is indexed, archived, and searchable, the issue of security and privacy within social media becomes increasingly salient. There is certainly much to be said about the issue of security online in general, as well as security issues specifically related to social media (for some good guides, see Flynn, 2012 and Solove, 2011). For now, we will briefly and broadly explore concerns that may be immediate to you as an individual and as you enter the corporate workforce.

Personal Information

Using a social medium is inherently a trade-off. You provide the social medium personal information and data, and in exchange for releasing that information you can use the social medium's services. A common feature of many social media is a profile: places on the medium where users can present information about themselves. Examining what information individuals self-disclosed on their Facebook profiles, Lampe and colleagues (2007) found:

- 87% of users disclosed their high school
- 83% disclosed their hometown
- 45% provided information about their current residence
- 80% identified their favorite movies, 78% their favorite music, and 47% their favorite TV shows
- 83% provided their birthday, including month/day/year
- 14% disclosed their home mailing address

Though these numbers have likely changed since the initial survey, that individuals volunteer such information is of interest. We likewise, and to the same effect, still see frequent threads and quizzes on contemporary social media about our hometowns, favorite teachers, and parents. What makes individuals disclose these types of personal information, and what are the risks of such disclosures?

Reasons for Disclosing Personal Information Individuals may disclose personal, identifying information for many reasons. Perhaps one of the most salient reasons for such disclosures is to connect more deeply with others. Social penetration theory, a hallmark of communication science, posits that individuals learn more about each other as the relationship between them deepens and intensifies (Altman & Taylor, 1973). Given that many of the individuals with whom we interact via SNSs are already friends, providing information about ourselves on sites such as TikTok and Instagram allows others to learn more about us, thereby creating more meaningful relationships. For example, many Facebook users who self-disclose their birthday online will find their inboxes and Timelines flooded on that day as friends, coworkers, family members, and acquaintances wish them a happy birthday. While some of these well-wishers may have remembered or had a note in their calendar about the birthday, many likely only thought to commemorate another trip around the sun after being reminded of the date by Facebook, which created a system-generated notification to the birthday lady's (or gent's) network based on her/his self-disclosed birthday (Viswanath et al., 2009). As individuals increasingly use social media like SNSs to learn about and maintain relationships with others, disclosing personal information on these sites serves the communicative role of facilitating relationship development and maintenance. Users may disclose personal information, from their birthday to their pets' names, to cognitively offload that information to the system rather than relying on an increasing number of connections to remember that information.

Another reason for disclosing personal information on social media is to achieve a strategic goal. For example, the professional networking site LinkedIn allows individuals to connect with others based on common career or vocational interests, and is often used by job seekers to identify new career opportunities. Likewise, many employers will turn to LinkedIn and other professional network sites to look up information about a job applicant before making a final hiring decision (Carr, 2016). Given LinkedIn's role (in part) as a way to connect employers and job candidates, many individuals post either elements of or an entire résumé on the site to maximize their exposure to potential employers (El Ouirdi et al., 2015). Just as in offline résumés, many LinkedIn users include contact and personal information in their LinkedIn accounts, either as profile elements or included in an attached résumé document, ultimately connecting the LinkedIn account with the user. Consequently, the very purpose of the social medium may encourage users to share personal, identifying information, from emails to mailing addresses, to help others reach them.

A third reason for disclosing personal information on social media is to add value to the medium or tool itself. Several online services collect and connect data about individual users and their on-site behaviors to increase the services the site can offer, and therefore add value for the individual. Though not necessarily social media, Web 3.0 tools like Amazon and Netflix utilize algorithms to examine members' purchasing and viewing behavior to make informed recommendations about other products or movies that users may also like (see Figure 10.10). If you have looked at several books by author J. K. Rowling, Amazon may recommend other young adult fiction such as the *Twilight* or *Hunger Games* series, based on what other users who searched for *Harry Potter* books also bought. Beyond these general recommendations, Amazon may also compare your search history to that of other users of similar demographics and realize your search is not normal behavior, functionally assuming you are seeking *Harry Potter* as a gift for a young cousin or sibling, and also recommend something for yourself as well—a new *Call of Duty* game—as when 20-somethings buy gifts for friends and family they often concurrently buy gifts for themselves. Services like Flickstr (for movies) and Pandora (for music) use similar algorithms and systems to increase the system's value (and concurrently the value to individual users) by essentially using personal information to learn more about users and better predict their behaviors. In many ways, these predictive services may be the future of social media. More directly, similar algorithms now also guide recommendations for connections on social media. By disclosing your network connections on Instagram and Twitter, the system may suggest other users or accounts with which you have something in common, even if you don't know what that shared characteristic or interest is. As such, disclosing your list of friends to a social medium may actually help the system find old or lost friends and allow you to get reacquainted (Ramirez et al., 2017), particularly as the simple act of adding a new friend on a social media platform can increase your attraction toward that newly added friend (Limperos et al., 2014). Ultimately, providing personal information may simply increase the quality and utility of the social medium tool itself.

A fourth reason for disclosing personal information on social media is to simply use the service. In many online applications today, terms of service require users to register in order to access the service. In these cases, personal information may not be disclosed to the broader user base of the service, and is simply disclosed to the firm administrating the service. For example, fantasy football pools, NCAA brackets for March Madness, and frequent customer cards for many retailers ask users to register their personal information so that they can use the service. For services with a financial aspect, such as a pay-to-play pool or customer rewards/refunds, you may even need to provide highly personal information such as a social security number, credit information, or a bank account or credit card number so that the service can process financial transactions. In all these cases, disclosing personal information is required simply to use the social medium, even if that use is not communicative or interactive in nature.

A fifth reason users may disclose personal information via social media is to express themselves. Even before social media as we think of them today, Turkle (1995) noted that CMC and online worlds can allow individuals to express themselves more freely and with less fear of personal retribution than offline, FtF disclosures. Given the highly identifiable features of many social media (e.g., your real name, photograph), these alternate or risky self-presentations may not occur as

FIGURE 10.10 You give up privacy to Netflix as you use its service, perhaps to increase its utility. Though Netflix may not share with others what you personally watch, the system keeps track of each show you watch, whether you binge the entire series or stop after a single episode, and your broad viewing habits. These bits of personal data are then used to recommend specific shows, genres, and movies that may interest you. We often do not think of this as being surveilled by Netflix, but the system is monitoring and recording our behaviors, and we must be willing to let Netflix know about our guilty and shameful reality TV pleasures in order to use the service—that's the trade-off we make.

frequently on social media, but the concept of disclosing via social media for self-expression remains solid. Research has shown a strong connection between Facebook use (including disclosure) and narcissism, so that individuals who are more prideful or ego-driven self-disclose on social media more (Jin, 2013; McCain & Campbell, 2018; Mehdizadeh, 2010; Ong et al., 2011; Ryan & Xenos, 2011; Scott et al., 2018). Social media like Tumblr and Snapchat can provide an outlet for self-expression, and those needing to self-disclose to others may find it therapeutic or a relief to engage in such disclosures, though they do so at the expense of spending a substantive amount of time and content thinking about and focusing on themselves, which is somewhat antithetical to the intention of social media.

A sixth and related reason to self-disclose personal information on social media is to use particular services or elements of the website. Even if the service itself is broadly free, some elements may require additional disclosures (either personal or financial) to function. For example, to send Facebook "gifts" (e.g., gift cards, digital icons posted to users' profiles often celebrating special achievements or events), you must provide Facebook your credit card information to process payment for the gift. Although gifts are not required of Facebook users, to utilize these features, users must provide personal information to Facebook. Though it may seem like you are just making a normal purchase or using these services as intended, it is important to remember that you are still providing personal information. Users may disclose personal information simply to utilize social media or social media services they perceive as valuable, but may not be aware they're doing so.

Take a moment and think of some of the social media you use. What information about yourself did you provide to these media? Why did you provide those data? Were you more comfortable sharing certain bits of information on some sites than others? If you have not explicitly thought about these issues for some services, it is highly likely you had some of these reasons in the back of your mind. Even if we are not highly aware of these issues, it is likely you've subconsciously worked through many of them. Disclosing personal information can provide you gratification as long-lost friends wish you a happy birthday after being reminded of the date via Instagram, save you search time and costs, recommend products you may have never thought of seeking out as Netflix recommends an old indie film based on your last search, or simply let you make a friendly wager with friends whether Michigan State University will make it to the Final Four this March. However, these benefits should be weighed against the potential risks of those same disclosures.

Risks of Disclosing Personal Information Though posting information online can have rewards, it can have risks too. Most of the time, these risks do not manifest. However, the *potential* for these consequences is very real, and users should carefully consider the risks they expose themselves to when posting identifying, personal information on social media.

One—perhaps the most substantial—risk in disclosing personal information online is the nature of the information often disclosed and others' potential misuse of that information. Recalling the types of information individuals typically include in their Facebook profile, does any of that information seem familiar? Many privacy experts have noted that the security questions banks and government agencies ask are the very bits of information many users provide in their SNS profiles (Nicholson, 2011). Think of the questions your bank's website uses when you log in or to verify your account if you forget your username/password. Questions likely include your favorite movie, mother's maiden name, address, or first pet. All of these data points are easily accessible by quickly checking your social media accounts (including both your profile and your network connections), and perhaps doing some cross-referencing with local records (see Figure 10.11). *Wired* author Matt Honan (2012) illustrated the ease of accessing a secure account, hacking into a friend's Apple account simply by providing Apple information from his friend's social media page, often giving several incorrect answers before hitting upon the correct response. Significant risk can be involved in putting your personal information online in general, as others (both individuals and organizations) may be able to access that information to do with as they will. Honan's example illustrates the ease with which an individual can be a victim of identity theft.

Even if others are not out to access your bank account, buy Justin Bieber albums on your Amazon account, or other immoral actions, organizations may still utilize your personal information in ways you neither anticipated nor desired. For example, the "Mood" discussion board of PatientsLikeMe. com was a message forum where users exchanged personal, detailed stories of their emotional disorders; and in 2010, its members saw its contents **scraped**. The board's

FIGURE 10.11 Making your personal information available online may allow others to access online accounts or other information about you to do so.

comments were automatically obtained and saved in a remote location (Angwin & Stecklow, 2010) by the Nielsen research and ratings firm, which was looking for information about potential markets for (among others) drug manufacturers. Upon finding their personal information and experiences had been captured and monetized, community members were outraged, feeling the ordeals about which they had posted to the forum had been stolen and their privacy violated. Although Nielsen was not out to engage in identity theft or bank fraud, the company's use of automated data gathering from a publicly visible social network and the ire that followed exemplify some of the concerns of self-disclosing personal information online. People can use information you provide in one social medium space for other purposes in other online venues.

A second potential risk to disclosing on social media is that information disclosed to the service may not stay private, even if you utilize rigorous privacy settings. When discussing the topic of privacy and safety in class, many of my students have stated something to the effect of "My information is secure. I use privacy settings." However, privacy settings can change, other users you know can distribute your information without your knowledge or permission, and ultimately the information you provide to many social media is no longer yours. Privacy settings are notoriously ethereal, changing from one month to the next as companies, economic and social structures, and technologies change. Facebook has been infamous for the many mercurial and ambiguous iterations of its privacy policies. Not only do the policies themselves change, but as the company updates its policies and technologies to facilitate these changes, major cracks have historically appeared in the armor of Facebook privacy settings. Perhaps most notable have been the repeated breaches of the privacy settings of Facebook creator and CEO Mark Zuckerberg and his family (Franceschi-Bicchierai, 2013; Hill, 2012b). (As a sidebar here, the Zuckerberg family's negative experiences with and reactions to privacy breaches on Facebook are particularly ironic given that Mr. Zuckerberg has been quoted as saying he doesn't believe in privacy [Van Buskirk, 2010].) Beyond circumventing or changing privacy policies, it is worth noting that often the information you post about yourself on a social media site is no longer your information. For example, Facebook's often-unread terms of service—the agreement users make with Facebook (https://www.facebook.com/policies/)—stipulates that anything you post to Facebook is the property of Facebook. Practically, this is because the text, images, videos, and interactions you put on Facebook are stored on its servers, so Facebook physically has a copy of the 0s and 1s that comprise your information. However, this also means that Facebook legally owns and can do with as it will the information you post, from messages to photographs (Silver, 2010). Indeed, much of the profit of most social media companies is derived from the user data they collect and sell to other organizations, including advertisers (Digital Information World, 2019). Knowing who talks about what to whom can be very valuable to marketers. Although many of my students indicate their information is private because they use privacy settings, I always am careful to remind them that: (a) privacy settings are not always effective, and that since (b) the information isn't theirs once posted, the settings do not really matter.

Balancing the Rewards and Risks of Disclosing Personal Information The previous two sections have raised several points about the help and harm of disclosing personal, identifying information on social media. Social media, from SNSs to discussion boards, often provide opportunities for individuals to disclose a lot of information about themselves. Sometimes, this disclosure can have benefits, such as social support from friends posting encouraging comments to your Instagram update that you aced your last exam. However, this disclosure can also have risks, such as an unwelcome stalker learning about your class schedule from the same post. How, then, do we make sense of social media and the double-edged sword of sharing personal information?

Every social medium presents rewards and risks from such disclosure. One way to begin making sense of information disclosure is to weigh the rewards you receive from using the service or providing specific information against the risks of others (including the system itself) learning about you. Another way is to consider if the service needs that personal information for you to use or obtain benefits from the site. Finally, consider how comfortable you would be to have the information you disclose posted publicly, even beyond the service to which you initially posted that information. Ask yourself, "Does the information this site wants me to disclose seem appropriate for what it does and for who else uses it, and would I be comfortable if that information were posted on a billboard in my grandma's neighborhood?" If the answer to all three of these is "Yes," then it may be okay to post that information. Just posting your name and the school you attend is pretty public information—most colleges and universities have searchable

FIGURE 10.12 When you provide information to a social medium, you're giving up some of your privacy, not only to other users but to the channel itself.

student directories that can publicly link the two datum anyway. Does eBay want your credit card to make an online purchase? If you trust the retailer and have read their privacy and information policies about what they'll do (and not) with your payment information, and think the benefit of buying through eBay offsets the risk of giving them your credit card number, then the information may be fine to share. However, if you are logging on to the newest SNS (Let's call it MyFlikBook [MFB]. Sure. Why not?) and MFB asks for your social security number to register because it wants to ensure everyone only has one validated account, does it matter how many of your friends use MFB? Your social security number is pretty secure and important: others could quickly put that information to unethical use by opening (and overspending) credit cards in your name or by using it to get other parts of your identity like your bank account number or school registration. As such, you may want to reconsider using MFB and stick with your Snapchat account—it's just fine for sharing pics and messages among friends. Ultimately, the decision to disclose personal information online, particularly via social media, should not be taken lightly.

Organizations Online

So far, we have looked deeply into how and why individuals must balance privacy and usability in social media. But these same concerns should extend to other entities as they seek to enter into and utilize social media. Organizations increasingly find themselves entwined in the rich world of social media tools, sometimes of their own choosing and other times because they are thrust into the medium without any say. Regardless of how an organization finds itself in a social medium, organizations should seek to carefully manage their presence in social media.

Companies often try to use social media to enhance their brands and connect with audiences, and they increasingly seem to have to manage their presence across multiple social media to do so. As of 2019, 99% of Fortune 500 companies had a presence on LinkedIn, 96% were on Twitter, followed closely by 95% on Facebook, 90% on YouTube, and 73% on Instagram (see Figure 10.13; Barnes et al., 2020). These figures represent a drastic year-over-year increase in the use of all platforms by these companies: just six years prior, only about three-fourths of Fortune 500 companies had an active Twitter and Facebook account (Slegg, 2013). Large corporations are turning to social media, in part to communicate directly with consumers and to allow the corporations' community of stakeholders to communicate with each other (Laroche et al., 2012; Lima,

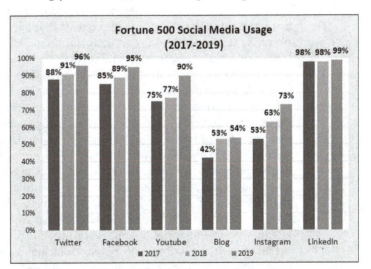

FIGURE 10.13 Fortune 500 companies continue to employ online tools—especially social media—to provide information to and interact with their publics. Though Fortune 500 companies have almost universally adopted most of the biggest social media tools, they also use less ubiquitous social media channels to reach specific markets, including channels like Discord, Caffeine, and Steemit, based on the social medium's user base and the organization's target. For example, organizations may create or engage on gaming streaming platforms like Twitch or Discord to target 18–35-year-old males, which accounts for a large component of site users. https://www.umassd.edu/cmr/research/2019-fortune-500.html

Irigaray, & Lourenco, 2019; Wang & Yang, 2020). Having an Instagram account can allow a company to send messages to consumers who have self-identified as interested in the company and its goods and services. This type of interaction may initially seem similar to the one-way mass communication afforded by billboards and television ads. However, the same Instagram page can allow users to engage with the company directly as well as to talk amongst themselves, creating rich dialogues about the company and its goods and services, and providing feedback about current and desired offerings (Harris & Dennis, 2011). Because of social media's ability to increase engagement about and with brands and organizations, channels like Twitter and Instagram have become a rich, interactive media for companies. Although this interaction requires careful monitoring, it allows new opportunities for organizations seeking a more personal connection with individual consumers, facilitated through masspersonal communication. As users feel like they can quickly and effectively communicate with organizations, their attitudes toward the organization can become more positive (Harris & Dennis, 2011; Hayes & Carr, 2015).

A benefit social media can offer over traditional static online Web 1.0 tools like webpages and emails is increased opportunities for user engagement. *Engagement* is a notoriously fuzzy term, and exactly what it means can change based on who is using it and in what context. Sometimes engagement for social media can mean "stickiness," or how long a user spends on a page or profile either reading or engaging with content. Other understandings of engagement refer to specific social media metrics, such as retweets or click-throughs on links to redirect users to other sites. One of the more common definitions of **engagement** with respect to social media is user-initiated action leading to the co-creation of value, often with behavioral, cognitive, and emotional dimensions (Khan, 2017). *Behaviorally*, engagement can manifest as clicks, creation of comments or other messages, or even just time spent reading or reviewing an organization's social media content—including the content about the organization other social media users have created. *Cognitively*, engagement can manifest as thinking about content—either content that has been posted or about content the user may create—or about the organization, which likely increases as a user spends longer on an organization's social media content or in brand communities. Finally, *emotionally*, engagement can manifest as emotions that arise from the online content, either in changes in emotional state due to persuasive messages, new information about the organization or its members, or even general emotions raised due to being on the organization's social media page. Engagement is a critical goal for many organizations' use of social media, as such behavioral, cognitive, and emotional engagement can result in positive outcomes for the organization. Specifically, individuals who engage more with an organization's social media tend to have increased trust, affective commitment, brand or organizational identification, and even purchase or donation intentions (Farrow & Yuan, 2011; Helme-Guizon & Magnoni, 2019; Jin et al., 2017; Vivek, Beatty, & Morgan, 2012). In other words, engagement can directly translate into greater financial outcomes for organizations as well as more connection on behalf of users, both of which organizations highly desire. This engagement, however, really calls for organizations to utilize and embrace the user-generated nature of social media, being comfortable giving up some control over their image and allowing others to share and interact with them online. Many organizations seek to guide social media content toward user engagement, holding contests, asking questions, and otherwise seeking to guide users' engagement with the organization's content.

It is worth noting here that organizations may often find themselves immersed in social media even if they do not actively create and manage accounts. For example, while a local restaurant may not have the resources to or interest in developing and maintaining a presence in social media, they likely have a profile created for them on food or restaurant review social media services like Yelp or Zomato. These sites allow users to comment on and share dining recommendations unsanctioned by individual establishments. Similarly, employer review sites, including GlassDoor and Indeed, allow current organizational members to review their employer and provide information to perspective applicants about the actual working conditions and environment. Companies should take care to continually monitor the Web and social media for reviews and references to their own services. Oftentimes, reviews may be positive and highlight the competitive strengths of the company. Besides, everyone likes hearing how well they are doing. Other times, reviews may be critical of services or staff, but even negative or critical reviews can provide opportunities for the organization to learn about itself and improve (Gossett & Kilker, 2006).

FIGURE 10.14 Online reviews can be difficult for organizations to manage, presenting both challenges and opportunities for the target of the review. Favorable reviews may be desirable to influence and affect others' impressions, but even critical reviews can be a chance to identify and correct problems.
KenWiedemann

Whether organizations venture into social media themselves or find themselves thrust into them, how organizations handle their online presence is critical. An unfortunate number of case studies exist where companies ignore the common online axiom of "Don't feed the trolls" and try to lash out against online critics. In the opening vignette, Applebee's attempts to combat online naysayers resulted in negative sentiment against the company (see Figure 10.1), and it took significant time and resources to restore perceptions of goodwill. Indeed, seeking to create good impressions of the organization can help the organization more than seeking to minimize or respond to others' negative comments (Hayes & Carr, 2021). Consequently, organizations can utilize social media to enhance their organization. Some organizations monitor their online word of mouth to be aware of organizational concerns not brought up through formal channels. Beyond simply using social media to monitor negative sentiment, some organizations have humorously embraced negative sentiment online and sought to turn it into a positive, using it either in marketing campaigns or to create interest in the company (see Figure 10.14). In both cases, it is important to remember that, just like an individual, organizations must be careful about how social media are used and how they facilitate self-presentation. Though much more can be (and has been) said about organizations monitoring their presence online, for now it is sufficient to say organizations should be careful to select appropriate social media and monitor vigorously to ensure the company portrayed via social media is the company as it wishes to be seen, and to manage its identity carefully.

Concluding Social Media

With a history dating back almost 30 years, social media are not necessarily new, and are certainly not revolutionary to current generations. Many of their features are appropriated and convergent from earlier forms of mediated and FtF communication: social media allow direct interpersonal messaging and broad mass communication, facilitate interactions with others, and foster socioemotionally rich relationships. However, their increased presence in society and their focus on self-presentation and networked inter-action have penetrated deep into our social fabric and collective consciousness. Many of us cannot go through our day without checking what our networks are saying or doing. However, as social media become more deeply rooted into the consciousness of the in-dustrialized world, more questions continue to emerge about the use and effects of social media. Questions arise about the way we present ourselves to others via social media as compared to FtF self-presentation and how we see ourselves as a result, how social media impact and are impacted by political campaigns, and even how the ability to visualize and traverse our social networks changes our relationships with others and our network's structure itself. Although many may see social media as omnipresent, ubiquitous, and well understood, issues like privacy concerns and social support through these emergent tools will continue to develop, and scholars, practitioners, and designers will need to continue to explore how human communication changes alongside ever-evolving social media.

Key Terms

social media, 186

disentrained, 187

cyberbullying, 188

parasocial, 189

viral, 190

social networking sites, 193

strong ties, 195

weak ties, 195

absent ties, 196

social support, 197

scraped, 203

engagement, 206

Review Questions

1. In what ways are social media different from static websites? What's the differ-ence in the form, the function, and the communication on your school's web-site versus social media page?

2. In social media, the value is derived from the interaction of users. "Value" can mean more than stock price or the valuation of the social medium's parent company. How may users "value" using a particular social me-dium—what gratification do they re-ceive? How does the form of users' engagement affect how they receive gratification from that interaction?

3. What is the difference between an SNS and a social networking site? After making that distinction, iden-tify one of each service that you use. What features do they have in common? In what ways do they look or operate differently?

4. Using SIPT or the hyperpersonal model, address how social support may be effectively (or even more ef-fectively) provided via social media than face-to-face. Are some forms of social support more readily trans-mitted via social media than offline?

5. Using your favorite social medium, what is the cost of using the me-dium (e.g., loss of privacy, selling of information, etc.), and what is the cost of *not* using that medium (e.g., lack of awareness, absence of social connections)? How do you weigh those costs? Are there things you would not be able to do were you not an active user of that medium

Influence Online

Let's begin this chapter with a little experiment. Do the following four things one step at a time, completing each step before reading the next:

1. Raise your *left* hand.
2. Say your name aloud. Be sure to enunciate clearly.
3. Now raise your *right* hand.
4. High-five yourself.

Did you do them? Good. How did that make you feel? Were any of the steps more challenging to accomplish than the others? If you did the four steps in a public place, how did others around you react? And, most importantly, why did you do what a book told you to? This is a textbook, not a cop. As silly as this activity was, it hopefully illustrates that influence can occur through mediated tools, even if from your doofus of an author via your textbook. Influence can be as explicit as a sender telling you what to do through a mediated message and you complying with them. Influence can also be more discreet and harder to identify. How do the channels we use to communicate affect how we seek to influence others? How do we tell when someone online isn't being truthful or accurate? And how are marketers and organizations adapting to and using newer digital channels to affect how we see the organization or their good/service?

In this chapter, we answer these questions by exploring social influence online from several perspectives, including online deception and deception detection, memes as tools for communication and persuasion, and more structured persuasive efforts in the form of advertising and public relations. Many online channels can be utilized to persuade others and ourselves, and though we'll spend a little bit of time looking at specific channels—such as social media ads and billboards in video games—even more critical is understanding the human communication processes

209

that uniquely occur or play out online. When we understand how individuals are influenced online, we can then take that understanding and apply it to whatever channel or context we find ourselves considering.

Social Influence

In this chapter, we'll discuss social influence and persuasion somewhat interchangeably, so we should begin by acknowledging their differences. **Social influence** refers to "symbolic efforts designed (a) to preserve or change the behavior of another individual or (b) to maintain or modify aspects of another individual that are proximal to behavior, such as cognitions, emotions, and identities" (Dillard, Anderson, & Knobloch, 2002, 426). Alternately, **persuasion** has been defined as an "intentional effort at influencing another's mental state through communication in a circumstance in which the persuadee has some measure of freedom" (O'Keefe, 2002, 5). Distinctions between these two terms can quickly be identified. For example, persuasion emphasizes the intentionality of efforts, and social influence considers behavioral changes as potential outcomes. For our purposes, though, we will focus on the shared elements of each: how individuals use communication to affect, influence, and persuade people's behaviors and perceptions.

One way such efforts can occur online is through the interactions and social connections we have online. Cialdini and Trost (1998) identified three major forms of social influence: social norms, conformity, and compliance. Though these forms of social influence were developed to articulate and describe social influence processes offline, in the following subsections, we explore how these forms of social influence and persuasion can manifest through CMC and affect users online.

Social Norms

Defining and Developing Social Norms Online Social norms refers to "rules and standards that are understood by members of a group, and that guide and/or constrain social behavior without the force of laws" (Cialdini & Trost, 1998, 152). Any time your thoughts or actions are guided by the implicit norms or habits of a group of which you see yourself a member—whether that's your family, your school, your team, or your circle of friends—you are being influenced by social norms. Offline, such norms are quickly evident once you know what you're looking for. After Elisha Grey pioneered the invention of the telephone, Alexander Graham Bell gained notoriety for its popularization. Interestingly, Bell advocated the use of the nautical greeting "Ahoy" when answering the telephone (Krulwich, 2011). Why, then, do you likely answer the phone with "Hello" rather than the intended "Ahoy"? The answer is likely social norms: you do it because that's just what everyone else you know does, and it would feel weird not to do the same. It's not that others would be mad at you or kick you out of a group because you answered with "Ahoy" or "Yo," but answering the phone with "Hello" is just what the people around you do. (It is worth an aside to note that a job applicant my manager once called to schedule an interview did not help his position when he answered the phone by yelling, "Who 'dat?!" into the receiver. The job was not with the New Orleans Saints, and thus this greeting violated social norms.) Social norms are the things your relevant social network seems to understand and agree to as sufficient and valid patterns of acceptable attitudes and behaviors.

Online social norms can be just as evident as their offline counterparts, particularly as individuals and groups are given time for these norms to establish and guide the community's behaviors, whether that refers broadly to the community of all online users or just to the networks or groups to which you belong. One of the most evident of these social norms is the online convention that TYPING IN ALL CAPS IS TO BE READ AS YELLING. (There's a good chance you just read that sentence in a different, louder voice than the rest of the paragraph.) Adapting to the lack of vocalics online, and consistent with SIPT, users have come up with social norms like "all caps" to create meaning. In this example, users may not intentionally attempt to persuade others through the paralinguistics or argument made. It may not even matter if the receiver understands or uses all caps to denote yelling in written communication. Instead, the influence is in the creation of the norms itself: whether or not you adhere to it, the collective culture of CMC users has adopted this convention as a communicative norm. A similar social norm in CMC has been the use of the phrase "Hello, world" as the first output of any new program, code, or social network (Trikha, 2015). The social norms of the sites, communities, or CMC tools you use and associate with can therefore influence how you communicate—both the topics and the very construction of your messages.

The Influence of Others A second way social norms can exert influence online is indirectly through perceptions. In an interesting phenomenon, the people with whom we associate online can influence how others see us, simply by the nature of our public connection. Because CMC tools can limit the information we have about people, people may attend more to the small or limited cues about a person that *are* available online to make inferences about them. One of these cues can be the people you know or with whom you associate, commonly accessible SNSs (Ellison & Boyd, 2013). Whom you associate with online can influence others' perceptions of you, in line with their assumptions of the others. For example, individuals with more physically attractive friends on social media are perceived as more physically and socially attractive themselves, even if no picture is visible or when holding the profile photo constant (Walther et al., 2008). (It's worth noting a similar process, sometimes dubbed the *cheerleader effect*, occurs offline as well, when a group of people is holistically perceived as more attractive than the average of its individual members [Van Osch et al., 2015].) The simple implication here is that if you quickly want to make yourself more physically attractive online, delete your more homely network connections. A more complex—and less superficial—implication is that influence online can occur indirectly: others' impressions of us are indirectly affected by the company we keep or the behaviors of others, rather than directly by our own actions. How we're seen is a function not only of our own communicative interactions but also of those with whom we associate (Figure 11.1).

Another way others online can influence your attitudes is through the desire to believe or hold a view held by a relevant social group, like peers and friends. For example, when students read instructor reviews (see Figure 11.2), their view of the instructor and the course are influenced by the reviews their peers leave of the instructor, regardless of the objective quality of education the instructor provides (Edwards et al., 2007). A group of students will have different perceptions of a class instructor, different attitudes about the course and learning, and even perform differently in the class based on whether they've seen different online peer ratings and reviews of an instructor, even though they're all in the same class session with the same instructor (Edwards et al., 2009). This is because we are influenced by the perceptions of others whom we see as similar to ourselves.

Many spaces online allow peer ratings and group feedback that can subsequently influence our attitudes about a target, whether it's an instructor, movie, or public service announcement. In such cases, the sender of the message—and whether

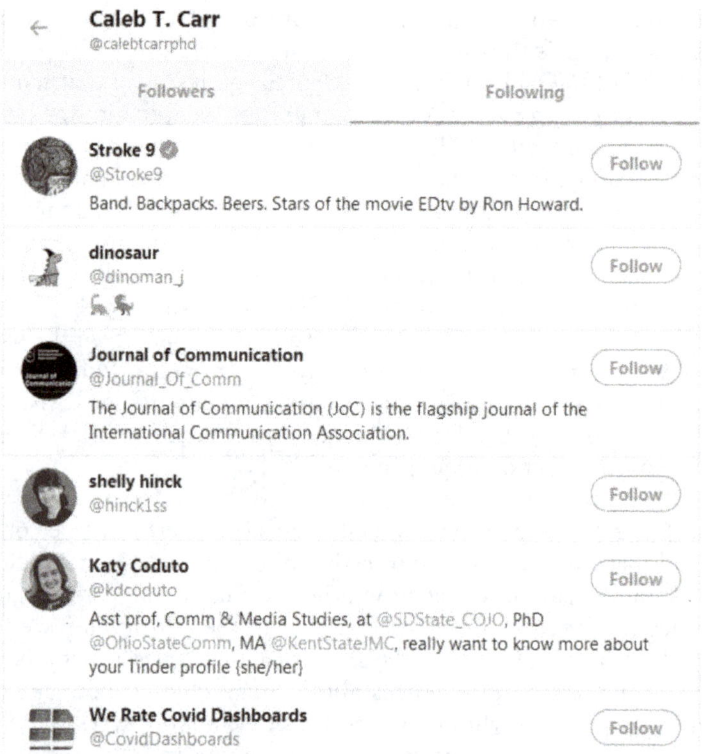

FIGURE 11.1 Part of how we are perceived online is based on with whom we associate. My Twitter associations may be more influential in your perception of who I am than what I may say about myself. What do my Twitter associations say about me? How would I strategically alter how I'm perceived by adding to or deleting from those I follow?

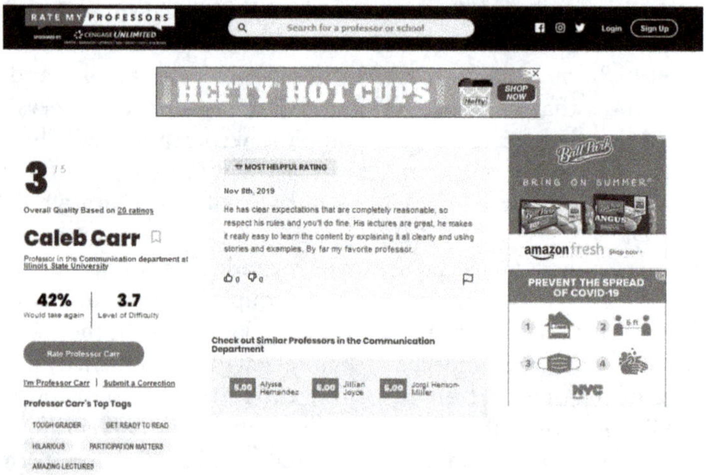

FIGURE 11.2 Peer influence can affect perceptions online, as your own views and perspectives are influenced by the views and perspectives of others online that you believe to be similar to you. How do online ratings affect your view of an instructor, movie, or product? Do you think your view of me as an instructor would be different if you hadn't seen this online rating from RateMyProfessor.com? How would the influence of the rating differ if you thought the ratings were from people very different from yourself (e.g., bad students, nonmajors)?

we see ourselves as similar to that sender—is critical to whether that message influences us (Walther, DeAndrea, et al., 2010). If we see the sender as someone like us, that sender's message will likely influence how we see the target. If the sender is unlike us, however, their message may have no effect, or may even boomerang to influence us counter to their message. As an example, after watching a public service announcement on YouTube informing you of the dangers of ingesting Tide laundry pods, you are more likely to be convinced not to eat detergent if you see the same comment, "Yeah! The Tide Pod Challenge is dumb. Don't eat pods!" from a fellow awesome college student rather than an older, boring adult who has nothing in common with you. This normative social influence can explain why not all messages we encounter online are persuasive. The effects of others' social influence are particularly important as we can increasingly access multiple social groups online.

As online networks and groups develop, their social norms naturally emerge as the members set informal (and sometimes formal) standards of behavior and interaction. You likely won't be jailed or fined for violating these norms, but the pressure to follow the normative behaviors or ideas of online communities is strong and substantively influences how we act and communicate online (Yee et al., 2007). Additionally, the violation of social norms can negatively influence how others perceive you, as nonconformity may make you seem like less of an ingroup member, creating social distance between you and the social group.

Social Norms and Self-Presentation A final way social influence can influence our behaviors is by affecting how we present ourselves online. As CMC tools let us connect across our offline social networks, social norms can be challenging, as individuals may struggle to understand which sets of behaviors should be prioritized. For example, if you and your friends do not consider drug and alcohol abuse taboo, there may not be a social norm against their use. However, when you post a picture online of yourself using alcohol and drugs, and that picture is accessible to a potential employer (whom may have different sets of social norms about alcohol/ drug use), the conflict between your norms and the organization's norms can lead the employer to sense a lack of fit between you and the employer (Bohnert & Ross, 2010; Rui, 2020). Likewise, signing off on an email with a religious scripture may seem like a normal articulation of your faith, but it can actually make receivers' perceptions of you more negative if they do not share your religious beliefs (Carr, 2017) or if religion is not relevant to the topic of the email.

The blurring of boundaries and ease of access across diverse social networks and groups online influences how we present ourselves and interact online. Particularly in social media, which frequently allow traversal of numerous social groups, the identity we seek to present is what Hogan (2010) calls the **least common denominator self**, or the self that is the least objectionable to the broadest swaths of one's various social networks. Said another way, the least common denominator self is the persona that is acceptable or consistent with the social norms of the largest number of people or groups with whom you expect to interact. This is not to say you lie about yourself—you instead only present the parts of yourself that comply with the social norms that span your diverse groups. You present the version of you that would be consistent with how your grandmother, your clergy, your softball team, your boss, *and* your friends would all see you, all without violating any of the social norms of their various groups. Being constrained by multiple social norms at any given time means our attitudes and behaviors are influenced by the expectations of the many individuals and groups with which we associate. This may persuade us to be less risk-taking or to more strongly conform to broad social categories online than we would offline.

Conformity

Conformity is when real or perceived pressure from others "causes us to act differently from how we would act if alone," including when we do *not* act in a certain way because the group would disapprove (Cialdini & Trost, 1998, 162). In other words, conformity refers to a desire to adhere to direct and explicated group norms, for fear of sanctions from group members or loss of one's group membership. Importantly, conformity does not need actual punishment or disapproval from others to work. Instead, individuals may behave in a way perceived as valued by or consistent with the group's articulated norms, because the individuals merely worry about what group members may think. For example, when interacting in online chat rooms, users may change their language style to be more or less aggressive in their messages based on the aggressiveness of other contributors in the chat room (Rösner & Krämer, 2016). This conformity goes beyond mere disinhibition. Instead, conformity occurs when individuals feel implicit or explicit pressure to follow a social convention they may not hold.

Conformity is often addressed as a negative, antisocial form of influence, as the broader group seeks to cajole individuals into adhering to a group norm. Underage drinking, social smoking, and listening to country music are all unfortunate behaviors that may arise as individuals seek to conform to the behaviors and attitudes they perceive as normative to their peers. But conformity can be used for good, prosocial outcomes as well. One way conformity has been used to positive effect is to motivate individuals to become organ donors. Building on the rivalry between the University of Michigan (Wolverines) and Ohio State University (Buckeyes), researchers (Smith et al., 2016) targeted Facebook ads to students enrolled at the two schools and portrayed organ donation as a contest, activating the desire for conformity (or outgroup antagonism) to donate blood. By noting that "Wolverines like you are becoming organ donors" and "Buckeyes are currently leading in organ donation signups," ads appealed to viewers' desires to conform to the norms of their social categories: Wolverines and Buckeyes. Encouraging students to register as organ donors relied on their desire to conform. The implicit pressure to act similarly to others of their social group increased individuals' organ donor registration far beyond typical donation registrations without the appeals to normative group behavior.

Compliance

Finally, **compliance** is acquiescing to an implicit or explicit request (Cialdini & Trost, 1998). Compliance may manifest as simply following a sender's request to click a link in an email. Even with something as common as spam email, just under half of users have opened an email suspected to be spam, and 11% of email users—often older adults—have clicked on a link in a spam email because the emailer asked or directed them to (Messaging Anti-abuse Working Group, 2010). To answer the perennial question of why you receive so many emails from deceased Nigerian uncles you never knew about, it's because *many* people fall prey to these scams and comply with the request, at least enough to make it worthwhile to scammers. Even a small percentage like 11% represents millions of users falling victim to online scams each year. The old adage that it never hurts to ask seems to apply online as well: with a low cost to send messages or request users to behave in a certain way, a simple call to action costs little, and even a fraction of receivers complying can be scaled to large numbers of people.

Individuals may comply with online requests, which are less nefarious than spam email, for several reasons. First, compliance may simply serve their interest or be of little cost. A common refrain at the end of many vloggers' episodes and Twitch streams is "Click the Like button and subscribe." This simple request costs viewers little time and few resources, as viewers are already on the page and Likes and follows are not fixed or scarce resources. Compliance may also come from the desire for *social validation*: the principle that individuals act a certain way after seeing others successfully acting that way (Cialdini, 2009). In other words, once others have demonstrated that complying will be rewarded or viewed favorably, they are more inclined to act that way themselves. When you book a trip or hotel room online, and the site pops up to note "2 other travelers have booked in the last 24 hours," it is an appeal to compliance (as well as possibly resource scarcity). Noting that other travelers like you have successfully booked this trip makes it look more favorable and normative for you to do so yourself, and therefore increases the likelihood that you too will book that trip or room. In a very practical example, learning that other participants have cheated in an online study or course can exert peer pressure to do the same (Kroher & Wolbring, 2015). Unlike offline compliance, however, online compliance does not seem to be a function of the requester's likability: whether you find the person making a request of you socially attractive or likable doesn't influence whether you actually comply (Guadagno et al., 2013). Ultimately, compliance online is likely still a cost-reward calculation for many users, as they consider what resources complying with the request would take and what potential benefits (or lack of consequences) may come from acquiescing.

Memes

The prior section has broadly addressed some of the social and theoretical processes underlying influence online. With that knowledge in mind, we can now explore the forms of persuasion online, starting with a topic you may have expected to have seen by now: memes. Memes as we know and use them are slightly different than how the term originated. *Meme* originated as a biological term describing genetic replication (Dawkins, 1976). In social interactions, the term *meme* was initially used to mean "an actor meaning structure that is capable of replication, which means imitation" (Spitzberg, 2014, 312). In other words, memes could be mimicked or riffed

on from the original version and used in a new way. That meaning is much closer to how we often talk of online **memes** now, as "units of culture that spread from person to person by means of copying or imitation" (Shifman, 2011, 188). These cultural units are typically either images or audiovisual clips altered from their original source to convey new content or ideas. Early memes on YouTube often shared a subset of characteristics: portraying "ordinary" people, presenting flawed masculinity, use of humor, simplicity, light and whimsical content, and some degree of repetitiveness (Shifman, 2012). The focus and portrayals of memes have continued to grow and diversify alongside their wider adoption, and now memes are drawn from movies, television shows, comics, current events, and other media. Users then overlay new text and/or audio onto the original in order to create new content. As more new iterations are made from the original content (see Figure 11.3), it is said to have

FIGURE 11.3 Memes take images, videos, and audio and replicate and modify them to create new meaning. The above meme references a popular *Game of Thrones* quotation—"Brace yourselves. Winter is coming."—and alters it to address the upcoming discussion of memes. Though the meme can stand alone, much of its value is in the knowledge of the source material or quotation. Unfortunately, like with a good joke, I've now made the meme remarkably less funny by explaining it.

become "mimetic," in that it is replicable and sharable, demonstrating the properties of a meme.

As communicative tools, memes have expanded to almost all domains, and which meme may be popular on any given day is hard to predict. Users of sites like Reddit and Imgur often use memes to convey information, humor, news, and relationships, and sites like Imgflip and Kapwing host templates to allow users to readily create and distribute their own memes. Regardless of their source, memes now often rely on a generally simple and replicable image or message that other users can co-opt and alter. A basic understanding of the original source of the meme's development helps other users adapt the meme for new purposes, but often with the same underlying structure. Because memes are easily edited and relatively low cost to try out, new iterations can be quickly created and shared among networks, with those resonating with or being well received by the online community shared and additionally altered (Spitzburg, 2014). By nature of being humorous, quickly created and disseminated, and tapping into collective understanding, memes have become a prominent source of both entertainment and—as we'll discuss in a moment—influence.

Memes in and as a Form of CMC

Online, memes began as forms of humor and geek culture, which were later slowly adopted into the mainstream (Denisova, 2019). Early memes were often silly non sequiturs, such as the "I can haz cheeseburger" meme, in which a cat looked longingly at a cheeseburger, and the "All your base are belong to us" meme, in which a character from the *Zero Wing* video game utters an ominous (and poorly translated) update on the player's campaign. Online discussion board and bulletin board systems (BBSs) co-opted these (and other) images from popular culture and continued to adapt and appropriate them. Later memes have employed cartoons, stock photos, classic artworks, and numerous other multimedia to express ideas.

As a form of CMC, memes are very versatile, allowing users to encode and transmit large amounts of information through a relatively short message. The juxtaposition of novel copy (i.e., text or words) over an established image can build on the assumed or prior knowledge about a meme (such as the implication that the man and woman on the left of the "Distracted Boyfriend" meme (see Figure 11.4) are holding hands and in a committed relationship, and that the man's facial expression denotes sexual attraction or interest at the expense of his girlfriend. To then label the three characters as two things that may present competing interests but without explicitly discussing the presented figures' relationships presents an opportunity to convey complex information relatively simply. Because of their simplicity and interactivity (encouraging others to also adopt and recreate the meme), memes have become a prominent feature in CMC.

Memes as Persuasive Communication

Increasingly, memes are used as persuasive tools. One reason memes have become effective persuasive tools is that their editability and shareability quickly diffuse information among like-minded users. Because memes can be fast to make and are rapidly disseminated via social media, email, and other online tools, users can quickly create and distribute messages that reinforce their opinions and attitudes or that disparage dissimilar opinions and attitudes. In looking at the sharing behavior of memetic videos, Guadagno and colleagues (2013) found that individuals were more likely to share and spread a meme—to make it "go viral," in keeping with the biological origins of memes—when it evoked a strong emotional response, often

FIGURE 11.4 The "Distracted Boyfriend" meme used a stock photo of a boyfriend ogling a woman other than his girlfriend (top left). It became a meme to articulate when someone was paying attention to something other than what they were supposed to. In these memes of the original, we see references to *Daily Bugle* editor J. Jonah Jamison's grudge against Spider-Man (top-right), a reference to the Greek god Zeus's infamous promiscuous infidelity (bottom left), and the desire to take a nap rather than doing anything else (bottom right) which just describes this book's author. In these three examples, we see references to comics, Greek history, and personal shortcomings, all demonstrating the versatility of a good meme. What message could and would you convey using the "Distracted Boyfriend" meme, and who would understand, appreciate, and be influenced by its humor?

based on social identities. In other words, a meme strongly against an outgroup or for the ingroup was the most likely to be shared. This ability to speak to ingroups or against outgroups may be one reason memes have been heavily adopted for trolling and the spread of misinformation or conspiracy theories. Particularly among groups with strong ideologies, memes—including Pepé the Frog by the alt-right and Russian trolls' creation of "snowflake" recruitment posters to dissuade people from voting for Hillary Clinton (Donovan, 2019; Nuzzi, 2016)—can be powerful tools to spread false claims or information among like-minded network members (Mocanu et al., 2015) without using traditional news channels that would censor or correct the misinformation. Because of their ability to reinforce previously held beliefs and social identities, persuasive memes may be most heavily circulated in ideological silos (see Chapter 13).

Memes have also become popular persuasive tools as means of discourse or critique about otherwise charged or sensitive issues, such as politics or social issues. Much like political cartoons, addressing taboo or delicate subjects indirectly through memes allows senders to avoid direct resistance to messages and receivers to consider a perspective different from their own (Kumar, 2015). For example, if you were to be directly told that your favorite show was terrible, you'd likely not pay much attention. But the same person sharing an image from the show with a deprecating caption may get past your biases and make you think for just a moment about whether the acting in that show is actually good. Similarly, memes can be

TABLE 11.1 **Properties of Advertising and Public Relations.**

Advertising	Public Relations
Paid	Earned
Builds exposure	Builds trust
Audience is skeptical	Media gives third-party validation
Guaranteed placement	No guarantee, must persuade media
Complete creative control	Media controls final version
Ads are more visual	PR uses language
More expensive	Less expensive
"Buy this product"	"This is important"

The goal of both advertising and public relations may be to influence consumer attitudes and behavior, but how that influence manifests and the processes by which it occurs differ widely.

Reprinted from Wynne (2014)

used to persuasively delegitimize or minimize others' persuasive efforts. For example, Greenpeace created a campaign that satirized a persuasive campaign by Shell Oil, altering Shell's messaging to mock and challenge Shell Oil's identity and trivializing and reducing the persuasive appeal of Shell's campaign (Davis, Glantz, & Novak, 2016). By mocking, satirizing, or otherwise making a joke out of a persuasive message, creators and their messages can reduce the original message's persuasive appeal or effectiveness.

Advertising and Public Relations Online

Influence online can also occur more formally, such as when organizations seek to influence consumer behavior in the form of advertising or public relations. **Advertising** refers to the "nonpersonal communication of information usually paid for and usually persuasive in nature about products, services or ideas by identified sponsors through the various media" (Bovee & Arens, 1986, 5). Alternately, **public relations** is "a strategic communication process that builds mutually beneficial relationships between organizations and their publics," and as such focuses on building relationships with key stakeholders to "shape and frame public perceptions of an organization" (Public Relations Society of America, 2019). Notable in these definitions is that advertising is nonpersonal (i.e., one-way) communication to sell a product or service, whereas public relations consists of two-way communications seeking to build relationships and trust. Additional distinctions are presented in Table 11.1. Organizations and groups may use one or both of these persuasion techniques in their marketing mix, seeking to persuade their publics to either buy goods and services (advertising) or view the organization or their relationship with it more favorably (public relations).

One-Way Advertising Online

Advertising is typically considered one-way communication transmitted from the organization to the audience or members of the audience. We see traditional advertisements on television or billboards and in magazines or newspapers, receive them in our mailboxes, and hear them on the radio. In many cases, online advertising is not substantively different from offline advertising, and advertisers simply take established advertising techniques and apply them to digital tools. Direct mail advertisements have become emailed ads delivered directly to our inboxes. Television and radio ads are video and sound files that precede or interrupt our YouTube videos, streamed shows, or Pandora and Spotify playlists. And print advertisements in our newspapers and magazines are now digital picture ads inserted into the sidebars of our emails or interspersed among our social media feeds. In these cases, online advertisements are very

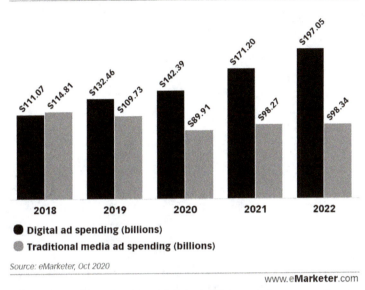

Digital vs. Traditional Ad Spending
United States, 2018-2022

- Digital ad spending (billions)
- Traditional media ad spending (billions)

Values shown: 2018: $111.07, $114.81; 2019: $132.46, $109.73; 2020: $142.39, $89.91; 2021: $171.20, $98.27; 2022: $197.05, $98.34

Source: eMarketer, Oct 2020

www.e**Marketer**.com

FIGURE 11.5 Money spent on digital advertising, whether on emails, social media advertisements, or in-game content, is expected to continue to increase and even to exceed spending on traditional media advertising.

similar to their offline analogs, simply broadcasting a message to buy a product or use a service to an undifferentiated set of receivers. As individuals spend more time using CMC tools, it is likely unsurprising that advertisers are spending more of their ad dollars on digital ad campaigns, sometimes reducing their ad buys in traditional channels to do so (see Figure 11.5).

But digital tools and CMC also give advertisers new channels and opportunities through which to sell their products and services. One opportunity CMC tools provide advertisers is greater control over the audience or subset of the audience that receives or can access their advertisement. The advertising concept of *narrowcasting* suggests that it may be a better investment to spend a little more money per viewer on an advertisement if that audience member is the right demographic or psychographic target for the ad, and as such targeting a narrower audience more deliberately can be more effective. The copious amount of information about us available online makes for a narrowcasting advertiser's dream, as ads can be targeted and displayed to a very small and particular subset of receivers. Additionally, targeted ads typically have higher click-through rates than broad, undifferentiated ads (Yan et al., 2009), suggesting their higher cost is worth it and provides a higher return on investment (ROI). Many social media and digital platforms' primary income streams come via advertising revenue, and advertisers have developed complex (but easy-to-use) tools to target specific audiences and users (see Figure 11.6), and therefore narrowcast much more effectively than they could offline or in traditional media.

Virtual worlds also present new channels and opportunities for advertising. From open worlds like *SecondLife* and *Sansar* to video games like *Need for Speed* and *PUBG*, virtual worlds present advertisers new audiences and placements for advertising content (see Figure 11.7). Advertising in virtual worlds has several benefits. First, as audiences become more averse to the interruptive advertisements they encounter elsewhere, ads in video games do not need to be repetitive or intrusive (Kim, 2008). More diverse ads that do not distract players can help advertisers get

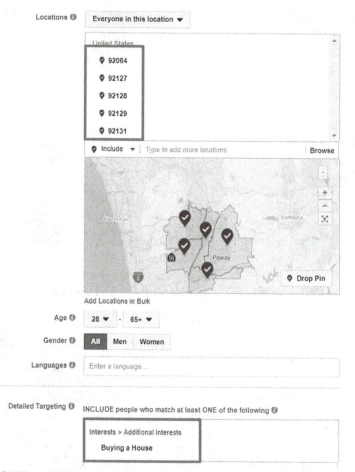

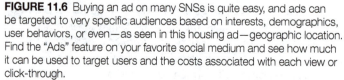

FIGURE 11.6 Buying an ad on many SNSs is quite easy, and ads can be targeted to very specific audiences based on interests, demographics, user behaviors, or even—as seen in this housing ad—geographic location. Find the "Ads" feature on your favorite social medium and see how much it can be used to target users and the costs associated with each view or click-through.

past mental or perceptual filters to reach out to audiences in new spaces. This benefit will likely wane as individuals become more aware of and accustomed to ads in new virtual environments, but for now the sheer novelty of discreet advertisements in online spaces can help advertisers more effectively shill their wares.

A second benefit to embedding advertising is that many users do not mind and may even appreciate the gameplay more. Players *can* react negatively if the advertisement or product placement does not fit with the tone or theme of the game (e.g., a Mountain Dew ad in *World of Warcraft*'s fantasy world, Azaroth). However, if the ad fits with the context of the game, gamers do not mind the ad placement and experience high levels of recall (Molesworth, 2006), in part because it can increase the realism of the experience. In other words, if you play a racing game and drive past billboards, it actually makes the game and experience more believable when the billboards have advertisements. In this case, players as well as advertisers may benefit from making the gaming experience more realistic and immersive.

A third and final benefit to advertisers in video games is that because ads are entirely digital, they are relatively cheap to update, substitute, or modify. Unlike a physical billboard, which must be climbed and to which a new ad must be physically adhered, the code of a digital billboard in *Animal Crossing* can simply be replaced with the code of a new billboard. This means that online advertisements can be updated much more often than offline ads. For example, digital advertisements may not only promote a television program but also be updated every seven days to reflect that week's specific episode. Ads can also be customized and tailored to the individual player. Given that greater personalization of ads online can increase the ads' effectiveness (Walrave et al., 2018), this ability to quickly update and

FIGURE 11.7 Wherever you go, ads may be inescapable. Advertisers seek to buy spots wherever their market is—even in video games. And since digital ads can be quickly edited or targeted to audiences, you may even see ads more personalized to you online than in traditional mass media.

tailor ads online and in games can increase their effectiveness.

One challenge—both ethical and procedural—to advertising via CMC may be the increased blurring of boundaries between advertising and other forms of communication—it can be increasingly difficult to identify an advertisement amid other media. One of the most evident of these boundary blurrers is **advergames**: advertisements in the form of games. Advergames may be as simple as changing the protagonist to the advertised product, as Coca-Cola did for its *Pepsi Invaders* game, which was like *Space Invaders* but with a Coke can fending off invading cans of Coke's competitor, Pepsi. Advergames can also simply include the product in the sidebar or frame of another game. But advergames can also include more direct and involved incorporation of the brand into the game, such as when Doritos chips created the *Doritos VR Battle* game, in which players compete in a virtual world to collect Doritos. *Doritos VR Battle* may seem like any other game, as its premise of chasing and collecting some McGuffins is not dissimilar from games like *Pac-Man* or *Sonic the Hedgehog*. However, you can readily see how competing to get Doritos chips' tasty, orange-hued goodness is really an advertisement. Particularly when targeted at children, advergames can become challenging to regulate and monitor as these adverts can label themselves or masquerade as "games," and thus avoid regulation on advertising that targets children (Montgomery & Chester, 2009). Kids can have a hard time distinguishing between an escapist pastime and an advertiser's attempt to influence their behavior (see Figure 11.8), and most advergames do not clearly identify themselves as advertisements (An & Kang, 2014). Though digital advertising may be effective, both advertisers and consumers should consider the ethical and social implications of advertising via CMC.

Dialogic Communication and Public Relations

Unlike traditional mass media channels like television and radio, CMC tools—especially social media—typically provide environments facilitating interactivity. This

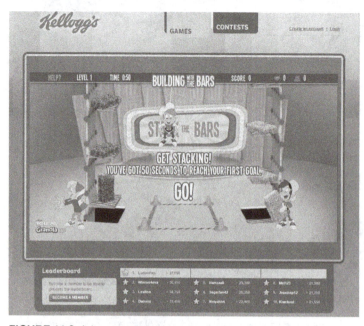

FIGURE 11.8 Advergames provide a means for brands to have customers interact with them, providing exposure to the brand without explicitly asking consumers to buy a product or service. The challenge of stacking your Rice Krispies Treats™ belies the fact that you're spending a lot of time engaging with the brand, albeit through the Snap, Crackle, and Pop of an online game.

interactivity means public relations practitioners increasingly engage in spaces where they are both the sender and receiver of messages. This has necessitated a shift in how public relations is conducted online, toward a two-way model of communication. A **dialogic** approach to public relations online shifts the emphasis of communication as a tool to cocreate and negotiate (rather than manage) relationships with publics and stakeholders (Kent & Taylor, 2002). The practitioner may create an initial campaign or message, but once it is online, the public is free to engage with, respond to, and further share it. Because users can interact with the organization, message, or each other, online public relations efforts are more collaborative than offline public relations efforts, as organizations and their public are equal contributors to the design and success of persuasive messaging.

Because public relations emphasizes the development of relationships, it takes time and interaction to develop such relationships via CMC PR efforts, consistent with SIPT. Consequently, two-way exchanges—or at least the potential for them—between organizations, brands, or practitioners and the publics they are seeking to influence are important to foster that perception of relational development. The benefit of dialogic public relations efforts is that, when successfully applied, they can increase the sense of closeness, immediacy, and/or consumer behavior intentions of the public. Examining Fortune 500 companies' use of Twitter to engage (or not) with Twitter users, Rybalko and Seltzer (2010) found that companies who responded to users via Twitter were almost twice as likely to have users interact with the company later on Twitter. Not only is approaching online public relations dialogically appropriate for social media and relationship management with publics, doing so can also lead to practical, monetary benefits for the company.

Dialogue with the Organization: Company-Consumer Interaction Particularly in social spaces online, the ability to interact with the company and peers about a message can influence the persuasive nature of that message. Seeking to prevent negative or counter-brand messages (as well as just off-topic feuds among users), some organizations disable comments on their online platforms. Though preventing others from critiquing or discussing the brand's posted content can dissuade trolls or unwanted comments, doing so can also decrease the collaborative spirit of being online and the ability to engage with the brand. Those seeking to promote or influence others about a brand or cause online who disable commenting or other social features can actually find a loss of trust, negative attitudes toward the brand, and lowered purchasing intentions as a result (Hayes & Carr, 2015). Alternately, greater engagement from a brand or organization with publics can lead to stronger brand-public relationships (Seltzer & Mitrook, 2007). By preventing others from speaking ill of a company online, the brand may also prevent others from saying good things about the brand online or limit the brand's interactions with its publics.

That brand-public interaction is the power of CMC. Unlike traditional media, CMC tools can provide opportunities for relationship-building dialogue between publics and practitioners. Rather than just broadcasting messages via mass media, public relations practitioners can now take advantage of CMC tools' interpersonal and masspersonal abilities to nurture interactivity with their publics. One opportunity in creating this dialogue is for companies to invite users to create or share content, particularly that which may favor the organization or reinforce its commitment to its brands. Sharing a story or a photograph with a brand represents a directional flow of communication very different from traditional channels, as the brand solicits feedback from consumers or seeks to motivate them to initiate contact with the brand (Tsimonis & Dimitriadis, 2014). Doing so can have substantive benefits for the brand, including perceived immediacy with its consumers and its incorporation into consumers' social identity. Individuals talking about or sharing favorable experiences with a brand—whether via posting a Yelp review, writing a blog post about the brand, or sharing a photograph with the brand—can increase individuals' social identification with the brand (Carr & Hayes, 2019). In other words, those times you post a status or email a friend to share your glee that "It's #PumpkinSpiceLatte season again at @Starbucks!!!" may have the unintended consequence of making you identify more strongly with Starbucks (see Figure 11.9).

Social media play a dominant role in dialogic communication with publics. Whereas static webpages are primarily used as mass media by organizations to broadcast information about new projects or campaigns, by their nature social media provide opportunities for interaction with the brand and with other users. Simply following a brand on a social medium can increase the perception of a relationship with that brand (Labrecque, 2014). Even if you never send a message directly to the brand, you *could* and you see others like you do so. Perhaps, then, it is unsurprising that the most common reason users follow a brand on a social medium is not to learn about the brand, but rather to interact—either to send messages to the brand's corporation or to talk with others about the brand (see Figure 11.10; Nisar & Whitehead, 2016).

← **Tweet**

BIGGBY® COFFEE ✓
@BIGGBYCOFFEE

We've found them. We found the cutest hot cocoas. Don't forget that tomorrow is the last day to enter our Instagram photo contest! For contest rules, please visit our highlights on our Instagram page at Instagram.com /Biggbycoffee

2:00 PM · Oct 30, 2019 · Hootsuite Inc.

FIGURE 11.9 Use of tools such as photo contests can help publics send messages to brands and others, increasing interaction with the brand and potentially reinforcing their identification with it. Here, a photo of kids dressed as Biggby coffee cups was submitted to an annual photo contest held by Biggby Coffee in which users shared their best photos (involving the Biggby brand, of course) via Instagram.

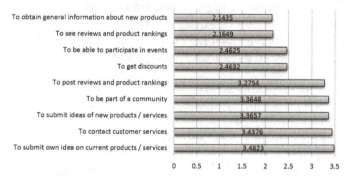

Reasons for following brands via social media

To obtain general information about new products	2.1435
To see reviews and product rankings	2.1649
To be able to participate in events	2.4625
To get discounts	2.4632
To post reviews and product rankings	3.2754
To be part of a community	3.3648
To submit ideas of new products / services	3.3657
To contact customer services	3.4376
To submit own idea on current products / services	3.4823

0   0.5   1   1.5   2   2.5   3   3.5   4

FIGURE 11.10 When asked why they follow a brand on social media, users' biggest motivation was communicating with or about the brand, more so than receiving messages from the brand. What does this say about social media as channels for dialogic communication and relational development?
Reprinted from Nisar and Whitehead (2016)

Dialogue among Users: Brand Communities A final dialogic opportunity for brands online is the establishment, maintenance, and observation of brand communities. A **brand community** is "a specialized, non-geographically bound community, based on a structured set of social relations among admirers of a brand" (Muñiz & O'Guinn, 2001, 412). In other words, brand communities are groups of people brought together by their shared love of and social identification with a particular brand. Because they bring together those who feel affinity with the brand, brand communities can be valuable to brands and their public relations efforts.

Sometimes, brands may seek to develop the community, either by establishing a space online for brand aficionados to interact or by encouraging or seeding a community online. For example, by sponsoring popular blogger "J Crew Aficionada," the J Crew apparel brand helped legitimize the blogger and the blog's followers as a semiofficial community of the J Crew brand (Farris et al., 2014). Brands can also take advantage of brand communities by helping maintain communities and community members' association with the brand. Brand communities can serve as a place for individuals who share a social category—fans and supporters of the brand—to come together, maintaining the salience and relevance of that social identity. Because they can keep members' brand identities active, there is a strong association between engagement in a brand community and subsequent attitudes and (purchasing) intentions toward the brand (Zhou et al., 2012) that brands seek to nurture. By creating a "community forum" or fan page, brands can maintain individuals' active associations with the brand. Finally, brands can benefit from monitoring brand communities, which are often the source of new ideas or thoughtful critiques that may advance the brand. For example, community members following "J Crew Aficionada" identified J Crew's need for a maternity line (Farris et al., 2014). Lego similarly frequently draws ideas for new building block projects from designs constructed by fans and brand community members in Lego's virtual online sandbox (Antorini, Muñiz, & Askildsen, 2012). Even if not directly communicating with brand communities, these associations of publics can provide valuable messages to brands online if the brands watch and listen.

Digital Deception

All of these new or modified forms of persuasion present new opportunities for communicators to be dishonest, misrepresenting (or not representing at all) their intentions or persuasive efforts. Such dishonesty and misrepresentation are in some ways made easier online. The selective self-presentation enabled by being online and disentrained allows senders a lot of latitude in what they present online, only giving off cues or information they see as central to the self they want to present. As the classic *New Yorker* cartoon goes, "On the Internet, nobody knows you're a dog"

(Figure 11.11). If anyone can be anyone via CMC, does that mean deception runs rampant online and nobody is who they seem? Such concerns should be approached cautiously, as though deception does occur online, it may be less frequent or problematic than first assumed.

Frequency of Deception Online

Because users can say whatever they want online about themselves or others, one might think that CMC leads to deceptive communication. Yet the data tell us otherwise: individuals are actually *less* likely to be deceptive in most mediated channels. Tracking college students' interactions with others and the number of lies told or deceptive statements made, Hancock, Thom-Santelli, and Ritchie (2004) found lies to be most frequent when communicators were face-to-face, followed by via telephone. The frequency of dishonest statements was lower in traditional CMC channels, so that instant messages (IMs) and particularly emails typically contained the

"On the Internet, nobody knows you're a dog."

FIGURE 11.11 By having more control over what we do and do not present online, we can exert control over how others perceive us. This is perhaps best summarized by Peter Steiner's famous *New Yorker* cartoon. Cartoonist: Steiner, Peter

fewest deceptive messages (Figure 11.12). Why did this happen? What motivated people to be more honest in CMC channels, particularly given our earlier discussions of disinhibition effects online?

Exploring their results, Hancock and colleagues noted the CMC distinction may not have been as critical as the verbal/written distinction in predicting deceptive communication. Written channels innately create and preserve a log of the conversation that can be later referenced. Fearing a communication partner may "check the transcript," communicators via IM and email may be less likely to lie or deceive, as future contradictions or alternate information can be checked against the text record. Alternately, verbal channels like face-to-face and telephone do not retain a persistent

record of the interactions, perhaps allowing users to feel more able to get away with small deceptions. Someone referencing a prior phone call or in-person exchange can be more easily convinced of mishearing or misremembering a deceptive statement when there is no ability to go back and verify the prior exchange.

Communicators were also more frequently deceptive in phone calls than in person. The CFO approach (Chapter 3) provides an explanation for this difference: phone calls

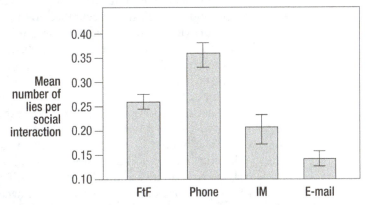

FIGURE 11.12 Rates of deception by communication medium (Hancock, Thom-Santelli, & Ritchie, 2004). Notably, individuals are actually less likely to lie via CMC channels (i.e., instant messaging and email). Where do you think teleconferences (e.g., Skype, Hangouts) would fall here—more or less deceptive—and why?

lack visual cues that could be used to verify information or look for confirmatory body language. When face-to-face, we can check to see if our communication partner is maintaining eye contact, looking nervous or twitchy, smiling or grimacing while saying they like our new shoes, or is even just generally attentive. All of these cues may raise concerns that our partner is lying in person, and communicators may be less likely to lie in person than when those many nonverbal cues are filtered out by the nature of a phone call. While phone conversations maintain vocalic cues, many of the other nonverbal cues we would use to detect deception are prevented by the channel. Given these findings, it would be easy to extrapolate that deception may be more common in audiovisual CMC (e.g., Zoom, Skype) or ephemeral channels (e.g., Snapchat, Bolt) where there are no persistent records.

Catfishing and Astroturfing

Although deception may be rarer online within interpersonal conversations between known partners, when the communicators do not have prior knowledge of each other or do not necessarily anticipate interacting again in the future, deception may be more frequent. Such deception may be most evident through individuals' and organizations' attempts to influence initial attributions about a target, whether those attributions are about whom to date or where to eat.

Catfishing Catfishing refers to presenting alternate personal characteristics or pretending to be someone else entirely in an effort to establish a—often romantic—relationship. Catfishing need not be malicious in nature and can be used to simply create a relationship under an alternate identity. But often, particularly in popular culture including MTV's television documentary series *Catfish*, catfishing has become synonymous with a duplicitous attempt to create a false relationship through mediated interaction. Catfishing draws on the hyperpersonal model to work, as instead of presenting an idealized self, senders present an alternate self believed to be particularly attractive or of interest to the target individual. Seeing this person online who appears to be giving off attractive character traits, the target begins to idealize the catfisher and sends reinforcing feedback to further intensify the relationship.

In an early and popularly discussed example of the practice, Notre Dame college student and football player Manti Te'o was understandably upset when the girlfriend he had met and established a romantic relationship with online (and later text messages and phone calls) succumbed to leukemia and passed away. Except there was no girlfriend. The relationship was initiated by a California man who knew Te'o, who created social media profiles (using pictures of a female obtained online) to perpetuate the ruse, and otherwise pretended to be Ms. Lennay Kekua (Curry, James, & Harris, 2013). Though a fabrication, the mediated nature of their relationship made her "death" real and impactful to Mr. Te'o. Such a relational deception would not be as possible offline, where upon the first meeting suitors would be able to discern physical characteristics and Mr. Te'o would realize his paramour was not actually Ms. Kekua. But online, where the perpetrator could articulate any identity for Ms. Kekua, deception could be used to persuade a target like Mr. Te'o, both emotionally and behaviorally. Other examples involve individuals creating false personas online in order to gain access to either personal information or interpersonal trust, ultimately coercing others to send financial or emotional resources (Nowak & Fox, 2018; Van Gelder, 1996; Walther & Whitty, 2021).

Catfishing may not be an entirely online phenomenon. As *Catfish* host Nev Schulman (2014) notes, the historical figure of Cyrano de Bergerac engaged in a type of catfishing. With a nose he believed made him unlovable, Cyrano whispered lines to his attractive friend Christian so that Christian could seduce Roxane. Wooed by Cyrano's words uttered by Christian, Roxane experienced a form of catfishing. (See also *Roxanne* [1987], starring Steve Martin and Daryl Hannah.) Though catfishing can occur offline, the ability to construct personas online and to develop and maintain relationships entirely via CMC makes the online environment particularly well suited for the practice.

Astroturfing/Crowdturfing A second way CMC has been used to perpetrate digital identity deception is by misrepresenting others, either their identities or their attitudes, in an effort to manage a target's identity. If catfishing is presenting yourself deceptively, astroturfing is manipulating others to present you deceptively. Most commonly a concern in politics, astroturfing has been defined as "a form of manufactured, deceptive and strategic top-down activity on the Internet initiated by political actors that mimics bottom-up activity by autonomous individuals" (Kovic et al., 2018, 71). As this definition suggests, the most common form of astroturfing is the false generation of statements, reviews, or comments by members of the public to sway perceived sentiment. These deceptive statements can take several forms (see Table 11.2), but ultimately are used to influence public opinion about a politician or issue. More broadly, **astroturfing** refers to the use of messages that members of the public at large or an engaged audience perceive as in support of a position, but that are actually messages authored or influenced by the target the messages address. As

TABLE 11.2   Understanding the Types of Deceptive Information Online.

Disinformation	*False information spread with the intent to deceive.* Examples:
	• "I went to Paris for spring break" posted to Facebook, when you spent it at home.
	• Claiming you work for a prestigious law firm on a LinkedIn profile when you don't.
	• Using the "delay send" function of your email to send an email you write at 2 p.m. to your boss at 10 p.m. so that the email's time stamp makes it appear you are working much longer hours than you are.
Misinformation	*Incorrect information spread with the intent to harm.* Examples:
	• "Barack Obama is not a US citizen," which is not accurate, but is claimed to sow discontent or negatively influence potential voters.
	• Making untrue, negative claims on GlassDoor about a law firm's discriminatory hiring practices, not because you (or anyone you know) experienced them, but because they didn't hire you for a full-time job after your internship there.
	• Giving a restaurant a poor rating on Yelp, not because the food or service was bad, but because you personally dislike the server and are trying to reduce her/his tips.
Malinformation	*Strategic dissemination (and often misrepresentation) of true information with negative intent.* Examples:
	• An online dating profile that overstates your "love of the outdoors and camping" (You only went once!) because you think potential suitors will perceive those profile elements as attractive.
	• Claiming you worked for a prestigious law firm on a LinkedIn profile, when you actually only did a day of job shadowing there.
	• You had a singular bad experience at a restaurant that you frequent (and otherwise always get good service and food), and so you write a really negative review to make it sound like the bad service is a constant problem rather than an isolated incident.

The types of inaccurate or deceptive information communicated can be categorized in several ways, often based on the persuasive intent of the source. Can you think of times you've seen or used these strategies online? Reprinted from Keller and colleagues (2020)

such, astroturfing goes beyond politics and has been used online by advertising and public relations firms, organizations, religious groups, and other agencies to influence public attitudes or behaviors.

In politics, astroturfing has taken several forms. A more nefarious application of astroturfing has been the recent rise in disinformation campaigns via social media, authored by governments or campaigns or individuals working on their behalf. For example, during the 2016 US national election, Russian operatives used paid, strategically placed advertisements on popular social media, as well as more obfuscated efforts such as generating false Twitter accounts (run by either automated bot programs or paid operatives), to create the perception of public sentiment or to give traction to conspiracy theories (McKew, 2019). Examining the social media ads purchased by the Russian Internet Agency (IRA), al-Rawi and Rahman (2020) found that much of the geopolitical astroturfing the IRA conducted in Facebook and Instagram ads—seeking to influence US politics, the UK's Brexit vote, and depictions of refugees seeking asylum in Europe—focused on microtargeting select demographics and exploiting their fear of outgroups. Through divisive ads and unfounded claims, astroturfers like the IRA have influenced the global attitudes people hold about politicians, local issues, current events, and the nature of science.

Astroturfing is not merely a concern of international politics, and local political groups have used similar tactics to influence local issues. The South Korean election of President Park Geun-hye in 2012 was scandalized after it was revealed the South Korean National Information Service had created false Twitter accounts to spread misinformation about Park's rival (Keller et al., 2020). Organizations can also use astroturfing to influence public sentiment about political issues, creating websites, hiring influencers, and creating posts online in favor of the company's interests and seeking to sway public attitudes (Cho et al., 2011). Kraemer, Whiteman, and Banerjee (2013) offer an example of astroturfing in their case study of a local effort in eastern India to prevent a company from using open-pit mining to excavate bauxite, which would lead to environmental harm in the region. After strong early support in the community, the anti-mining effort was stymied when the organization hired one of the social movement's leaders to provide pro-mining messages about the same site. This apparent change in position from one of the movement's leaders led to an erosion of public confidence in the mining effort, and many were unaware the mining company had actually hired the leader to engage in astroturfing. Astroturfing has been used as a technique to influence others in other domains as well, including advertising and marketing.

Another form of astroturfing is the creation of fake accounts or reviews to influence consumer behavior, often called **crowdturfing**. One of the most common examples of crowdturfing is the creation of false Yelp reviews—restaurant or business owners create accounts to artificially inflate the positivity of reviews (Yao et al., 2017). Microtask sites like Fiverr and Amazon's MTurk have even been used to hire individuals to go online and author favorable reviews of places they've never patronized (Lee, Webb, & Ge, 2014), essentially allowing owners to buy an online reputation. Online product and restaurant reviews matter a great deal in the contemporary business environment because—in line with SIDE—we believe these third-party reviews are written by users like us (Walther et al., 2012), and therefore they can significantly influence how we perceive a business (Luca, 2016).

Detecting Deceptive Presentations

As selective self-presentation and the filtering out of certain cues can make deception easier via online communication, how can we have any confidence in the reviews or people we encounter online? First, take some solace in that people still deceive in FtF communication—sometimes more than online. Ask your neighbor how his dad's new restaurant is, and of course you'll be told the restaurant's food is the best in the world. Remembering that people you already know may be less likely to be deceptive in their online communication (Hancock et al., 2004) should be a point of reassurance. However, when your target is unknown, detecting deception may be more challenging. Even then, there are still ways CMC tools and ideas can be used to vet the information and claims found online.

Identifying Astroturfing As astroturfing seeks to influence us by affecting our perceptions of how the public—or at least others similar to us—views something, how then can we identify astroturfing? One way is through careful and critical media literacy, being mindful not just of what's posted but also by whom and how. Analyzing the South Korean National Information Service's astroturfing efforts during the 2012 election, Keller and colleagues (2020) found several telling practices to detect astroturfing, at least at scale:

- **Accounts posting on regular schedules.** Astroturfing accounts posted at regular intervals (e.g., on the hour, every 90 minutes), and primarily during the workweek (i.e,. Monday through Friday between 9 a.m. and 4 p.m.). These patterns do not follow the less-scheduled social media user behavior of casual users, and instead suggest the posters are doing it as part of a job or coordinated effort.
- **Clusters of retweets or signal boosts.** By exploring how user accounts interact with each other, social network analyses can identify astroturfing accounts by the nature of their frequent interaction with each other, particularly to reinforce or rebroadcast information provided by others in the network. We all have our close friends or social media network ties, but when most of an account's content comes from retweeting or reposting content provided by a very small subset of the account's network, it may suggest astroturfing.
- **Clustered account identifiers or user codes.** Systems that assign users some form of identifier based on the order in which they create their accounts can identify astroturfing by exploring the patterns of identifications of users posting about a topic. As astroturfing accounts are all created about the same time, the identifiers should be much closer together. Just as seeing a series of dollar bills with similar serial numbers can tip you off they were all printed and issued about the same time, when a large number of comments or discussion posts come from a small cluster of user identifiers, you are likely seeing astroturfing.
- **Hyper-focused or limited content.** An account or group of accounts having a singularly myopic focus may indicate astroturfing. We all have multiple interests, and we like to discuss them. Unless you are deliberately using social media for a specific purpose (e.g., an organizational account just to discuss your academics), most accounts typically have broad posts and contributions in topic and in target. Astroturfing accounts often focus on just one topic (e.g., healthcare reform, a political candidate) to the exclusion of all others. Another manifestation of this may be creating an account just to provide one review—a five-star review from a Yelp account opened a year ago and only reviewing that one restaurant may suggest it is a friend or family member of the restaurant owner more than it suggests the user created that account simply to review the one restaurant (Zhang et al., 2016). An account with an unusually small social network, unnaturally limited focus, or little activity beyond one area may be a sign of astroturfing.

Tell Me Lies, Tell Me Sweet Little Lies

It is likely that sometime recently, prior to making a purchase online, you looked at peer reviews of the product, service, or company. Whether buying shoes or finding a new doctor, we now often look to peer-generated reviews to inform and influence our behaviors. After all, any restaurant can put out a sign reading, "Best coffee in New York," but that doesn't make their coffee taste any better, because they have control over their sign and claim—and no restaurant would claim, "Worst cuppa' in Los Angeles." So we turn to online reviews to guide many of our consumer behaviors. But how do we make sense of online reviews?

DeAndrea et al. (2018) set out to understand how users make sense of online reviews. After asking half of their 123 college student participants to read a *Bloomberg Businessweek* article about business use of astroturfing or spamming fake Yelp reviews, the researchers asked all participants to look at reviews of a fabricated local pizzeria. In some conditions, reviews were formatted as a fake Yelp review, as if reviews were made by previous customers and unfiltered, and in other conditions, reviews were formatted as part of the pizzeria's website under the header, "Look what people on Yelp are saying about us," as if the restaurant had selected reviews to use. In both conditions, all three reviews were identical.

Although all participants saw the same reviews of the same restaurant, the format of the reviews—and thus the participants' belief in the bias or curation of those reviews—and their knowledge of Yelp review spamming affected the influence those three reviews had on their attitudes toward the business. Individuals who saw the Yelp page (rather than the website) were more influenced by the review and thought more favorably of the restaurant. Additionally, participants who read the Yelp review spamming article before seeing the review were more cautious, perceiving the reviews as less genuine.

Taken together, these findings indicate that as users believe the reviewed restaurant, product, or company has less control over the reviews—either through selecting which reviews to highlight or by populating a site with false reviews—the more they are influenced by the reviews. Said another way, people are influenced by the experiences and reviews of others they believe to be like them (e.g., other Yelp reviewers), but efforts to manipulate reviews can reduce the effectiveness of that peer influence, even when the reviews are exactly the same. For businesses or producers, the lesson here can be one that's difficult to learn: if you are going to ask others to review your products online, you may need to let go of control and not attempt to micromanage reviews (or at least make it look like you are), because if you do, you may actually harm the influence other users can generate.

Reference

DeAndrea, D. C., Van der Heide, B., Vendemia, M. A., & Vang, M. H. (2018). How people evaluate online reviews. *Communication Research, 45*(5), 719–736. https://doi.org/10.1177/0093650215573862

Warranting Identities Online Another type of deception we may want to detect is identity deception, as individuals can use mediated tools to construct and present a false self online. This type of deception may go farther than a slight or strategic discrepancy between a true self and an ideal self as we discussed in Chapter 9, as online we can present ourselves however we want. Sometimes this deception can be mild and strategic. Someone with a stigma (think a large birthmark, burn scar, or other physical characteristic that may be obvious on first meeting and unnecessarily distract from the communicator) can use CMC to act as themselves but not be known or defined by that stigmatizing characteristic (Turkle, 1995). Other times, though, incomplete or inaccurate online self-presentation may be more nefarious as a communicator seeks to alter our attitudes and behaviors toward them, such as via catfishing. How, then, can we determine whether someone is who they say they are online?

One way to vet someone's online claim is through the warranting value of that claim. Developed by Walther and Parks (2002), **warranting theory** refers to the strength of a connection between an online claim and the perceiver's belief the online target truly possesses the espoused characteristics. Online information thought to strongly connect the offline individual to the online persona is said to be a "high warrant claim" with **high warranting value**, and information thought to be easily fabricated—and thus not reflect the target offline—is said to be a "low warrant claim" with **low warranting value**. When making confident attributions about a person based on online information is important—such as for a dating partner, a job candidate, or choosing a dentist—we look to high warrant claims.

Parks (2011) claimed that for warranting to occur, three conditions must be met: "First, the source must make an identity claim and, second, a third party must comment on that claim in a way that others can observe. And finally, it must be possible for observers to compare the claim and comment in practical and meaningful ways" (pp. 559–560). Later research demonstrated that, with respect to Parks's second condition, the mere *ability* for a third party to comment on the identity claim is sufficient to activate warranting, even if no claim is actually made (Hayes & Carr, 2015). Ultimately, warranting occurs when individuals make self-claims which can then be assessed by other sources.

We can determine a claim to be high in warranting value in several ways. One of the easiest ways to do so is to find third-party claims. Though the person herself/himself may deceptively self-present to strategically achieve a personal goal (e.g., get a date, be hired for a job), others usually do not have the same goals and are therefore more likely to provide accurate information online about the target's offline self. Consequently, others' statements either confirming or refuting a claim are said to have high warranting value, as the target has less control over the nature of that other person's claim than her/his own self-claim (see Figure 11.13). For example, an individual can claim she is outgoing and gregarious. Someone else following up to agree by pointing out how fun Saturday night was getting everyone together would solidify a perceiver's thought that the target is actually as outgoing offline as claimed online. However, should someone instead follow up to challenge that claim by pointing out they stayed at home every weekend this month, perceivers would likely put more trust and impression-formation value in the others' claim, and ultimately believe the target is more quiet and isolated than her initial post wanted you to believe (Walther et al., 2009). In another example, anyone can simply state online that s/he is in a new relationship, but social media like Facebook can increase the warranting value of that claim

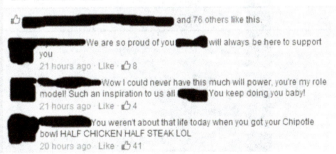

I have decided to announce that I am officially a Vegan! After seeing recent videos on Facebook regarding animal cruelty I have decided that I HAVE HAD IT. Meat and anything dealing with animals is just so not for me. I am SO PROUD of my decision and I feel as if it is a factor of growing up!

Like · Comment · Share

and 76 others like this.

We are so proud of you will always be here to support you
21 hours ago · Like · 8

Wow I could never have this much will power, you're my role model! Such an inspiration to us all You keep doing you baby!
21 hours ago · Like · 4

You weren't about that life today when you got your Chipotle bowl HALF CHICKEN HALF STEAK LOL
20 hours ago · Like · 41

FIGURE 11.13 Warranting theory says the harder it is to make, manipulate, or influence a claim, the stronger the connection between the online identity and the offline self. What forms the stronger impression of the original poster's offline dietary choices: the initial post claiming to be a vegan or the subsequent reply from a friend noting the poster had just eaten meat today?

by forcing the other party to recognize and legitimize the claim to become "Facebook official," in which the system displays the formal relationship between the two. This act of becoming "Facebook official" actually has such strong warranting value that couples often do not switch their profiles right away upon beginning to date, as the act of becoming "Facebook official" is a signal of a strong and meaningful offline attachment beyond simply going on a date or two because it requires the third party to verify the relationship status and significance (Lane, Piercy, & Carr, 2016).

Other humans' statements are not the only sources of high warrant claims online. Systems themselves can also generate information about others to help us be sure people are offline who they say they are online. For example, most SNSs keep and display a running log of all of an individual's connections on that network, whether they are called Friends, Followers, or something else. Though it takes little effort for someone to simply claim to be very popular with a lot of friends, that claim can be confirmed or refuted by the Friend count in the system in which the claim is made. Indeed, someone claiming to be very popular with few network ties is ultimately perceived as less social and lonelier offline (Lane, 2018; Tong et al., 2008; Utz, 2010). Similarly, devices themselves often send information about an online user, from time stamps to the type of device itself, which can warrant the individual's online claim (Carr & Stefaniak, 2012). For example, employers can use the time and date stamp of an email to increase confidence the worker is at her/his desk during assigned work hours and promptly responding to work emails (Paczkowski & Kuruzovich, 2016). A restaurant can claim to be "Best in Town," but if the amalgamated reviews of 563 Yelp users give it only one of five stars, the high warrant cue star rating will likely lead you to think poorly of that restaurant and to eat at a different restaurant. Consequently, computer systems themselves can generate or report other-generated information to serve as a high warrant cue (Figure 11.14).

FIGURE 11.14 Warrants can come from systems themselves, including time and date stamps, user activity logs, or even the device used to generate the message.

Another way to determine if a claim has high warranting value is to assess the degree of control the target has over the claims about her/him. Information that is readily manipulated or influenced by the person to whom it pertains is thought to be low in warranting value (DeAndrea & Carpenter, 2018). An organization may seek to create a flattering image of itself online by noting all of the good outreach efforts it does, only to find that others refute those claims. What should the organization do? In many online channels, including the organization's website or social media profile, the organization can control what third-party comments are displayed, deleting or hiding comments that may refute or work against the image claim the organization initially made. However, deleting or editing a user's message could leave traces, including system-generated notes that prior messages were deleted, other users saving the message before it was deleted and subsequently calling out the organization's activity, or a system acknowledgment that the organization has control over what third-party messages are presented through modification or deletion. Even the perception the organization is moderating third-party comments can increase users' concerns about the objectivity of claims made in social media. Such traces or suggestions of deletion of comments may result in users perceiving information on that service as lower in warranting value, as although people can make whatever comments they want about the organization, the organization has the power to remove or promote those comments to manipulate the claim in its favor. This ability of the target to delete or suppress unfavorable or inconsistent information reduces the warranting value of any claims it and others may make (DeAndrea et al., 2018).

Ultimately, warranting theory can help us understand how claims online connect back to the offline person or organization they describe. Claims (a) from others that support or refute an individual's self-claim, (b) that can be vetted and confirmed by others, and/or (c) that are less likely manipulated by the target described are said to have greater warranting value, and therefore provide a stronger link between the online claim and the offline target. So how can you tell that someone—either a potential romantic partner or an employer—is offline who they claim to be online? One answer to this question is to check what others say or to check to see if there are objective ways to verify online claims.

Concluding Persuasion via CMC

As we conclude Chapter 11, give yourself a pat on the back. Now high-five yourself. Now smile at your study partner. Did you do all that? If you did, it's hopefully because you wanted to, not because a book told you to and you conformed. Once again, this is a textbook, not a cop. But if you did, perhaps it's because you felt like other students or members of your study group would too, and you didn't want to feel left out (conformity). Perhaps it's because you assumed all the other students in the class would be doing so too (social norm). Or perhaps you were just looking to make your roommate or study buddy laugh (strategic influence). Many of these same processes and goals play out online. But given the billions of mediated interactions occurring each day, sometimes without direct intention (such as an email signature line), influence is a large part of the online communicative experience.

In this chapter, we have explored how individuals seek to persuade and influence others online. Sometimes, these persuasive efforts reflect their offline counterparts: someone may present their best self on an online dating profile just as they may stay on their best behavior on a first date. But CMC gives us new opportunities and challenges with respect to persuasion. Because social influence can be stronger online through the activation of social identities and the depersonalization of online actions, social norms and desire for conformity can be stronger online than offline in many instances. And in line with the additional means and mechanisms to be deceptive in various CMC channels, new ways have emerged of vetting claims to determine whether online statements accurately reflect reality. Even organizations are engaged in new means of influencing publics, finding novel ways to persuade consumers and stakeholders to purchase goods or services, or providing new and more interactive venues. Although the nature and function of influence remains stable online, the exponential growth of channels and means of communication means the ways we are influencing ourselves and each other are also expanding, and they require constant vigilance to be aware of when and how others seek to influence us.

Key Terms

social influence, 210

persuasion, 210

social norms, 210

least common denominator self, 213

conformity, 213

compliance, 214

memes, 215

advertising, 216

public relations, 218

advergames, 220

dialogic, 221

brand community, 223

catfishing, 225

astroturfing, 226

crowdturfing, 228

warranting (theory), 229

high/low warranting value, 229

Review Questions

1. What social norms are unique to online contexts and guide your behaviors when interacting via CMC?

2. How do ingroups online influence the persuasiveness of online communication, whether simple claims, complex behaviors, or the structure and content of memes?

3. What makes memes so pervasive and impactful in our contemporary communication? Consider both system factors and social influences.

4. In what digital channels are individuals least likely to be deceptive? What about those channels makes communicators less deceptive than in other channels?

CHAPTER 12

Politics via Computer-Mediated Communication

LEARNING OBJECTIVES

After reading this chapter, you should be able to …

- Address how people use CMC to talk about politics—both among themselves and with politicians and their campaigns.

- Consider how some of the old means of political activity have been moved online, and new ways to engage politically online as well.

- Reflect on the communication between politicians and constituents through CMC tools, and how these new forms of interaction influence perceptions of politicians.

- Identify how political campaigns are integrating new communication channels into their efforts and their effect on the campaigning process.

The use of media for political communication is not new. When citizens of a young American nation were debating the proposed US Constitution, politicians published and circulated *The Federalist Papers* to persuade the newly independent American people of the need for a federal government to oversee the 13 states. President Warren Harding's 1922 dedication of the memorial for Francis Scott Key, the composer of the "Star-Spangled Banner," was the first time a president's voice was broadcast via radio. Dwight D. Eisenhower was the first presidential candidate to air televised political ads, which he did in his 1964 "Eisenhower Answers America" campaign. The visuals of Richard M. Nixon and John F. Kennedy in the first televised presidential debate are largely credited with JFK's subsequent win in 1960 (although this long-held position has been challenged; Druckman, 2003). And you can thank Tony Inocentes for automated calls to your home or cell phone, as he was the first to use automated dialing and prerecorded messages for political purposes when he announced his run for California's 57th Assembly District in 1983. The incorporation of CMC tools into political communication seemed inevitable. We now see political communication all around us online. Politics and discussion of politics appear in our emails, our social media, and our websites, and in almost all facets of our online experiences. But what part of this is truly new?

Digital tools now give individuals means of interacting with and about politics in ways that weren't practical offline or through other mediated channels. The most significant way digital tools have changed how people talk about politics is the scale on which communication occurs. Online websites, forums, discussion groups, and social media now give individuals venues where it is not only appropriate but expected to talk about politics. In bringing together those who are politically involved, CMC tools can provide a place online for individuals to discuss politics.

The slow erosion of **third spaces**, locations where we congregate socially beyond home and work (Putnam, 2001), has been a large concern for those worried about political engagement at the national level, and even more so at the local level. As individuals are less likely to frequent bowling alleys, Elks' clubs, community centers, and other third spaces where they may casually discuss politics, they are also less likely to be aware of and engaged in political discussions and activities. Computer-mediated communication can provide numerous spaces specifically devoted to communication of and about political issues: podcasts, online news aggregators (e.g., Flipboard, Feedly), political groups, websites, and social media threads all serve as new, virtual third spaces for individuals to engage with others casually and informally about politics. Numerous spaces online are devoted to political communication: Reddit threads (e.g., https://www.reddit.com/r/politics), blogs and websites researching and discussing politics (e.g., https://fivethirtyeight.com/politics/), and emails and text messages increasingly are used to relay information about political issues. For those who are interested, large corners of the Internet are devoted to discussing and engaging in politics.

Computer-mediated communication tools also provide environments not devoted to politics, but in which political communication can inevitably occur. Increasingly, it seems like we are inundated with political messages online, as our inboxes, feeds, and timelines are filled with advertisements, political statements, and people sharing information related to politics. It may therefore be surprising to learn that only a minority of the general discussion in social media is about politics. Additionally, although about 15% of online content is about politics, that content is generated by about 10% of social media users (see Figure 12.1; Hughes, 2019). This tells us that a lot of people spend a little time talking about politics online, but also that a few people spend a lot of time talking about politics online, and is yet another example of the "Long Tail."

With all of these technologies and tools available, it is therefore worth some time to

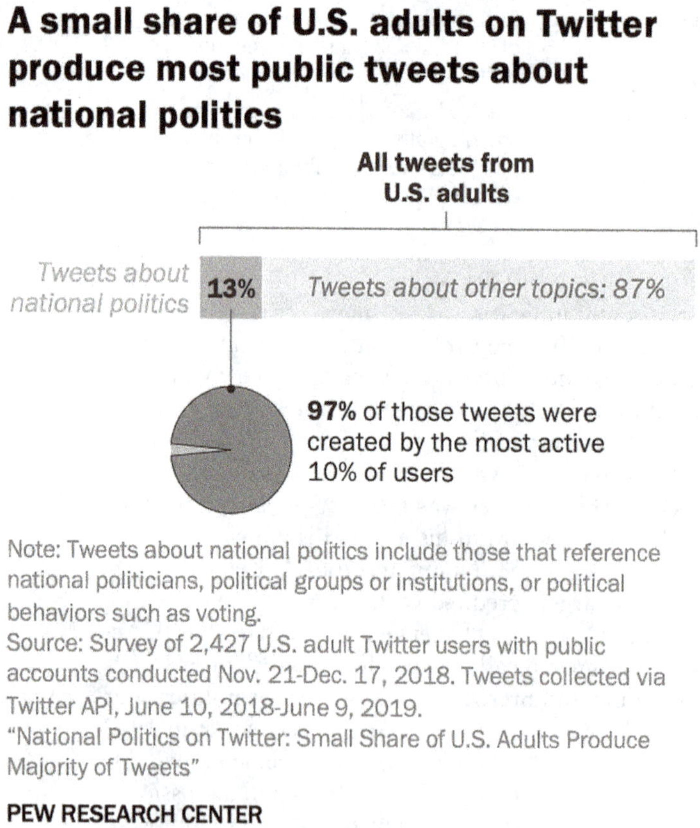

A small share of U.S. adults on Twitter produce most public tweets about national politics

All tweets from U.S. adults

Tweets about national politics **13%** — Tweets about other topics: 87%

97% of those tweets were created by the most active 10% of users

Note: Tweets about national politics include those that reference national politicians, political groups or institutions, or political behaviors such as voting.
Source: Survey of 2,427 U.S. adult Twitter users with public accounts conducted Nov. 21-Dec. 17, 2018. Tweets collected via Twitter API, June 10, 2018-June 9, 2019.
"National Politics on Twitter: Small Share of U.S. Adults Produce Majority of Tweets"

PEW RESEARCH CENTER

FIGURE 12.1 Though it can sometimes seem like all people talk about online is politics, discussion of politics actually makes up only a small fraction of online communication. Additionally, much of that political communication is sent by a very small subset of politically active users.
Pew

consider how CMC is used for political communication. In this chapter, we explore politics via CMC from three perspectives: how individuals communicate with each other online about politics, how constituents and publics communicate with politicians and political issues through digital channels, and how politicians and advocacy groups have used CMC to communicate with constituents and publics.

Communicating about Politics

Though politics pervades our lives, it is a topic we don't often talk about face-to-face. We may shy away from discussing politics with friends and family for fear that our attitudes and perspectives will differ from theirs, leading to hurt feelings or strained relationships. Because our self-presentation is innately linked to our identities when across the dinner table, we are hesitant to share our political ideas or beliefs at family dinner for fear that crazy Uncle Mitch will have a different idea, leading to a shouting match about Referendum 34 and ruining Thanksgiving dinner *again*. But online, we can distance ourselves from our political communication to varying degrees. Taking advantage of how we identify ourselves, with whom we talk, and the topics with which we engage, CMC can sometimes make people freer with their political communication, for better or for worse.

Before discussing these factors, it is important to note that not all online political communication is inherently problematic. Even in discussions of highly polarized topics like the 2014 Scottish independence referendum to cede from the United Kingdom, obscene or inciteful communication typically constitutes only a small proportion of all online discussion about the topic (Quinlan, Shephard, & Paterson, 2015). Often, online tools are simply bigger places for the rational sharing and discussion of political ideas, basically digital versions of the ancient Greek agoras. Several of the subsections that follow address how CMC can sometimes facilitate conditions that make communicating about politics more problematic, but it is important to remember these same processes may also make discussions more honest, frank, and productive.

Disinhibition and Ingroups

One thing about political CMC that may be of immediate note—particularly in contemporary online discussions—is the common perception that CMC discussions are increasingly uncivil, toxic, vitriolic, and otherwise simply not nice (Duggan & Smith, 2016). Politics is not always comfortable or easy to discuss (Zompetti, 2017) because of individuals' closely held and often immutable political views. And yet online discussions can be even more negative or aggressive than offline discussions (Figure 12.2). What makes normal people jerks when talking online, especially about politics?

Disinhibition and Flaming One reason for this lack of civility may be that individuals discussing politics online (rather than face-to-face) feel more **disinhibited**: freer to communicate without fear of personal or social repercussions. Online disinhibition is a product of several factors co-occurring online, including the ability to communicate anonymously, minimal presence of authority to enforce social norms, and asynchronous discussions to further distance posters from their messages (Suler, 2004). Often, the deindividuation fostered by many online tools—sometimes even tools in which users are identifiable—can lead to this disinhibition, as individuals' suppressed sense of personal identity makes them less concerned for the consequences of their actions (see Chapter 5). When disinhibited, individuals are less concerned about the reactions to their messages and thus may engage in more negative or antisocial posting. This is why some online news sites have closed off discussions following posted news stories,

Many users see social media as an especially negative venue for political discussion, but others see these sites as simply "more of the same"

% of social media users who say their political discussions are more or less _____ compared with other places people might discuss politics

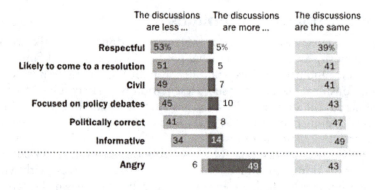

	The discussions are less ...	The discussions are more ...	The discussions are the same
Respectful	53%	5%	39%
Likely to come to a resolution	51	5	41
Civil	49	7	41
Focused on policy debates	45	10	43
Politically correct	41	8	47
Informative	34	14	49
Angry	6	49	43

Source: Survey conducted July 12-Aug. 08, 2016.
"The Political Environment on Social Media"

PEW RESEARCH CENTER

FIGURE 12.2 Social media users report decreasing civility in online discussions of politics. How may the nature of being online make discussions less respectful or civil?
Pew

or at least made it so that some form of identifiability is required to post: anonymous comments tend to be more uncivil, off topic, and comprised of ad hominem attacks on other posters (Rowe, 2015).

Online, disinhibition may lead to **flaming**, or "hostile intentions characterized by words of profanity, obscenity, and insults that inflict harm to a person or an organization resulting from uninhibited behavior" (Freelon, 2010, 1179), particularly in discussions about politics. The prevalence of flaming in political discussions may stem from the closely held nature of political beliefs central to the core of individuals' social identities. Individuals online engage in more flaming when their beliefs are directly challenged (Hutchens, Cicchirillo, & Hmielowski, 2015), so that raising alternate political views or counterpoints to their strongly held beliefs can increase their flaming behaviors. Individuals flaming via CMC may seek to release their tension or internal conflict regarding politics or opinions that may be unpopular or have undesirable consequences if expressed offline. For example, saying "Abraham Lincoln was a bad president" in person may make others think negatively of you or lead to conflict, but the same statement can be made anonymously online so that others do not know you and cannot subsequently treat you differently (see Figure 12.3).

Another reason CMC may make political communication less civil and communicators more disinhibited is because of the asynchronous nature of many CMC tools. When you have a political conversation with someone in person or over the phone, the rapid and real-time exchange of messages and feedback keeps the conversation flowing naturally and lets individuals understand how communication partners have interpreted their messages. However, in online chat forums and discussion boards, messages may be posted with no immediate response, resulting in a

FIGURE 12.3 Without fear of identifiability or repercussions from their comments online, individuals can feel disinhibited and engage in more inflammatory or inciteful conversation than they would offline.
Cartoonist: Cantú and Castellanos

loss of conversational coherence or thinking the worst of someone while awaiting a reply (Stromer-Galley, Bryant, & Bimber, 2015). Basically, because people lose track of the discussion or do not hear back quickly from others, they begin to be more aggressive or less polite in their discussions online. The very nature of being online in a public forum may reduce individuals' willingness to engage in civil discussion or feel concerned about the consequences of their comments.

Ingroup Influence Another reason online discussions of politics may be less civil than their offline counterparts relates to the group rather than intrapersonal influences of CMC. Just like talk radio and cable news, online tools can help people find politically homophilous others—individuals who share their political attitudes, leanings, and views (Sobieraj & Berry, 2011). The social identity model of deindividuation effects (SIDE; Reicher, Spears, & Postme, 1995; see Chapter 5) would predict that when individuals are in an online environment where group identities are salient and activated, group processes drive interactions so that individuals associate with and conform to norms of the ingroup while disassociating from and being antagonistic toward members of outgroups. Politics makes ingroups/outgroups particularly distinct, as individuals can quickly identify and associate along party lines (e.g., Republican, Democrat, Green, Socialist, Labour, Tory, Social Democratic Party of Germany), and thus be guided by ingroup processes, especially affiliation with ingroup members and disaffiliation from outgroup members. Consequently, individuals may be even more antagonistic online toward other political views or opinions than they would be online, as their social—rather than personal—identity is guiding the interaction. Rather than calmly discussing politics with your lovely Aunt Lynn (who happens to be left-leaning), online you identify as politically right-leaning and consequently may lash out or insult any left-leaning individual you encounter (even if they happen to be Aunt Lynn).

Another consequence of these ingroup effects is when individuals perceive their form of political discourse as normative to the ingroup. Flaming behaviors are particularly susceptible to this form of normalization, as individuals who view aggressive online communication patterns as normal and acceptable are more likely to engage in flaming themselves (Hmielowski et al., 2014). Important here, the flaming does not have to be viewed as normative to the broad group discussing politics. The individual must merely see flaming as normal with respect to the ingroup, even if that ingroup is a minority social category in the larger discussion. This normalization of flaming among small, homophilous groups online then explains some of the antisocial and offensive behaviors of some groups online, from those simply engaging in trolling or flaming, to the legions of *Fortnite* players threatening bodily harm on others after they lose a match—it's because they think or see others like them also make such statements.

This application of SIDE also suggests online incivility may be less attributable to individual factors, and that incivility may simply be a manifestation of the polarizing nature of politics. As blogs, discussion fora, Discord servers, and other CMC tools often identify a political leaning or position (e.g., right- or left-wing), the communication in and from individuals using that tool may be innately polarized, favoring individuals and attitudes consistent with that shared by the group and antagonistic and mean toward those with different views (Lawrence, Sides, & Farrell, 2010). From this perspective, CMC tools are providing us new and more persistent channels devoted to political communication. Whereas the upcoming election may be an occasional topic in the break room, in the "Sophomores for Socialism" Facebook group, listserv, or Amino group, communication may predominantly be about broad local or regional political topics. Communication is therefore guided by the relevant political social identity and its members, all of which could lead toward more contentious discussions when encountering individuals with other political attitudes or beliefs.

Political Activity: Engagement or Atrophy?

Another way CMC has been influencing political communication is by giving us new forms and outlets for political activity. Offline, one can engage in many forms of political engagement: attending a political rally, going door to door to canvass for a candidate or issue, calling an elected official to voice your opinion, or simply going out to lunch with friends to talk about the upcoming election. Many of these activities have online counterparts: watching a political rally streamed online, creating emails to send out to friends in support of a candidate or issue, emailing or tweeting an elected official to voice your opinion, or simply chatting with a friend about the upcoming election. Digital tools have provided several online equivalents for traditional political activity, but they have also created new opportunities for individuals to engage with and communicate about politics (Bennett, 2008). Individuals can now post status updates about politics, interact realistically with others' political behaviors, or publicly share photos of politicians or political events.

Intensifying Online communication may just be a new way or set of channels for those already politically engaged and active to continue to be so, ultimately intensifying their political communication and behavior. Exploring college students' political activities online, Vitak and colleagues (2011) found that some of the new ways to be politically active online were really only used by those who were already politically active offline. Follow or friend a politician on social media? Good chance you've also attended a political rally. Talk about politics in your WhatsApp group? You're also likely willing to talk about politics around the dinner table. One implication of these newer forms of political activity may therefore be a "rich get richer" argument, whereby those already predisposed to talk about and engage with politics offline supplement their offline political activity with newer online channels.

One of the ways this intensification of offline behaviors is most evident is through the consumption of political news and information. Individuals who seek political information offline through reading newspapers, listening to news radio, or watching televised news programs are more likely to seek out additional sources of political information online via blogs, podcasts, news aggregation sites, and politically oriented social media accounts (Gil de Zúñiga, Puig-I-Abril, & Rojas, 2009). Computer-mediated communication may be giving those already interested in learning about politics additional new channels through which to do so. This is perhaps why several studies have noted that many of the predictors of *offline* political activity—socioeconomic status, gender, age, political self-efficacy—also predict *online* political activity, so that those who are older, better educated, and believe themselves more involved and able to influence political issues are more politically active online (Di Gennaro & Dutton, 2006).

Digital communication tools can also engage new sets of users, particularly those with lower socioeconomic status or those who are politically active offline (Di Gennaro & Dutton, 2006). Poorer individuals may not be able to take time off work or travel to attend political events, subscribe to news programming, or have the social capital to engage in political activity offline. These individuals can find that the barriers to entry in political participation are lower online. Once you have Internet access, it is functionally free to watch a political event streamed online, find free news podcasts or programs, or to talk to others online about politics. Similarly, because CMC can transcend physical space, CMC tools may help more rural individuals be more active in politics as rather than driving long distances to attend an event or interact with others, they can do so from their home or their public library. Consequently, CMC tools may be giving politically interested individuals new and more practical ways to communicate about politics.

Finally, CMC may actually engage more individuals in political communication by exposing them to more political information. The **news finds me** (NFM) (Gil de Zúñiga, Weeks, & Ardèvol-Abreu, 2017) perspective suggests individuals who do not actively attend to professional news outlets (e.g., newspapers, magazines, or television broadcasts) will remain politically informed through peer networks via social media. In other words, even if you do not watch the news or read *Time*, you will still be exposed to current events as they are shared by others on social media, discussed in chat rooms, and otherwise pushed to you by those you know who *are* politically active. Problematically, however, those who take this NFM perspective typically find themselves less politically active (including voting) and politically knowledgeable than those actively seeking out news (Gil de Zúñiga & Diehl, 2019; Lee & Xenos, in press), and are also less likely to begin using more traditional news outlets (Gil de Zúñiga et al., 2017) as they become increasingly comfortable letting news find them. As CMC now can thrust news upon us and expose us passively to current events, some individuals may be less likely to actively seek out political news and discussion.

For those individuals who *do* engage in political CMC, engaging politically online can intensify subsequent offline behaviors. Online groups and discussion fora about political issues and candidates give like-minded individuals a space to discuss politics without having to be in the same room at the same time. Online political groups and discussions are means of finding other ingroup members based on a shared political social identity. Immersion in the discussions of these like-minded group members can strengthen individuals' social identification with a political cause and their motivation to find ways to engage that social identity offline. As there is a strong correlation between being a member of a political group page online and political engagement (Conroy, Feezell, & Guerrero, 2012; Rojas & Puig-I-Abril, 2009), it seems that being a member of a political group may intensify that political social identity, leading to additional offline political communication and engagement. It is notable, however, that the relationship does not hold for actual discussion in the online group (Conroy et al., 2012). Consequently, online group membership itself appears sufficient to intensify offline political communication and activity, more so than actually communicating with members of that group.

Although new and emerging CMC tools give us new ways to engage politically, do we necessarily use them to that end? Comparing college students' political behaviors on Facebook leading up to the 2008 and 2012 US presidential elections, Carr and colleagues (2016) found that many new voters used several Facebook tools less in 2012 than 2008 and for political behaviors, but reported more political behavior on Facebook from their network members. One potential explanation of this difference is that Facebook was a new tool in the 2008 election, still used primarily by college students, so young voters were still getting a sense of how Facebook could be used for politics. By 2012, college students just didn't see Facebook as a useful or appropriate place to talk about politics (Hayes, Smock, & Carr, 2015; Kruse, Norris, & Flinchum, 2018), partially because Facebook had been adopted by other users with whom college students did not identify. Students' parents, grandparents, and the broader user base of Facebook in 2012 *did* see it as a place to discuss politics, resulting in college students seeing more political activity from their Facebook Friends, and simply not wanting to see or engage in political communication with these broader social networks. It may be that as specific channels grow and users' demographics change, how the channel is used also changes, consistent with SIPM (see Chapter 2; Fulk et al., 1987).

A second potential explanation for these findings is that perhaps there is just a new meaning and understanding of what it means to be "politically active" online. Now, political activity may not be as substantive as volunteering for or donating to a campaign and may instead include simply following a politician on Twitter, observing a Reddit AMA (ask me anything), or just being passively exposed to political information from trending topics. Digital natives may increasingly find novel ways to engage politically online, through tools and in ways older pollsters have never thought of. In one unplanned and unconventional form of political engagement stemming from online communities, a 2020 political rally anticipated a standing-room-only crowd to fill the 19,000-seat arena in Tulsa, Oklahoma, after more than a million people requested a ticket online. The campaign heavily bragged about the anticipated turnout—particularly amid the ongoing COVID-19 global pandemic—as a sign of the public's support for the candidate. However, only about 6,200 people actually showed up to claim their tickets and attend the event in person. What happened to the 993,800 unclaimed tickets? The culprit was a grassroots effort by TikTok users—especially fans of Korean pop (K-pop) groups—who registered for hundreds of thousands of tickets as a prank, substantively throwing off the campaign's estimate of attendees (Lorenz, 2020). Registering for a campaign event you don't plan to attend—and using TikTok to convince others to do so as well—simply to create confusion and embarrassment for a campaign is surely not a "political activity" as it's been considered and measured previously. But apparently it should have been. Consequently, we may need to expand our understanding of what it means to be "politically active" online, going beyond following candidates or attending virtual rallies (see Table 12.1) to consider less conventional forms of political activity like engaging in a guerrilla flash mob in order to overbook events or creating political memes.

Slacktivism Sometimes, the political activities we undertake online may not seem terribly impactful or resource-intensive. It's easy to click a Like on a political post, but does that like really do anything? Some have claimed CMC has given rise to online political slacktivism (McCafferty, 2011). **Slacktivism** (sometimes also called *clicktivism*) refers to political "activities that are easily performed, but they are considered more effective in making the participants feel good about themselves than to achieve the stated political goals" (Christensen, 2011). Online political slacktivism can include low-cost acts like changing a profile picture, liking a political post, or joining an online group about politics. While these activities are all political in

TABLE 12.1 Political Activity on Facebook and Exposure to Networks' Political Activity on Facebook in 2012 Election (Carr et al., 2016) and Comparison to Political Activity and Network Exposure on Facebook in 2008 (Vitak et al., 2011)

In the past week, which of the following have you done in Facebook/ seen in your news feed?	% of Carr et al. (2016) performing this behavior	% of Carr et al. (2016) observing this behavior being performed by others	% of Vitak et al. (2011) performing this behavior	% of Vitak et al. (2011) observing this behavior being performed by others
Added or deleted political information from their Facebook profile	7.8	37.1*	5.8	26.8
Added or deleted an application that deals with politics	7.8	27.5*	3.8	19.8
Became a "fan" of a political candidate or group	10.2	71.9‡	8.8	51.0
Discussed political information in a Facebook message	6.0	n/a	8.9	n/a
Discussed political information using Facebook's instant messaging system	6.6	n/a	6.9	n/a
Joined or left a group about politics	10.8	52.1	13.8	51.2
Posted a status update that mentions politics	14.4	91.6‡	18.4	70.0
Posted a photo that has something to do with politics	7.8	79.6‡	10	49.3
Posted a photo of someone at a political event	1.8‡	50.9	9.6	48.4
Posted a wall comment about politics	8.4‡	71.9‡	20.4	43.2
Posted a link about politics	7.8	72.5	6.1	41.9
Posted a Facebook note that has something to do with politics	1.8‡	30.5	3.6	35.5
RSVPed for a political event	3	45.5	13.8	42.5
Took a quiz about politics	4.8	21.6†	2.7	11.1

Significant differences (*$p \leq 0.05$, †$p \leq 0.01$, ‡$p \leq 0.001$, two-tailed) between 2008 and 2012 samples as determined by paired z-tests

Reprinted from Carr et al., 2016.

nature, given their low cost and lack of direct action there has been a concern over whether online political slacktivism constitutes meaningful political communication or activity. Many would argue, probably rightly so, that Liking a post by someone running for political office is not as meaningful or impactful as actually going out and voting for the candidate (Halupka, 2014; Morozov, 2009). Likewise, changing your social media profile photo's frame to indicate your preferred candidate may be functionally very similar to wearing a campaign button, taking little effort and being more performative than functional. However, whereas a photo frame can only be viewed by those who see your profile, the button can be seen by anyone who passes by regardless of whether or not you know them, and therefore the button may be a more effective political activity to raise awareness or change opinions about that candidate.

And yet evidence is beginning to mount that political slacktivism may actually have demonstrable effects. For example, individuals who shared a video about a social cause on Facebook were subsequently more willing to volunteer for a cause similar to the issue they shared about (Lane & Dal Cin, 2018). One reason a low-effort act like sharing a video may actually influence someone's attitudes or behaviors is the concept of identity shift (see Chapter 9). Sharing a video, Liking a politician or political group, or even seeing yourself type a post about an upcoming election are all forms of mediated public self-presentation, which can influence how an individual see herself/himself. Although slacktivism may not have direct effects similar to raising money for a campaign or canvassing for signatures on an upcoming ballot initiative, slacktivism may change how individuals see themselves with respect to politics or political issues. That change in self-concept can then drive more substantive political communication in the future. For example, Liking a politician's Instagram Story now may make you more likely to vote, vote for that politician, and/or talk about that politician with others in the future. Though much has been said to demonize online political slacktivism, perhaps there are some important indirect effects of the time we spend engaging in lightweight political communication online.

Social Media: Bubbles or Bridges?

One of the largest and most-contested recent questions regarding the impacts of CMC on political communication is how social media may influence the way we talk about politics and with whom. On one hand, social media let us extend far beyond our close friends and family members to encounter new ideas and perspectives, particularly with regard to politics. On the other hand, social media can also let us find like-minded others, exposing us only to ideas consistent with those we already hold. Perhaps because social media use—especially for political communication—is still developing and normalizing, there does not seem to be a clear consensus in the literature about which effect social media is having, so it's worth talking a little about both.

Polarizing Political Bubbles Early thinking about the potential of CMC for political discussion, deliberation, and debate was optimistic (e.g., Hague & Loader, 1999): By providing a venue for all voices and perspectives, the Internet would be a tool for democracy. Yet early findings of what CMC tools—especially social media—actually did to political communication pointed toward the creation of political silos in which individuals found and limited their interaction to those who shared their own political views. Democrats talked with Democrats, Tories followed Tories, and centrists Friended centrists. In many ways, this reflected our homophilous offline networks—we tend to associate with people who act, think, and behave similarly to us. But without being forced to accidentally encounter oppositional viewpoints or information, users found social media to be political silos or **echo chambers**: closed networks in which individuals with similar ideologies share information, perspectives, and attitudes that reinforce their existing views.

On Twitter, Reddit, Tumblr, and other SNSs, political discussion is typically clustered by partisanship (see Figure 12.4). Users reinforce their ingroup and outgroup distinctions and associations by spending most of their time communicating with others who share their political ideology and not interacting with those of different political perspectives (Conover et al., 2011). Because people are associating and talking primarily with others who think as they themselves do, it reinforces their previously held beliefs. If everybody I talk with online shares my view, my view must be correct, and therefore I should hold that belief more strongly. This feedback loop reinforcing social media users' political identities and further polarizing them and their interactions is particularly notable as conversations turn more political. As politics among online users contains more messages about politics, ingroup and outgroup memberships become evident and members with dissenting or minority perspectives may leave, resulting in more political messages by the remaining members to reinforce their shared political values. This political siloing is further intensified as users share polarizing or one-sided news and political stories (Conover et al., 2011). As users speak and listen primarily to each other, and as cross-talk among political categories is often antagonistic, these political discussions online become separate groups based on political affiliations, resulting in structural silos of like-minded individuals. Consequently, on many social media, users are simply never exposed to political views or ideas that differ from their own (Himelboim, McCreery, & Smith, 2013).

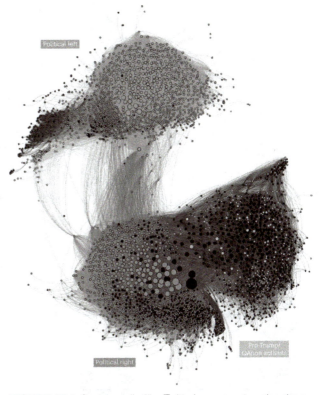

FIGURE 12.4 Social media (like Twitter) may create echo chambers or silos of users where individuals interact only with others who share their political (or other) ideas, reinforcing their previously held beliefs and reducing perspective taking. This map of Twitter networks shows strong clusters of online discussions in which individuals interact in tight, dense networks of like-minded others, with few accounts spanning the political left and political right.

A second factor that may cause echo chambers is the proliferation of outlets presenting news or political stories, particularly outlets based online. New digital tools now enable almost anyone to host a podcast, tweet political stories and information, blog about politics, and otherwise present political "news" and commentary, but without the rigorous journalistic standards and gatekeeping role of traditional news media. **Gatekeeping** refers to the "editorial intervention and validation of authorship" (McQuail, 2010, 139), including the ability of channels to regulate the flow of messages. Traditional news outlets—including newspapers and television or radio broadcasts—serve as gatekeepers, limiting the type or quantity of messages transmitted to their publics by selecting what messages to broadcast. But as the number of outlets proliferates, new outlets have to slice out smaller or more extreme niches to attract an audience. This has led to the polarization of many "news" outlets (see Figure 12.5) and the political communication they facilitate, often at the expense of journalistic standards and the reporting of newsworthy stories (Baum & Groeling, 2008). Troublingly, individuals attending only to polarized media purporting to be objective news actually find themselves less politically knowledgeable than if they had consumed no media at all (Cassino, Woolley, & Jenkins, 2012).

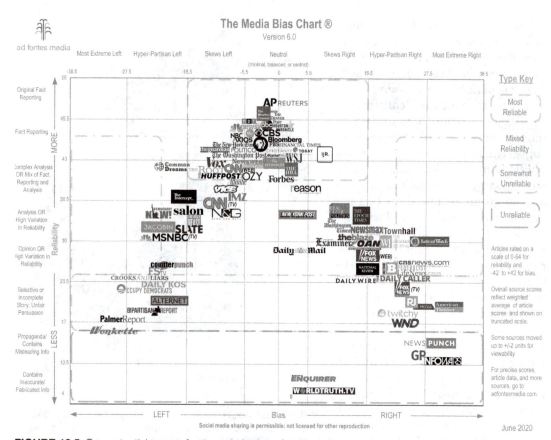

The Media Bias Chart ®
Version 6.0

FIGURE 12.5 One potential reason for the polarization of political discussions may be the polarization of news sources that have emerged online. How may the news sources users predominantly consume influence how they perceive political issues?
Ad Fontes

The polarization of "news" online is then echoed as individuals select and share stories from sources that share their political beliefs and perspectives. Right-leaning individuals can find increasingly right-leaning stories to share with others, and likewise see increasingly extreme stories that reinforce their worldview. Similarly, left-leaning individuals may find their political views supported and intensified as their friends share news stories from the extreme left. The same process happens even with the online information sources themselves, as blogs and other online stories typically reference and link to other online blogs and stories that share their political ideology (Adamic & Glance, 2005; Hargittai, Gallo, & Kane, 2008). The recursive and insular sourcing of political "news" sources and lack of gatekeepers to political information may be polarizing the political information to which we have access, further segmenting and distancing political groups and discussions.

The increasing polarization of the political information available online is compounded by the segmented networks users immerse themselves in within social media structures. Smaller circles of users share increasingly biased political information, reinforcing their own views and isolating them from other perspectives. Following this logic, a large amount of work initially concluded that online deliberation—particularly within social media—was resulting in political bubbles and ingroup discussions of politics (e.g., Gilbert, Bergstrom, & Karahalios, 2009; Goldie et al., 2014; Quattrociocchi, Scala, & Sunstein, 2016). These online political bubbles would make people more insular, being only exposed to political views and messages that are similar to their own previously held beliefs, resulting in less politically informed individuals and more polarized discussions about politics online in the rare instances it happens.

Political Bridges Recent research has been more optimistic about the potential of CMC to spur political discussion and dialogue, more reflective of the hopeful early thinking about the Internet's possibilities. From this perspective, tools like social media are living up to their potential to let us expand beyond our own immediate physical networks and social connections to interact with new perspectives. For example, studies have shown that individuals are more likely to encounter political information and views that differ from those they hold via online networks rather than in their FtF social networks (Bakshy, Messing, & Adamic, 2015; Barberá et al., 2015; Barnidge, 2017). Because our online networks—especially social media—can span more different groups than we could in a single FtF setting, they may allow us to see political ideas that differ from our own more readily than we do offline.

Political discussions that include multiple perspectives have several benefits. For example, engaging in political discussions online can encourage users to seek out additional news sources to learn more about the new information and perspectives they encountered (Klofstad, 2009; Levendusky, Druckman, & McLain, 2016; Levitan & Wronski, 2014). Exposure to different political views and information can also increase individuals' political engagement, as they are motivated to become more politically active following rigorous and civil political discussions (Klofstad, 2009; Levendusky et al., 2016; Levitan & Wronski, 2014). And, perhaps guided by the contact hypothesis, interacting with other political ideologies online can also increase individuals' political tolerance, making them more aware of and informed about oppositional views and rationale (Mutz, 2002), even if the individuals do not necessarily change to share those perspectives.

Does CMC Bubble or Bridge? Scholars are still struggling to determine whether CMC creates political bubbles or bridges for several reasons. First, these phenomena are relatively new and still emerging. Prior to CMC, the places people interacted about politics—work and third spaces like bowling alleys or social clubs—tended to be homophilous and social conventions constrained the conversation. Online, our discussions can be much more diverse and less constrained, so in some ways political CMC is letting us communicate about politics more freely. Second, it seems many individual factors influence political CMC processes. For example, Munson and Resnick (2010) found that while some people seek out ideas and viewpoints that may challenge those they already hold, others purposely avoid information that challenges what they already believe. This means much of how any individual uses CMC for politics may be more idiosyncratic than generally predictable. Ultimately, more research is still needed, and there's no clear and definitive answer. Computer-mediated communication may silo and insulate us within our existing political ideologies, it may bridge and diversify our exposure to and interactions with other ideologies and perspectives, or it may do both depending on our individual attributes. So far, it seems political communication online may be generally guided by group processes, but continued work is necessary to determine when, how, and to what degree group processes (rather than interpersonal, intrapersonal, organizational, or other communicative processes) drive our (in)ability to cross political lines online. The answer to this question is likely "It depends," and the more interesting question will be "Upon what does it depend?"

Communicating with Politicians

Computer-mediated communication tools represent some of the newer and more exciting ways for individuals to communicate with politicians, engaging more rigorously and completely in their democratic or representative governments. Prior to CMC, opportunities to interact with politicians were limited, particularly for those holding larger offices. Attending a political rally or campaign event can be costly—both in time and in money—and does not guarantee attendees a chance to meaningfully talk with the politician. You may be able to schedule a meeting with your local mayor or run into your town councilperson in a coffee shop, but it can be challenging to talk with—either in phone or in person—a state representative or federal official, as scheduling a verbal conversation can take time. And though you can mail a letter to these individuals, once the letter is put in the postbox, you have little assurance it will be received or read by the politician directly. Computer-mediated tools can facilitate the public's communication—or at least the *perception* of communication—with politicians.

Upward Communication and Direct Interaction

An immediate way that CMC can make upward communication from constituents to politicians easier is through reducing barriers to access. Whereas making an appointment to meet with an elected official can sometimes take weeks (if they are willing to meet with you at all), sending an email or posting a message on their social media profile is immediate, and can take advantage of CMC's disentrainment by allowing you to compose and send the message whenever it is ready and regardless of your physical location. Even if your email or tweet is not read, the knowledge your message was sent and received may be sufficient to feel as if you've been heard as a constituent or voter. Moreover, using social media to attempt to communicate upward with politicians can create masspersonal messages visible to others, helping bring marginalized groups or issues (Bekafigo & McBride, 2013) to the attention of the broader public and (hopefully) the politician.

Even more engaging as a constituent is when the politician *replies* to you. This direct interaction may be in regards to political matters, such as an upcoming vote by the official, the politician's views on a political issue, or suggesting a new issue for the politician to consider. Some channels may be more likely than others to garner replies from politicians, likely based on the norms and intended function of the channel. For example, Kalsnes, Larsson, and Enli (2017) found that politicians were more likely to reply to a Facebook post than a Twitter tweet, possibly because politicians thought their engaged constituents were more likely to be on Facebook than Twitter. Other politicians have turned to online communities such as Reddit to host an "Ask Me Anything" (AMA) in which the politician sets aside time for community members to ask questions and to respond directly. President Barack Obama's AMA (https://www.reddit.com/r/IAmA/comments/z1c9z/i_am_barack_obama_president_of_the_united_states/) involved users asking him questions about student loan debt, small business initiatives, net neutrality, and even his basketball icons. This type of dyadic (i.e., two-way) interaction online can humanize politicians, allow them to interact with constituents and receive feedback about the concerns of their voters and electorate, and increase voters' likelihood of supporting that candidate in upcoming elections. However, such interactions may not always be genuine: Exploring the habits of politicians in Queensland, Australia, in 2012, Bruns and Highfield (2013) found that replies from politicians to public queries and comments came to an almost complete stop following the election, suggesting interactions were merely another tool in the politicians' campaigns rather than genuine opportunities for voters to interact upward with their elected officials.

Computer-mediated communication tools can also be used to reach out to politicians outside of their official capacities as fellow citizens. Sighting a local politician such as Calgary mayor Naheed Nenshi in person and reaching out on Twitter, FourSquare, or other CMC tool can provide a chance to reach out to the politician and elicit a reply, either in the same channel or in person (see Figure 12.6). As channels continue to emerge and

FIGURE 12.6 Calgary, AB, mayor Naheed Nenshi has been notable in his use of Twitter to interact and engage with constituents on a personal level. How would you feel if your mayor, alderman, or representative directly contacted you online? What if you just saw that politician interacting with someone like you (like Ms. Straub)? Would that change your attitude about the politician?

decline and as users and usage habits evolve, where politicians are most likely to be responsive will likely change. But email, social media, and even online communities remain channels in which individuals can feel empowered to reach out directly to politicians.

Another way constituents' political communication can manifest via CMC is not attempts to communicate with the politician directly, per se. Rather, constituents may use CMC to communicate *about* the politician. Particularly in social media, individuals can post comments to or about politicians. Even if the politician does not see the message or reply, this user-generated content can be used to shape others' views about or attempt to engage the politician. Political and satirical cartoons, for example, can be messages about politicians that can serve as constituents' feedback or perspective about the politician's actions or performance as well as to raise awareness about issues or public sentiment. Because online political cartoons can be shared, directed toward specific intended receivers, and reappropriated (e.g., memes), their influence can go far beyond a flyer posted in a local coffee shop or a local newspaper message (Terblanche, 2011).

Online petitions and polls are another way political messages may be directed upward, though not directly to a specific politician—at least immediately. For example, in 2011, the White House established the "We the People" petitions site (https://petitions.whitehouse.gov/), which allows any individual to create a formal petition for consideration by the United States' executive branch. These petitions can then be circulated and support sought from other individuals to indicate their agreement and sign the petition. Any petition receiving 100,000 signatures within 30 days then will be read and receive a formal reply from White House staff within 60 days. Whether the issue is repealing a bill, honoring a servicemember, or changing the national mascot, the "We the People" site expedites communication about political concerns and upward communication from interested constituents to the United States' executive branch of government. Whether it's mentioning a politician in a tweet or starting a formal petition for a new law, CMC tools make communication from the public to their elected and appointed officials easier, more visible, and more accessible than many traditional forms of offline or mediated communication.

Parasocial Interaction

Another unique opportunity for constituents to communicate with politicians is the vicarious effects they may experience when they see *other* constituents interact with politicians. As masspersonal channels, social media allow for interpersonal interactions to occur in a way that is visible to those beyond the two interactants. When you post a message to a politician on a social medium and the politician responds, not only do you see that interaction, but other people can see that interaction—particularly those in your networks. Being able to observe and vicariously experience others' interpersonal interactions with politicians is a different experience than simply listening to a politician's mass message as part of the broad and undifferentiated audience or watching the politician interact with a newscaster.

Watching others similar to yourself interact with a politician (similar to how you would) can change your perceptions of and attitudes toward the politician just as if you had interacted directly, serving as a form of parasocial interaction. **Parasocial interactions** occur when members of an audience develop the perception of friendship or personal relationship with a personality with whom they have never interacted directly. Early work into parasocial relationships focused on viewers' perceived relationships with and attachment to soap opera characters and newscasters (Horton & Wohl, 1956), whom the viewers frequently saw and could imagine interacting with but never actually interacted with (because that's not how television works). Later, parasocial interactions became more possible with social media, as individuals could observe interactions between other social media users and celebrities or media personalities and—as other social media users are considered like them—imagine the interaction as if it had happened with them (see Dibble, Hartmann, & Rosaen, 2016). These parasocial interactions can occur with politicians, as individuals can see politicians interact with others online and, provided they see the others as similar to themselves, feel closer to them (see again Figure 12.6). If you see a politician interacting favorably and personally (e.g., using their name, addressing their issue directly and explicitly) with another constituent, you will like that politician more (Lee & Oh, 2012; Lee & Shin, 2012).

Functionally, this parasocial interaction may also be a chance to vicariously get information or answers to your questions. As there are more constituents than politicians, officials may not be able to individually respond to every tweet, email, and text they receive. Using masspersonal channels like social media or wikis may help both politicians and their constituents by providing a repository of information for all. Imagine if you had a question about your state representative's position on spending for higher education. You may be able to find the answer to the same question asked by someone else on the politician's social media profile, saving you from having to ask the question and the politician

Somebody Like Me

How do you influence someone politically without actually communicating with them? One answer may be to interact online with someone else who is similar to that person and who can act as a surrogate or stand-in, in a way the person you're trying to influence can see. Because conversations on Facebook, Twitter, and other social media are often masspersonal, influencing one person may be like influencing thousands of others.

Exploring the effect of observing interaction online, Lee and Jang (2013) assigned 100 South Koreans to view either a social media profile or a static web page of a political candidate. The politician could be seen interacting with constituents directly in the social media condition; the same information was provided as additional information in the website. Individuals assigned to the SNS condition, in which they observed the politician interacting with other potential constituents (like themselves), perceived themselves as psychologically closer to the politician, felt more of a relationship with the politician, and were more likely to vote for that candidate.

This effect is caused by parasocial interaction—the illusion of intimacy or direct interaction with a media figure with whom we cannot or do not interact. As it may be impractical for a politician to interact with *everyone*—either online or offline—using social media as a means of publicly interacting with people can persuade and influence even more constituents. Other people can act as proxies for you online, and statements or replies made to others can be taken as if they were made directly to you. Experiencing this vicarious interaction with a politician (or other public figure or organization) would feel as if they were communicating directly with you, and their messages may influence your attitudes about the politician, your relationship with the politician, and perhaps even your future attitudes and behaviors.

Reference

Lee, E.-J., & Jang, J.-w. (2013). Not so imaginary interpersonal contact with public figures on social network sites: How affiliative tendency moderates its effects. *Communication Research, 40*(1), 27–51. https://doi.org/10.1177/0093650211431579

from having to answer the same question another time. In these ways, masspersonal channels provide opportunities for parasocial interactions among constituents and candidates as well as ways for politicians to vicariously answer your questions or provide you information by communicating with other constituents who had similar information needs. Consequently, even this indirect interaction can be helpful both perceptually and functionally, making politicians feel more accessible and making information and prior interactions more visible than in traditional channels.

Communicating with Constituents

Computer-mediated communication tools can also be effective ways for political communication to flow from politicians and campaigns down to constituents and publics. This type of communication may take two forms. First, elected and appointed individuals may communicate from their official office with constituents. Second, political communication may come from campaigns—organizations seeking to influence voters to vote for a politician's (re)election or political issue during an upcoming election. In the following two subsections, we'll address each of these forms of communication from politicians or campaigns to their audiences in turn, and help make sense of how and when each of the various tools may be most appropriate or effective for communication through the lens of media richness theory (see Chapter 3; Daft & Lengel, 1986).

Politicians Communicating with Constituents

When an individual is elected or appointed to office, that politician may use several CMC tools to communicate with her/his constituency—those individuals whom the politician has been elected or appointed to serve or represent. In what may be considered downward-flowing communication, this allows the politician to address constituents and publics in the capacity of the office s/he serves, whether that role is that of a prime minister or an alderperson.

Theoretically, a politician could use any CMC channel to communicate with constituents, and each election and political cycle is a chance to find new uses and new channels for politics. For example, during the 2020 US election, the Biden-Harris campaign introduced signage and merchandise in the *Animal Crossing: New Horizons* video game (see Figure 12.7). However, there are some limits to where politicians and constituents may intersect in practice, both because the number of channels continues to increase and because of the inappropriateness of some channels for attempting to connect politicians and constituents. For example, it may be weird or inappropriate for your county coroner to reach out to you through your personal Snapchat account. Consequently, we will focus on some of the most common CMC channels politicians and political offices use to communicate with constituents, but with the understanding that the same principles can usually guide and inform communication in other digital channels as well.

Websites Politicians can use websites to communicate with constituents and interested parties. Many political offices maintain websites for the individual holding that office to communicate relevant information to constituents, provide information about the official, and give updates on what that office is doing. Many websites are designed as push mechanisms, in that they provide (i.e., "push") information to individuals who were already interested in the individual or office and sought out the website. In other words, websites are often resources for constituents who already have some degree of interest in the individual or political position and want to learn more. Additionally, websites are typically **static** in that the information on the sites does not frequently change and is authored only by the person (or organization) whose website it is. Consequently, websites are good repositories for information, as they can contain pages devoted to different issues or audiences with relatively stable information (see Figure 12.8). Websites may therefore be considered lean media (Daft & Lengel, 1986), providing information to reduce uncertainty about a politician or the politician's office through a large and standardized set of information addressing some of the most common issues.

FIGURE 12.7 Politicians and campaigns are constantly seeking new channels through which to reach potential voters. In fall 2020, the Biden-Harris campaign introduced virtual signage in the Nintendo Switch's *Animal Crossing: New Horizons* video game, allowing players to put a political sign in their village, just as you would in front of your offline house.

Social Media In the past decade, social media have emerged as another critical form of CMC in politicians' tool kits for communicating

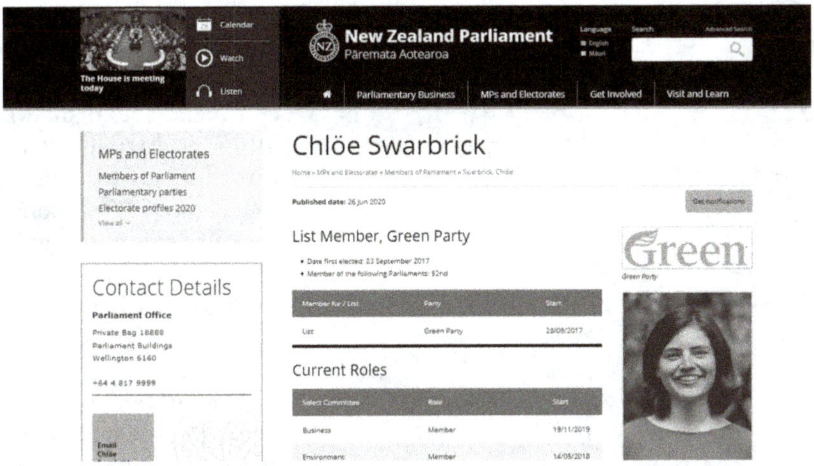

FIGURE 12.8 Like many politicians, New Zealand's parliamentarian Chlöe Swarbrick uses a website to keep constituents updated about federal issues that are of interest to her Maungakiekie constituency and to provide information about her perspective on contemporary political issues.

with constituents. The expectation that politicians—for both national and local positions—have a social media presence now parallels the expectation of a website, especially on some of the most-used social media like Facebook and Twitter (Larsson & Kalsnes, 2014). In some ways, politicians can use social media similarly to their use of static websites: presenting relevant information to voters and updating constituents on the politician's actions and efforts on their behalf. One of the most important opportunities of social media for politicians' communication with constituents and publics has been the further reduction of gatekeepers and access to large audiences. Politicians can use social media posts to disseminate content to publics unfiltered by gatekeepers like journalists or news media. However, this use of social media as a megaphone to speak to the public may be used without regard to the quantity, focus, quality, or truthfulness of the posted content.

As the prior statement acknowledges, the reduction of gatekeepers has mixed outcomes. On one hand, the reduction of gatekeepers can be prosocial and good for publics, allowing more messages or detail from politicians than media sound bites may allow. Likewise, reducing media gatekeepers allows for a greater diversity of opinions, positions, and views to be expressed, potentially giving the public a more detailed and accurate perspective of a politician than a sanitized or biased media outlet may provide. This reduction of gatekeepers can let both politicians and constituents feel as if the politician can talk to them more directly than when messages are filtered through traditional mass media channels (Bor, 2014). On the other hand, reduced gatekeepers can increase the amount of misinformation and disinformation politicians (and others) can communicate, leading to the spread of false information (Freelon & Wells, 2020; Valenzuela et al., 2019). Increasingly, politicians themselves have greater control over the information being disseminated about them and their political activities than traditional news media (Gainous & Wagner, 2013), which may be at the expense of the accuracy or veracity of information that has been fact-checked and vetted by formal news organizations, or without all of the standards and oversight that accompany marketing and advertising in traditional media. Consequently, simply as a means of mass communication, social media for politicians' communication is a bit of a mixed bag, allowing them faster access to publics and dissemination of information—at the cost of a loss of surety about the messages being put forward.

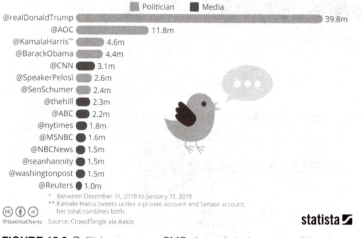

Politicians Have More Twitter Clout than Media
Total interactions on Twitter*

Politician   Media

@realDonaldTrump	39.8m
@AOC	11.8m
@KamalaHarris**	4.6m
@BarackObama	4.4m
@CNN	3.1m
@SpeakerPelosi	2.6m
@SenSchumer	2.4m
@thehill	2.3m
@ABC	2.2m
@nytimes	1.8m
@MSNBC	1.6m
@NBCNews	1.5m
@seanhannity	1.5m
@washingtonpost	1.5m
@Reuters	1.0m

* Between December 11, 2018 to January 11, 2019
** Kamala Harris tweets under a private account and Senate account,
    her total combines both.

@StatistaCharts   Source: CrowdTangle via Axios

statista

FIGURE 12.9 Politicians can use CMC channels to bypass traditional media and communicate directly with the public. But what does "interaction" mean? Here, "interactions" mean Likes and retweets of original tweets, not necessarily responses or engagement with or by the politicians or news outlets themselves.
Statista

Beyond allowing politicians to broadcast messages more readily than traditional channels, social media also allow politicians even more direct access to constituents (see Figure 12.9). Though politicians typically spend much of their social media messaging transmitting broad, non-personalized messages, when they do interact in social media, it tends to be with average citizens more often than with other politicians or heavy political influencers or commentators (Tromble, 2018). Consequently, it does seem like politicians are utilizing social media to reach out to constituents directly.

Where social media offer additional opportunities over websites is their user-generated and interactive nature (Carr & Hayes, 2015; see Chapter 10). Social media content is not meant to be static and one-to-many, and social media provide users opportunities to interact with messages presented within the social media channels (see Figure 12.10). That interaction may take the form of a simple PDA (e.g., Like, Favorite; see Chapter 8), or may be more communicatively involved, such as replying to the initial post. Users can agree with or affirm politicians' accurate statements and challenge or correct politicians' inaccurate or undesirable statements. For example, after President Donald Trump included Alabama in a tweet warning those who were about to be affected by the landfall of 2019's Hurricane Dorian, the National Weather Service tweeted to correct the inaccurate geography by

---

Thread

Tammy Duckworth ✓ @SenDuckworth · Nov 25, 2019

It's unacceptable that any Illinoisan—that any American—living in public housing finds themselves in units that are crumbling or unsafe. Kids & families rely on investments like these that help improve conditions, increase security & reduce homelessness.

☰ TAMMY ⬤ DUCKWORTH ☰
U.S. SENATOR FOR ILLINOIS

Duckworth, Durbin Introduce Bill to Address Public Housing Crisis in Eas...
The Official U.S. Senate website of Senator Tammy Duckworth of Illinois
🔗 duckworth.senate.gov

💬 2   🔁 41   ♡ 146   ⬆️

FIGURE 12.10 Politicians such as Senator Tammy Duckworth (Illinois) can use CMC to directly communicate policies, statements, and political news to constituents. Posting information directly or linking to static websites serve as means by which politicians can quickly communicate with the voters and communities they represent, but also by which constituents can react and respond to the politician. What would the 146 Likes to this tweet indicate to you? What do you think they indicate to Senator Duckworth?

noting that Alabama would not be impacted (Nace, 2019). By the nature of social media, politicians' contributions and messages on these platforms invite and encourage responses to and interaction with the politician as well as among users.

Politicians may use social media to communicate or evidence their connections and involvement with their constituency. Rather than being holed up in an office or government proceeding, politicians on social media can depict their actions in and with the communities they serve. Sharing photos of community events, blogging or tweeting about local happenings, and even taking pictures with constituents they can then share with others all can communicate to constituents that politicians are active and engaged civil servants (see Figure 12.11).

An underused tool of social media is the ability to create polls or surveys to seek input directly from constituents (potentially because of the challenge in limiting responses to just constituents). Using social media polls, emailed surveys, or other

← Tweet

Joan Burton ✔ @joanburton · Dec 8, 2015
Great to open @women4election review of 2015. Look forward to seeing more women in the Dail & around cabinet table.

💬 10   🔁 17   ♡ 22   ↑

FIGURE 12.11 Irish TD (similar to a congresswoman) Joan Burton often uses social media to evidence her involvement and engagement in social and political issues in her Dublin district.

digital panels, politicians can solicit feedback and thoughts from their constituents, using the channels as a pull rather than push mechanism, actively trying to pull in feedback, ideas, and perspectives from their publics (see Figure 12.12). Rather than waiting for a formal vote at the polls, social media or web polls can be used as tools to take informal, straw polls from constituents to inform and guide politicians' subsequent representation of their voters. One challenge of online polls, perhaps explaining why politicians have not broadly adopted them as a formal tool, is that they are subject to the biases of the platform, users, and networks. A Twitter poll may solicit quick feedback,

← Tweet

Rep. Eric Swalwell ✔
@RepSwalwell

Assault weapons have become the firearm of choice in too many mass shootings. We don't have to live this way — those guns only belong on battlefields. Do you support a ban AND buy-back of every weapon of war in America?

Yes!Australia already has	20.7%
No. We love Guns > Kids	79.3%

119,342 votes · Final results

8:39 PM · Jul 1, 2019 · Twitter for iPhone

11K Retweets and comments   **6K** Likes

💬   🔁   ♡   ↑

FIGURE 12.12 Representative Eric Swalwell (California) used a Twitter poll to seek feedback about gun control in the United States. (Notably, he does use some false equivocations in his poll, but it does exemplify asking the public for input.)

but it faces several challenges: it is difficult to constrain respondents to just those in a politician's district, social media polls innately prohibit responses from the many who may not want or be able to use them (see Chapter 2), and responses are more likely to come from the politician's followers and direct network ties and therefore are biased and not fully representative of the general public. Yet for smaller political activities, from school council elections to neighborhood associations, where the politician can be more confident that all users are aware of and contribute to the poll, social media polling can be a quick and cheap way to seek information, guidance, and voter sentiment.

It is worth mentioning here the underuse of online polls may be somewhat deliberate, as politicians learn about the potential pitfalls of crowdsourcing opinions and feedback via CMC tools. One such incident was a 2016 online poll by the British government to name a newly commissioned polar research ship. Eschewing the historic or commemorative options (e.g., *Endeavor* or *Shackleton*), the public—or at least online denizens taking the poll—quickly made "RSS Boaty McBoatface" the leading contender. Though "Boaty McBoatface" clearly won with almost a third of the vote, the National Environment Research Council (NERC) instead named the vessel after the fourth-place option, the RSS *David Attenborough*. The NERC did, however, make the consolatory gesture of naming one of the *Attenborough*'s remote submarines *Boaty McBoatface* (Rogers, 2016). Similar online naming contests have led to a zoo naming a newborn penguin Fluffy McFluffface, a Swedish train Trainy McTrainface, and informally christening an Australian ferry Ferry McFerryface. Finally, in a contest to replace Colonel Reb, the antebellum mascot of Ole Miss, students voted to replace their Rebel mascot—though not their name—with *Star Wars*'s rebel Admiral Ackbar in two separate votes, much to the chagrin of the Ole Miss administration, only to be stymied by usage rights from Lucasfilm (Malinowksi, 2010). These fun anecdotes demonstrate two things: first, CMC tools make this kind of polling readily available, and; second, allowing the masses to name something can lead to unanticipated or biased results. Just because you can poll voters does not mean the results will be desirable and reflective of the constituency.

Finally, social media are even more sharable than websites, with many social media encouraging the retransmission of posted content. Whether it's a Reddit or Twitter share or a Tumblr reblog, social media allow users to signal boost a politician's content, expanding its visibility. Additionally, when users rebroadcast a politician's message via social media, the message can reach new and different audiences, letting this broader audience see and potentially interact with the content, the politician, or other users. As illustrated in Figure 12.9, politicians' social media comments are frequently rebroadcast to new audiences and network connections, diffusing the politicians' ideology, platforms, and information along social—rather than political or traditional media—networks (Wallsten, 2014). Admittedly, the rebroadcasting of a politician's message may be to critique or admonish its contents, but even a negative framing of a political message boosts the message.

Though social media can spread politicians' messages to interested publics, sharing and retweeting can also introduce source effects beyond the original message source. In other words, my attitudes about the message are influenced not only by my attitudes about the politician but also by my attitudes toward the user who shared that post (Stieglitz & Dang-Xuan, 2012). Considering retweeting or sharing from an intergroup perspective, if you see a politician's message shared by someone from a social group or category with which you identify, you may view the shared message more positively, assuming that if another ingroup member shared the message it must be valuable, important, and/or accurate. Alternately, if you see a politician's message rebroadcast by an outgroup member—someone whose social group you do not identify with—you may innately distrust or diminish the message. In this way, we again see social media as a loss of control over a message: though the politician can control the initial post, once that post is online, it can be reproduced, stored, or shared in ways the politician may not have anticipated or desired.

Political Campaigning

A second process in which we have seen CMC play an increasingly dominant role is in **political campaigns**: organized efforts to influence public sentiment favorably toward a political candidate or issue in an upcoming election. These efforts can be simply persuading voters to vote for the campaign's candidate (or not vote for the candidate's rival), but can also include fundraising to maintain the campaign's structure and pay for staffers, advertising, and other campaign needs. Political campaigns have long used both FtF and mediated tools to reach out to voters, donors, and other publics. Campaign rallies and whistle-stop train tours have candidates deliver speeches to and meet with members of local communities. Phone calls connect with voters, raise money to continue the campaign, and encourage voting (Nickerson, 2006). Letters and donation forms can reach new or more rural publics. Ad campaigns on television, radio, and billboards overwhelm us every time we approach election season. These means of communication remain staples of political campaigns' communication, but CMC tools have complemented and extended the means and goals of campaigns' continued efforts.

Websites Political campaigns quickly adopted websites as means of fully articulating politicians' platforms and ideas in a level of detail television or newspapers couldn't afford. In 1996, Bill Clinton's and Bob Dole's campaigns were the first to create websites to host information about the candidates (Margolis, Resnick, & Tu, 1997). (Notably, these were also the first campaigns to have satirical websites created to look official.) Since then, it has become expected for a campaign to have a website (Druckman, Kifer, & Parkin, 2007), as not doing so may introduce a perception of illegitimacy. In addition to providing more information with greater detail than a pamphlet or public speech can offer, websites are also something individuals can quickly and easily share with others, forwarding and circulating URLs to help campaigns further diffuse and disseminate their messages.

Static websites can be great ways for political campaigns to communicate large amounts of information to broad audiences, exemplifying lean channels per MRT (Daft & Lengel, 1986). The same information can be transmitted to many receivers, reducing their uncertainty about local information, politicians' decisions or voting, or platforms for future issues. But websites do not typically tailor much information to individual users beyond providing pages devoted to specific topics, and the *interactivity* of these websites is typically operationally limited to how many pages a user can navigate to under the homepage (e.g., Kruikemeier et al., 2015). However, webpages can be used to increase involvement with a campaign by providing greater personal information about the candidate (i.e., to personalize and humanize the individual; Kruikemeier et al., 2013) or issue and soliciting user information like names and contact information, which can then be used to communicate more personally with users via other channels like telephone or email.

Email Though door-to-door and phone-based campaigns have long been staples of political communication, over the past two decades, political campaigns have heavily used email to communicate with voters. Unlike the impersonal mass messages of websites, emails can be more interpersonal. Email lists and databases can provide politicians and their campaigns complex data about potential voters and/or donors, allowing them to tailor messages based on voting histories, family composition, socioeconomic status, geographic location, hobbies, interests, or many other individual factors. Email databases and lists can consequently cost a political campaign almost $8 per email address because of the nuanced information in the profile of each email address and the subsequent ability to segment and tailor emails to subsets of email addresses who share a relevant trait.

Similar to a relationship marketing strategy (Jackson, 2005), email can be used to target specific messages to segments of a politician's public who may find the message particularly useful or engaging (see Figure 12.13). For example, drawing on a large email database of potential voters, a politician may send emails regarding her position on school funding to those working in the education field, separate emails detailing her position on agriculture to those living or working in rural or agrarian communities, and a third set of unique emails to those who frequently use public transportation about her proposed infrastructure legislation. Because these emails are received individually, they can be perceived as more interpersonal and direct than websites, even if their contents are broad-form letters received by many. Additionally, because they can be more tailored to individual receivers' interests and facilitate feedback in the form of a response email, emails are typically a richer form of communication with publics than websites. And finally, though politicians do not often have the time to individually reply to every email, they are more likely to respond to an email than a written letter (Jackson, 2005) and even the potential for a political figure to read and reply to an email still makes them widely used, especially among older adults (Xie & Jaeger, 2008). Even though we sometimes feel our inboxes are overloaded with political spam—and that politics is encroaching more and more into social media—email is still widely used (Doubek, 2015), especially for fundraising.

Social Media Most recently, political campaigns have leveraged social media tools to create grassroots campaigns, support, and advocacy. One of the most notable influences social media have had for political campaigns are the instances of political activism services like Twitter and Facebook offer. Founded and run by collectives of concerned individuals rather than established political organizations, this *hashtag activism* (based on the ability to use hashtags to identify and easily share and find related content; Augenbraun, 2011) has included notable and impactful political campaigns such as #BlackLivesMatter to address issues of police violence and racial inequality, #OccupyWallStreet to bring awareness to economic disparities, the #ALSIceBucketChallenge that brought awareness to

FIGURE 12.13 Campaigns can use emails to target potential donors who have expressed an interest in the politician, attended an event, or had their information purchased from a database. These emails are low-cost interpersonal tools to solicit funding or action, even by individuals who may not be residents local to the campaign. How are intergroup cues used in this email from Barack Obama's 2012 campaign to motivate campaign donations?

amyotrophic lateral sclerosis (i.e., Lou Gehrig's disease), and the antigovernment protests of 2010's #ArabSpring that began in Tunisia and spread through northern Africa and into the Middle East. By providing a channel for like-minded individuals to form groups, create organizations and structures, and then coordinate online and/or offline political activities, social media provide opportunities for everyone to construct and engage in a political campaign, whether it is altering broad political structures, providing means of resistance to or rebellion against authoritarian governments, or simply raising awareness of disenfranchised groups (Pearce, Freelon, & Kendzior, 2014; Thorpe & Ahmad, 2015; Tufekci, 2017).

The formal political campaigns of candidates, political organizations, and advocacy groups are also increasingly incorporating social media into their media mixes. Most often, political campaigns simply use social media as mass media channels, simply presenting information and propaganda (Klinger, 2013). But effective political campaigns are also using social media for their ability to foster dialogues, allowing campaign staff and the public to interact and discuss issues related to the campaign, like getting more information about an upcoming event or seeking clarification on one of the campaign's policy issues. Campaigns are also utilizing social media for their ability to mobilize groups formed in social media toward other political action, whether that's donating to the campaign, posting or sharing campaign information, or engaging in offline political activities. One reason social media can be so effective at mobilizing users to political engagement is that involvement in political campaigns' social media—whether it's simply following the campaign or actively engaging with the campaign and other users—relies on both interpersonal and intergroup phenomena. By providing individuals a place to interact with others who share their interest in the campaign's focus (i.e., a common-identity group), individuals can have their political categories activated and intensified, as campaigns use appeals to the social group to motivate action and beliefs. This is why campaigns typically try to engage groups or social categories online, using verbiage like "Together *we* can make a difference" or "*Us* young people are the change." These appeals to the social group can motivate individuals by activating their social identities and driving them to conform to the normative group behavior. Indeed, when campaigns try to individuate users online, not emphasizing their social identities, it can threaten the campaign as users feel disconnected from others and unable to enact change on their own (Fenton & Barassi, 2011).

Additionally, as users begin to form interpersonal relationships with other group members, the resultant ties make them even more likely to adopt attitudes, seek change, and otherwise engage in behaviors desired by the campaign (Parigi & Gong, 2014). In other words, political campaigns can use social media to come at you from multiple sides: not only are other activists like yourself doing the desired behavior, social media can let you see that your friends and personal connections are doing it. Whether that's promoting the campaign, donating, or attending a protest, both intergroup and interpersonal factors can be activated by the campaign, energizing and engaging users politically.

Campaigns also continue to spend larger percentages of their finances on online advertising, including social media presence (Figure 12.14). Researchers anticipated that by 2020, more than a quarter of political campaigns' advertising dollars—an amount expected to exceed US$2billion—would be spent on digital advertising (Erdody, 2018). Notably, social media and online spending can lead to increased engagement, particularly among older voters, who are more likely than younger voters to click and follow an advertisement seen on social media (Tech for Campaigns, 2019). So whether it's a plea to volunteer at a local call center or

Digital Drive

Political campaigns increasingly focus their ad spending on social media and online platforms.

Political advertising spending

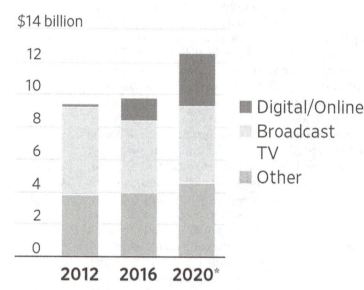

*Projections as of 2018

FIGURE 12.14 As social media have grown in use for our daily communication, political campaigns have put more of their resources into advertising in social and online media.
Borrell Associates

a targeted ad on Instagram, there is a good chance you'll be seeing even more from political campaigns during the next election cycle.

The Globalization of Local Politics

A final important way CMC has affected political communication is not necessarily how campaigns themselves communicate, but the influence on elections from beyond whatever voter bloc is influenced by an issue. Because CMC tools are not bound by space or time, emails, social media, websites, and other digital tools can communicate information, misinformation, and disinformation (see Chapter 11) about issues by parties that may not be directly affected by the political process or outcome, and yet influence the political process. For example, a city election for a county coroner may seem like a local issue and result in messages by the individuals and their campaigns running for coroner as well as local constituents seeking to influence neighbors and voters to vote for their preferred coroner. But as CMC tools make it easier to identify targeted publics, nonresidents may begin to seek to influence the coroner race. For example, a doctor in Toronto, Canada, may donate to the campaign of an old medical school friend running for coroner in Thurston County, Washington, USA (Carr, 2020b). Though a simple and seemingly benign example, this scenario illustrates a growing concern of more problematic or nefarious involvement in local issues from entities outside the locality.

In the past few years, campaigns have worked to communicate with individuals outside of the locality of the particular election to seek campaign donors and financing, particularly in close elections or key battleground districts. For example, the 2018 elections for US representatives and senators saw heavy uses of social media advertising to target out-of-district contributions, especially in districts likely to flip (i.e., vote for a candidate from a party different than the incumbent). North Dakota senator Heidi Heitkamp (D) saw 96% of her contributions come from out of state, Wisconsin representative Paul Ryan (R) received 95% of his campaign funding from out of state, and ~US$11M (42.7%) of Texas representative Beto O'Rourke's (D) state-record US$23.7M campaign came from out-of-district contributions (Center for Responsive Politics, 2019). In some ways, this makes sense: campaigns want to cast wide nets to seek the most funding they can for a candidate or ballot item, and donors may be willing to give to campaigns they cannot vote in if they perceive (a) their local election would not benefit from the same donation and/or (b) the distant campaign to which they are donating is a critical political issue. You may not think a donation to your local alderperson's campaign will make a difference, so instead you donate to an out-of-state campaign in the hope of giving your party an advantage in that close race. Aware of this, campaigns are increasingly sending emails, targeting social media advertisements, and using other online resources to communicate to individuals far beyond local geographic boundaries.

Another, much more concerning, trend is the influence of foreign nationals and/or government on domestic or local issues. The year 2016 saw several such problematic occurrences. For example, following Great Britain's very close vote to leave the European Union (i.e., Brexit), with 51.89% voting to leave, it was discovered that Russian operatives had engaged in an expansive disinformation campaign via social media, posting millions of statements on Twitter and Facebook to sway public opinion and influence Brits' Brexit vote (Wintour, 2018). Russian operatives similarly affected the 2016 US national elections, both by buying targeted ads on social media like Facebook and Twitter and by creating and spreading false narratives and information on social media (McKew, 2019). Russia or Russian nationals or hacker groups have also been linked to attempts to influence German and Dutch elections that did not directly deal with Russian relations or concerns (see Figure 12.15). And in 2019, the Chinese government began using public social media campaigns to discredit Hong Kong protests against an extradition bill proposed by the Hong Kong government and sway global public sentiment against Hong Kong and in China's favor (Wood, McMinn, & Feng, 2019).

Given the ability of CMC to help politicians, constituents, and the general public reach out to broad audiences, one substantive issue moving forward will be the globalization of local politics (Carr, 2020b). Politicians now can interact with constituents just as easily as individuals outside of their districts via email and social media, and online donation and payment systems (e.g., Kickstarter, Venmo) make

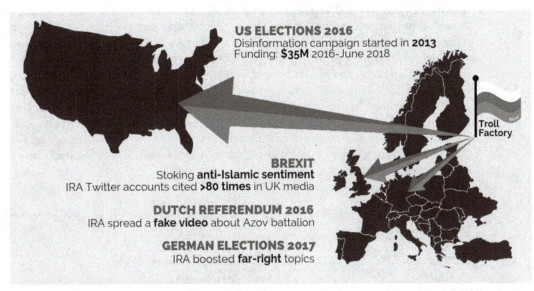

FIGURE 12.15 In the past few years, Russian nationals and agencies have sought to spread disinformation about local (i.e., non-Russian) elections and politics to influence key votes including elections and the Brexit referendum.
EUvsDisinfo.eu

campaign financing possible far beyond the geographic boundaries of an electorate. Geopolitical influence will grow as governments, groups, and individuals seek to influence elections that do not directly affect them but may be influenced to their benefit. As one of the first things a politician does when beginning a campaign is establish an online presence (i.e., webpage and social media profile), even local candidates may increasingly need to consider and communicate with publics far beyond their immediate constituents or electorate. Similarly, governments will increasingly need to consider how unwanted or disruptive involvement from outside of an election's locality may destabilize political communication and processes (Bennett & Livingston, 2018).

Concluding Politics via CMC

Political communication has found new homes beyond Federalist leaflets, televised fireside chats, and pool hall prattle. Email, blogs, chat rooms, online groups, social media, and other digital technologies give new ways to do old things and also provide new opportunities as humans engage in politics. We can still sign a petition to seek change in political processes, but we can now do so more easily via the "We the People" petition website than through physically signing a petition after running into a canvasser on a campus quad or town square. We can find people who share our political views just as we could in the bowling alley or knitting circle, but we can now more easily and naturally encounter oppositional perspectives that challenge and make us reconsider our own via social media and chat fora. And politicians and pundits still have traditional media outlet through which they can broadcast mass media messages, but the lack of gatekeepers online has led to a proliferation of political views and the ability of politicians and publics to talk at, to, and past each other in volumes we could not have imagined 50 years ago.

Perhaps the biggest challenges moving forward will be to sort the voices from the noise. How do we identify and engage with meaningful, faithful political dialogue online while avoiding disinformation, trolls and flamers, and malicious influence from external forces? One thing to bear in mind while wrestling with this important question is that as much as politics online may now seem like a common and normal occurrence, it is really still quite new and evolving. In 2009, we wrestled with the security of Senator Barack Obama's BlackBerry smartphone and whether the incoming president of the United States should have a personal and unsecured device to remind him of appointments or call his family (Choney, 2009), but just a decade later, US federal courts ruled that the president's Twitter account cannot block users because it serves as a public conduit to and from the president (Marimow, 2019). Even more than the technologies, the personal, social, and geopolitical processes of online politics are still developing and normalizing, and for now the best way to adapt may be simply to be mindful and critical communicators about politics online.

Key Terms

third spaces, 236

disinhibition, 237

flaming, 238

news finds me, 241

slacktivism, 242

echo chambers, 244

gatekeeping, 245

parasocial interactions, 250

static websites, 257

political campaigns, 257

Review Questions

1. What processes or features may make it easier to discuss politics online rather than offline?

2. Are following or liking a campaign or politician online "real" forms of political activity? What makes something a legitimate political engagement rather than simply slacktivism?

3. How could the hyperpersonal model or SIDE model explain your very positive or very negative perceptions of a politician from her/his online presence, without having ever met her/him?

4. As CMC and digital tools allow us to observe and act globally, what is the value of a local (whether city or region) political issue being online?

5. How can (a) dyadic, (b) interpersonal networks, and (c) social categories influence how you are exposed to or perceive politics online? When and how can these three influences intersect?

CHAPTER 13

The Future of Computer-Mediated Communication

LEARNING OBJECTIVES

After reading this chapter, you should be able to …

- Consider how new interfaces may change the way we communicate by providing communicators new channels to transmit and receive cues.
- Critically relate how new divides are emerging for digital media users based on skills and efficacy.
- Understand the potential surveillance possibilities and concerns from

continued integration and ubiquity of technologies.

- Discuss the potential benefits from continued integration and ubiquity of technologies.
- Explain the impacts of CMC on globalization and culture.

From their earliest room-sized forms in the 1940s, to their broadening adoption in the 1980s, to their networking in the 1990s, and their seemingly ubiquitous integration in the 2010s, computers have changed and been adopted in ways we would likely have not been able to fathom. Sometimes, even our best guesses as to their development are fundamentally wrong. In 1943, the president of IBM, Thomas Watson, predicted there may be a global market for—at *best*—five computers (which, at the time, were room-sized series of vacuum tubes with limited practical applications). Three decades later, predictions of computing's future were just as dire, as Ken Olson—who founded Digital Equipment Corporation, later acquired by Compaq Computers—warned, "There is no reason anyone would want a computer in their home." And in 1995, 3Com's founder, Robert Metcalfe, predicted the imminent and catastrophic collapse of the Internet.

These predictions are admittedly outliers among many more predictions of the success of computers. However, they do illustrate that it can be difficult to accurately predict how computing and humans' use of computers will change, even in the near future. The farther out we try to extrapolate, the more flawed our guesses become. Such predilection can lead to technodeterminism by presuming the world and the nature of human communication will inevitably change due to technological advances. But because some of these changes are growing quite near, and it takes some time to write or update a book, it's perhaps worth a chapter to consider some of the changes to technology and to our use of emerging CMC tools, and the potential impacts these will have on the processes by which humans communicate through mediated tools.

Let's begin by taking a look at some of the technologies just over the horizon—those that have been built and are undergoing testing and development, and whose broad commercial release or public adoption seems inevitable and quite near. Then we'll take a look at some of the applications—both actual and potential—of CMC tools and their continued diffusion and what it means for the human experience. While some of the potential applications and integration of CMC tools seem like they will benefit us, our work, and our life experience, other applications have more nefarious implications, suggesting caution and restraint as we move forward. In all of these discussions, though, remember that the technologies we'll explore are neither inherently good nor bad. Rather the utopian or dystopian potential of these tools is for us to decide based on what tools we adopt and for what purposes we use them.

Changes Just Over the Horizon

Some of the changes to digital communication tools and the way they influence human communication are just around the proverbial corner. We already see the early manifestations of these changes, which the next few years or decades will solidify as staples to our communicative ecosystems (Figure 13.1). Tools like smartwatches and smartglasses—watches and glasses connected to the Internet that can provide additional communicative tools such as sending and receiving messages and videos—are currently in use, but still seeking the "killer app" that will make them as indispensable as smartphones and broadly adopted beyond techies and younger users. Similarly, augmented and virtual reality experiences have become more accessible but have struggled to break into the mainstream as more than commercial kitsch.

Given that some of these devices are already entering the market, let's begin by focusing on some of the changes in CMC that are currently breaking. Rather than focusing on specific devices, let's explore the changes and opportunities you may already see around you that

FIGURE 13.1 Though some futurists' predictions (like a global demand of only five computers) are markedly incorrect, others have come surprisingly close to coming true. The form and function of today's smartwatches closely match (if not exceed) the watch worn in Dick Tracy comics by the eponymous detective, and the tricorder communication device in the original *Star Trek* was cited as an inspiration for early cellular flip phones. However, smart devices—from watches to glasses—are still looking for that feature that will make them must-have tools for communication, beyond the smart devices to which they're often tethered.
Bruno Vincent / Staff
Kim Kulish / Contributor

are still emerging or stabilizing. Some of the biggest changes in CMC that are currently emerging affect how we actually interact with and use our devices, as well as where our devices are. As everything becomes CMC, what will the experience of CMC and human interaction become?

Changing Interfaces

One of the most obvious changes in store for CMC in the near future is the way we interface and interact with computing devices, and therefore each other. No longer are "computers" simply monitors, keyboards, and laptops. New and emerging technologies are giving us ways to engage with our communication technologies and each other, new ways to experience CMC, and new senses we can transmit digitally. Excitingly, all of these technologies are in various stages of development or release, so it's not a matter of *if* we'll see these new CMC interfaces—it's just a matter of how soon we'll see them used and in what ways.

Virtual Reality

One way the interface of CMC has been changing is through the development and integration of immersive virtual environments. Though virtual reality (VR) has been commercially available since the 1980s, recent technological advances have made VR more cost-effective and accessible (see Chapter 9). The integration of motion-sensing technologies into smartphones and low-cost headpieces into which the phones can be placed to display three-dimensional images now makes it so that almost any smartphone can be turned into a VR headset for as low as five US dollars (see Figure 13.2).

Virtual reality will change the way we experience CMC. No longer will we have to see our interactions as outside of and distinct from ourselves, reading them on a defined screen or hearing them on a disembodied headphone. Instead, VR will let

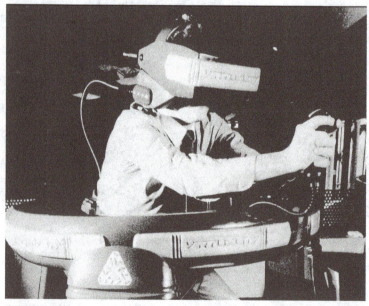

FIGURE 13.2 The VR headsets and equipment of the 1980s and 1990s (left) closely resemble contemporary VR setups (right). However, processing power has made the virtual environments much more realistic, and the reduced cost of technologies has made VR experiences as cheap as a five-dollar headset into which you place your smartphone.
Fairfax Media Archives / Contributor

us immerse ourselves into our communicative experiences. Consider, for example, when your roommate wants to share with you how amazing her alternate spring break experience is so far. She could email you a description of the trip, but email may be a lean medium and hard to really put the experience into words. She could send you a video file, but there's something about watching a two-dimensional image and a camera swinging wildly that just doesn't do the experience justice. Imagine, instead, if your roommate had a 360° camera to record the experience, and you could slip on a headset and actually walk through the site. Even better, imagine you could go a step beyond and virtually manipulate objects, "petting" a virtual llama at Machu Picchu or "hearing" the sound of Tibetan throat singers with her.

In another example, VR has the potential to change both education and journalism by increasing the immenseness of messages and experiences. Educators can take advantage of VR to have students experience historical events or engage in firsthand training (Dede, 2009). Medical students could use VR systems to practice surgeries firsthand without concern for danger to patients, just as tradesmen could learn the fundamentals of welding without fear of burns. Already, researchers are using VR environments to create and deliver health communication campaigns on the benefits of vaccinations (Nowak et al., 2020). Journalists could likewise use VR in news reporting to provide perspective-taking experiences, immersing you within the war zone, political protest, or local event being reported. These types of VR experiences make the news story more visceral, immediate, and impactful, enhancing the journalistic experience (de la Peña et al., 2010). Ultimately, VR provides the opportunity for richer, more immersive communicative experiences by which individuals go beyond merely reading or hearing a message and *experience* it.

As the costs of VR continue to decrease—particularly for the production of VR content—VR will become a more popular tool for complex, multisensory communication. These multisensory experiences currently focus on visual and audio cues to immerse VR users into the virtual world. But additional cues and channels are concurrently emerging that may be used independently or to supplement the multisensory experience of VR and that can even be applied to CMC more broadly.

Haptics

Haptic communication refers to the use of touch for communication. Handshakes, placing a reassuring hand on someone's shoulder, and high fives are all forms of haptic communication and play important roles in social processes. Touch allows us to detect "different types of stimuli (i.e., sub-modalities), such as pressure, vibration, pain, temperature, and position" (Haans & IJsselsteijn, 2006, p. 151). These stimuli are meaningful and substantive cues for conveying encouragement, social support, intimacy, tenderness, and other forms of personal immediacy (Jones & Yarbrough, 1985; Register & Henley, 1992), as well as to persuade others or gain compliance (Crusco & Wetzel, 1984). However, to date we've considered the inability to digitally transmit these haptic cues a limiting factor in CMC. Being able to convey haptic cues online could facilitate social processes and perceptions. New technologies are emerging to rise to this challenge, either physically replicating physical sensations (i.e., mechanical or kinesthetic feedback) or using small electrodes to simulate sensations. Both of these technical processes will convey nonverbal cues via CMC in the near future.

Again, as a form of nonverbal communication, haptic cues can help convey messages central to interpersonal perceptions, including encouragement, intimacy, and support (Jones & Yarbrough, 1985; Register & Henley, 1992). A sense of touch can complement the verbal messages exchanged online, supplementing the verbal message and making the interaction more socioemotionally rich (Chang et al., 2002;

FIGURE 13.3 Forthcoming haptic interfaces may allow you to "pet" a dinosaur, feeling the point of its horn or the texture of its skin. Such applications can go beyond games and displays and may soon let you physically receive a "hug" from your out-of-state grandmother.
Junko Kimura / Staff

Rovers & Van Essen, 2004). Additionally, touch can allow for more intimate interactions than may be possible by words alone (Haritaipan, Hayashi, & Mougenot, 2018; Mueller et al., 2005; Singhal et al., 2017). For example, receiving an email from your mom during a stressful final exam period is just not the same as receiving a physical and reassuring hug from her. Sometimes, physical contact and sensation can just mean more than words. Recently, devices for such haptic message exchanges have become available, such as a bracelet that vibrates when a partner taps his/her matching bracelet, providing a touch cue to indicate intimacy or awareness, just as laying one's hand on another's arm would. Finally, haptic cues can substitute for more laborious or implausible messages. For example, it can be difficult to accurately convey a sensation of pressure via text, but squeezing someone's fingers can convey the same information much more efficiently and accurately (see Figure 13.3). Likewise, when someone is concentrating on a cognitively demanding task, a light touch can be a less disruptive means of gaining the person's attention than a mediated message that requires an attention-dividing response (Rovers & Van Essen, 2005).

There are, of course, looming challenges to the inclusion of haptic communication via CMC. One of the most immediate and interpersonal will be the development of symbolic meaning to be associated with touch cues in new mediated interactions. Beyond the initial odd sensation of incorporating a haptic cue into existing CMC channels, users will also have to decide what those cues mean. For example, though running your hand across another's cheek is typically a very intimate form of touch offline, will it have the same meaning and intimacy if it's not another *person* directly touching our face, but rather an electrical pulse stimulating those nerve clusters? Users will also have to address the deliberateness of haptic cues. Someone bumping you may be determined to be accidental or aggressive when done in a shared space based on numerous other communicative cues, but users will need to negotiate the purposefulness of a haptic cue. Until common meanings are developed to associate with haptic CMC cues, users may struggle to appropriately encode and decode haptic cues transmitted digitally (Rovers & Van Essen, 2005). The formation of mutual meanings of haptic cues will likely follow the formation of mutual meanings of other nonverbal cues online, simply taking time and experience for norms to develop and be understood (Walther & Burgoon, 1992), first among pairs of individuals and eventually in broader social contexts. A final challenge will be incorporating the haptic interfaces naturally into our common experience. Currently, haptic interfaces are either highly specialized (e.g., telemedicine) or very complex (e.g., electrodes that must be adhered to one's skin at precise points), and neither is effective for broad adoption and usage. Surely we'll find a way to make the integration of haptic interfaces and channels more naturalistic experiences, just as we did text and audiovisual channels, but the widespread diffusion and adoption of haptic channels will take time.

Taste and Smell

The most challenging senses to transmit digitally and incorporate into our CMC have been smell and taste, which are closely linked—something's taste is largely influenced by its odor (Silver & Finger, 1987). One reason for the development of **gustatory** (taste) and **olfactory** (smell) cues to be transmitted online is the nature of flavors and odors. Though researchers have been trying for decades (Platt, 1999), it's been difficult to digitize the complex and nuanced chemical composi-

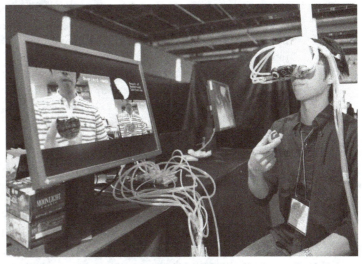

FIGURE 13.4 Capturing and replicating smell and taste can be challenging. At the Cyber Interface Lab (University of Tokyo), researchers are using devices like these to test the influence of smell on taste. You would still need a physical wafer to eat, but digital technologies like these may soon allow someone to transmit the smell and taste of that wafer, allowing you to share a five-star dining experience or relax and smell the roses. YOSHIKAZU TSUNO / Staff

tions of tastes and smells. In other words, the challenge of olfactory communication online may not be transmission, but digitization—actually turning taste and smell into 0s and 1s. Desired tastes and smells are certainly possible to reproduce, as predetermined chemicals or fragrances can be released to evoke a taste (think the flavor of a hamburger) or odor (think the smell of rain) desired by the sender (Figure 13.4). Numerous four-dimensional rides and amusement park experiences release chemicals to provide guests the desired olfactory experience to enhance the ride (SensoryCo, n.d.), like the smell of smoke during the burning of Rome on Disneyworld's Spaceship Earth ride. The biggest challenge to gustatory and olfactory cues in CMC will be the ability to capture and reproduce any flavor or odor, which will require complex input and output devices. Though research continues, it may be that we will soon indeed be able to taste Grandma's family recipe or smell the ocean air from our apartments. Additionally, the inclusion of these senses into the readily available channels for CMC could further widen the bandwidth of online interactions (Smeltzer, 1986), making CMC channels even richer and better able to facilitate perceptions of presence and socioemotional exchange among interactants.

Ubiquity and Integration

Beyond altering *how* we interface with digital communication technologies, an additional change whose foreshocks we are already experiencing is the growing ubiquity and perceived omnipresence of CMC. Recall that in the 1980s, "computers" (and therefore CMC) were limited to large, stationary desktop devices mostly placed in offices. Smartphones have now made it so that our computers go with us, transmitting information and making us accessible constantly. Though it may seem impossible, the ubiquity of CMC will continue to grow.

One factor increasing the ubiquity of CMC is the **Internet of Things**, which refers to the interconnection of networked everyday objects (Xia et al., 2012). Part of the implications for the Internet of Things (IoT) is the interconnection of objects themselves, allowing everyday items to "talk" with each other through embedded systems and artificial intelligence. For example, your interior lights could be connected to your garage door, so that as you arrive home and open the garage your lights automatically turn on so that you don't walk into a dark house. In this way, the IoT may not innately refer to CMC as we've defined it, when its communication is not between humans. However, the IoT can also be used and have implications for human-human mediated communication.

One CMC implication for the IoT is the perception of agency and immediacy the IoT may facilitate (Kim, 2016). For example, consider that on the morning of your birthday you wake up to find flowers delivered to your doorstep as you leave the house, receive a happy birthday text message at lunchtime, and rendezvous for a romantic dinner that evening at a restaurant. You may likely view these actions as thoughtful messages from your significant other. But would your perceptions change if you knew these actions were systems talking to each other: your significant other's calendar noted your birthday was imminent and last week automated an order for flower delivery to be charged to your account, their watch then sent a message automatically as they left the office for lunch when it also made a reservation for two at the restaurant of your first date (which was found stored on your calendar) and emailed invites to the both of you? Knowing this may make the same acts seem less meaningful or personal, actually reducing the romanticness of the gestures and resultant propinquity between you and the other person.

A second factor increasing the ubiquity of CMC will be the expanding capability of networks and digital telephony. Computers no longer tether us to our desks, and our smartphones will continue to allow us to take our CMC tools with us. As mobile network coverage expands and transmission speeds increase, this growth in infrastructure will parallel our ability to communicate through digital tools anywhere and everywhere, even as the number of communication tools we have increases. Moving beyond smartphones, smart devices—from tablets to toasters—enable us to use even more devices to communicate with others, and these smart devices are increasingly networked (see Figure 13.5). During 2021 alone, almost 7 billion new devices are expected to be connected to the Internet, resulting in a total of about 42.62 billion devices connected to the Internet (Bustamante, 2019), from phones to insulin monitors to dishwashers.

We are concurrently experiencing challenges in managing boundaries between the different domains of our lives, particularly between our personal and work communication. The ability to send and receive digital messages at any time means that work emails can be communicated at any time regardless of where we are physically or mentally (Mellner, 2016). The encroachment of work into our leisure time (imagine being at the beach on holiday only to get an email reminder from your professor about the term paper due next week) can be stressful, just as getting a message that your grandmother has fallen ill while you are in the middle of class can distract you from your professional obligations. The increased number of devices that are networked and facilitate communication will make the potential for accessibility ubiquitous, concurrently liberating and challenging us.

dorothy
@thankunext327

I do not know if this is going to tweet I am talking to my fridge what the heck my Mom confiscated all of my electronics again.

5:43 PM · Aug 8, 2019   LG Smart Refrigerator

15.6K Retweets   **67.2K** Likes

FIGURE 13.5 When everything is a computer, what's left to be CMC? In one notable example, a teen tweeted from her family's refrigerator after her other devices were taken away, possibly showing that ubiquitous computers and determination can enable CMC everywhere.

The Perils for Future CMC

As these new interfaces and forms of CMC continue to emerge, the future of CMC and human interaction is still in motion. The new devices, programs, and habits of individuals as we use technologies to communicate with each other will present us with new challenges and new opportunities, just as for the past 30 years new technologies have at times hindered and helped human communication. Of particular interest, two large perils loom in our near future: concerns of access and privacy.

Digital Divide

Back in Chapter 2, we discussed the digital divide: those who have meaningful access to CMC tools and those who do not. As we noted then, we have only recently surpassed 50% of the global population having regular access to CMC tools. Given that, in some ways the digital divide is closing. More individuals can access the Internet and the digital communication tools and channels it facilitates. But new manifestations of the divide will continue to split humans, separated not based on whether they have access to digital communication tools, but rather how they can use them. Beyond issues of mere access (first-level digital divide), we are beginning to see the second- and third-order effects of individuals' abilities and willingness to use online tools that will continue to divide people moving forward. To take a look at these changing divides, let's explore the three levels of the digital divide (see Table 13.1).

First-Level Digital Divide

The **first-level digital divide** refers to differences in access: who has a digital device (e.g., a home computer or smartphone) or meaningful access to one (e.g., a computer in a public library or shelter to be used without interruption) and who does not (Van Dijk, 2005). This is the manifestation of the digital divide we discussed back in Chapter 2. Simply being able to get online or use digital communication tools creates two classes of people, the haves and have-nots. The first-level digital divide is starting to close, but much progress and accessibility is still needed, as slightly less than 50% of the global population still has no form of Internet access at all (*The Economist*, 2018). The lowering cost of technology and the high adoption rates for mobile smartphones in particular have given many people basic Internet access. And yet parts of the divide remain to be conquered. The global south—including sub-Saharan Africa, Southeast Asia and northern Oceania, and parts of South America—is still disproportionately unlikely to have meaningful Internet access (see Figure 13.6). Rural regions may also be less likely to have Internet access than more developed urban or suburban areas with their established infrastructures. And

TABLE 13.1 Levels of the Digital Divide

Level	Focus	Factors Influencing the Divide
First-Level Divide	Access	Geographic location, infrastructure, global and regional socioeconomics
Second-Level Divide	Usage	Personal socioeconomic status, self-efficacy, personal preferences, perceived need to use CMC tools
Third-Level Divide	Engagement	General usage patterns, perceived value or outcome of use, discretionary time to use CMC tools

Adapted from Scheerder, van Deursen, & van Dijk (2017).

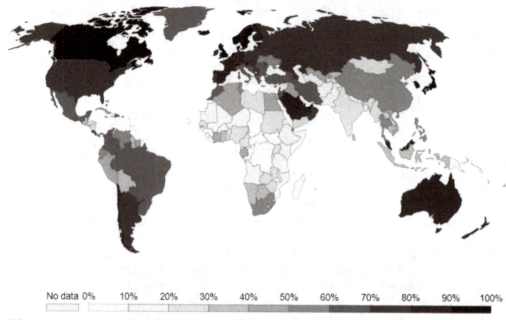

FIGURE 13.6 This map shows Internet use by country as a percentage of population. What countries or regions have the highest percentage of residents online? Which have the lowest adoption rates? What are the implications for a nation or region when its residents cannot get online?
WorldBank (CC BY 4.0)

though it can seem like everyone around you has abundant Internet connectivity, remember that may not be the case for others in your state, let alone worldwide. As the global COVID-19 pandemic made clear, when we do not have access to libraries, schools, and restaurants and coffee shops, even many Americans can suddenly struggle to maintain access to Internet connectivity (Auxier & Anderson, 2020). But again, this divide is closing. More infrastructure expenditures, lower costs of communication technologies, and increased social pressures to move online have begun to close the first-level digital divide, albeit slowly. However, as the first-level gap closes, the second- and third-level digital divides will continue to challenge us.

Second-Level Digital Divide

The second-level digital divide may be more challenging to overcome than the first. Whereas the first-level divide was mostly tied to physical and economic limitations, the second-level digital divide is more social and psychological in nature, addressing whether and how individuals *use* the digital tools they have available. The economic and social disparities of not having Internet access cannot be resolved if users do not *use* their Internet access. For example, the rapid adoption of smartphones has recently provided many individuals access to the Internet they previously lacked. But even with Internet access, many smartphone users without a computer still don't use their smartphones to engage in the online activities that would increase their communication or well-being (Pearce & Rice, 2013). This second-level digital divide reframes the concern away from merely having access to digital technologies—which are admittedly becoming more accessible—to consider if individuals who have digital technologies and Internet access can use those tools and be included in the communicative, educational, vocational, and other opportunities those digital tools provide (Livingstone & Helsper, 2007).

Much of the differences in the use of CMC technologies that make up the second-level digital divide are a result of individuals' skills and competencies (Hargittai, 2002). For example, one of the biggest factors influencing individuals' use of CMC tools is their **self-efficacy**: an individual's *belief* they can successfully use a tool or skill. Notably, self-efficacy is not necessarily related to actual skills and abilities, as individuals may not be confident in skills at which they are actually proficient. A lack of technological self-efficacy about their skills in using a computer or online tools (e.g., search engines, email) can reduce that individual's desire to use them (Hargittai, 2002).

This lack of technological self-efficacy is a critical reason older adults are less likely to use the CMC tools they have. If your grandparents are anything like mine (They are wonderful and amazing people, no matter what the rest of this example may suggest. Love you, Grandma!), they likely have a computer and smartphone in their home, and perhaps even a tablet or smart device. They have an email account, perhaps use Facebook, and even text from time to time. But just because they have these CMC tools does not mean they use them often or well. Older users tend to be less efficacious in their use of CMC tools, which reduces how often they use them (Cotton et al., 2014). Identifying needs for or benefits from using CMC tools (such as connecting with friends and family) or through supported use (such as when you show your grandmother how to snap a selfie) can increase an individual's self-efficacy and subsequently their use of digital tools for communication. It's important to note here that age alone is not the strongest predictor of using CMC tools, but rather personal factors such as cognitive function and income (Yu et al., 2016). It may not matter as much whether an individual is 55 or 95 years old to predict whether they use the technologies they have; instead it is their comfort and belief in their ability to use that technology that best predict its use (Figure 13.7).

Another way the second-level digital divide may present itself is through disproportionately high self-efficacy or desire to use online (rather than analog) communicative tools. Whereas having low confidence in one's use of digital tools may depress their use, having high confidence in one's digital tools may increase their use. For example, younger individuals can often have high self-efficacy about their computer and Internet use that exceeds their actual skill set (Sam, Othman, & Nordin, 2005). In a study of college students' Web search behaviors (of which they were highly self-efficacious), students rarely scrolled down in the initial search results screen and almost never went to the second page (Metzger, Flanagin, & Zwarun, 2003). It appears these users assumed that if a helpful search result is not displayed in the first few returned hits, no such information exists online, rather than considering

FIGURE 13.7 Even individuals who own a CMC device and have an Internet connection may have either too much or too little belief in their own skills at using them. It's easy, as in this FoxTrot comic, to point toward older users as victims of this second-level divide, but the same self-efficacy concerns can limit younger users in the same way.
FOXTROT ©1995 Bill Amend. Reprinted with permission of ANDREWS MCMEEL SYNDICATION. All rights reserved.

whether the search terms were the most applicable or how results may be displayed. I once had a situation in which overconfidence led to a shortcoming when a student asked me for help formatting a résumé. The formatting request was to help align the "Proficient at Microsoft Word" skill on the résumé, but that proficiency was demonstrably absent by nature of having to ask for help. Though confidence can be a beneficial attribute, overestimating skill with online tools may make users turn to online tools more often, recursively increasing their use (and self-efficacy with) of CMC tools, without actually increasing their skills.

That individuals with low self-efficacy will use CMC tools less while individuals with high self-efficacy will use CMC tools more suggests a broadening of the second-level divide, which may be harder to fix. It can be difficult to increase users' self-efficacy or demonstrate why they would benefit from using a technology, just as it can be difficult to decrease other users' self-efficacy or suggest why analog tools may be better. As more users have *access* to CMC tools, an emergent challenge will be the divide between those willing and able to use those tools. Because this divide is based more on individual differences (rather than infrastructure inequalities), redressing the second-level digital divide may require training, social demonstration, and greater support for certain demographics.

Third-Level Digital Divide

As the presence and impact of the second-level digital divide begin to be noticed, the third-level digital divide is beginning to make itself known as well. The **third-level digital divide** refers to individuals' engagement with and benefits from using digital communication technologies (Van Deursen & Helsper, 2015). If the second-level digital divide is about *whether* individuals use CMC tools, the third-level digital divide is about *how* and *for what* individuals use CMC tools. You may have already seen evidence of the third-level digital divide: if you use a smartphone app as a frequency card for your coffee, grocery, gas, or retail purchases, you're taking advantage of a digital tool that's not available to those not willing to use the app. Even users who have smartphones and are confident in their ability to use them may not be willing to use the digital loyalty program—due to either privacy concerns or a lack of interest in *another* app to have to install and monitor—and thereby take advantage of cost savings or freebies. Similarly, many individuals do not choose to participate in politics online (see Chapter 12), whether that's following politicians or signing digital forms, believing their online activities do not have demonstrable offline benefits or effects, or simply because they have grown more cynical about politics altogether (Lee, 2006).

Like the first- and second-level digital divides, the third-level digital divide seems to be strongly influenced by resources: individuals with greater financial and personal resources are more able to not only use CMC tools but also to derive benefits from their use (Van Deursen & Helsper, 2015). As more of our human interactions occur online, addressing this third-level digital divide will become increasingly important. It will not be sufficient that individuals have a computer, know how to use email, and feel comfortable and confident using email to communicate with others. As may seem self-evident, it will be important that individuals can send emails that elicit the desired response, because without that response there may be no reason or motivation to use the communicative tool again. Already we see many of our traditional offline benefits playing out online. For example, successful job seekers are increasingly incorporating digital tools into their job searches, including searching job boards and potential employers' websites, as well as asking others in their online networks about job openings and sharing that they are currently job seeking (Piercy & Lee, 2019). Individuals who can engage with online tools and obtain

benefits from their use will see social and offline returns, including jobs, social capital, and personal connections. Alternately, individuals who are unable to manage
CMC tools to elicit responses, feedback, and other benefits may not realize the same
personal and professional benefits.

This third-level digital divide may be the most complex of all to solve. Because
it addresses the benefits of using CMC tools, it is an innately social problem.
Whereas an individual can be given a laptop or trained to use email, interacting online or obtaining benefits from using online tools requires others—whether
the "other" be systems or other users—to interact with the individual. Because
its immediacy as a concern rests on overcoming the first two levels of the digital
divide, the third-level digital divide may not be the most pressing. But addressing
the social and systemic forces that allow some individuals to derive benefits from
interacting online while preventing others will require large sweeping—and likely
societal—changes to address. And until the gap between those who can effectively
engage with others via CMC and those who cannot has closed, CMC tools will
continue to divide society.

Surveillance

One of the largest potential threats of technology and CMC tools is the use of these
tools for surveillance as organizations, governments, and other users utilize communicative technologies to track our movements, behaviors, and actions. George
Orwell's classic *1984* is a cautionary tale of what happens when "Big Brother" is
watching over us. The ways we increasingly give up our privacy or have it taken
from us by technological tools are quickly approaching the Orwellian levels about
which we were warned.

Some of these surveillance concerns are already upon us, particularly as our
behaviors are tracked by businesses and governments using our identification, credit
cards, and frequency cards to analyze and predict our actions. In an early iteration of predictive marketing, a father stormed into his neighborhood Target and
derided the manager for Target mailing his teenage daughter coupons for cribs and
baby clothes. The manager apologized for the errant ads, and called a few days
later to follow up. On the call, the father admitted he'd only been recently told of
his daughter's activities and that she was due in August (Hill, 2012). Target's algorithms had accurately predicted the girl's pregnancy (even before her own father
became aware) based on small changes to her shopping patterns. Even had Target
not unintentionally disclosed the pregnancy to the girl's father, many are concerned
that such attempts to predictively and proactively market to consumers represent a
privacy threat.

A similar case of undisclosed monitoring of user behaviors occurred in 2017,
when the GPS company Stravia—used by numerous fitness trackers, including Fitbit and Jawbone—posted a map online that mapped 27 million users'
movements over the course of two years. Viewers quickly realized there were
high concentrations of activity in remote and war-torn areas. Stravia had just
unknowingly published the location of secret military bases in active conflict
zones, where soldiers used fitness trackers as they ran around their clandestine
installations (Sly, 2018). In this case, it was the aggregated surveillance of many,
rather than the surveillance of any given individual, that caused the unintended
breach in information.

Another way privacy can be reduced online is through the aggregation and
cross-referencing of multiple information sources. Individuals may use several strategies to manage their identities online, such as communicating anonymously, using

FIGURE 13.8 Increasingly, algorithms and users are connecting data and data patterns to connect otherwise discrete bits of information. When Facebook suggests a person to tag in a photo or Amazon suggests your next purchase, you are seeing that reduction in privacy in action.
facebook.com/Maxisdrawing

pseudonyms, or limiting personal disclosures across various channels. However, both humans and computer programs are getting better at making connections among the data points of personal information found online. For example, though you may not use your real name on your Instagram account, you also used the email address you used to sign up to Instagram to register your Xbox Live account, which has the same gamer tag you posted on Twitter last week. By following the information trail and consolidating information, curious individuals may be able to deanonymize you. Even more challenging are computer programs and algorithms that do the same, matching users' pictures, limited information releases, and metadata (including usage patterns and technical information) to identify users' otherwise anonymous or pseudonymous profiles (Su et al., 2017; Zafarani et al., 2015), manifesting the goal of Web 3.0 (see Figure 13.8). Consequently, in the near future, even information you thought was private may be discoverable, deanonymized, or connected to you.

An even more immediate and dystopian reality is the use of technology to monitor and evaluate individuals' social behaviors and then to assign scores. Social monitoring has been dramatized in the *Black Mirror* episode "Nosedive" and in *The Orville*'s "Majority Rule" episode, but reality is close on the heels of these fictional examples. The mobile app Peeple allows users to comment on and review other people based on professional, romantic, and personal relationships (Wattles, 2015). Going beyond a "RateMyDate" feature, Peeple allows aggregated user reviews of other people.

More widespread is China's nationwide "social credit system," which assigns each citizen a score (similar to a financial credit rating) based on behaviors. Behaviors perceived as prosocial or valued by the government, such as disposing of trash properly, walking your dog on a leash, and caring for the elderly (including those to whom you are not related) can increase your social credit score. Alternately, negative or undesirable behaviors like spending too much time playing video games, spending money "frivolously," and speaking ill of the government can lower your social credit score. Individuals' social credit scores can govern their behaviors and opportunities: those with high scores are eligible for better jobs, have higher Internet speeds, and can more freely access planes and trains for transit, while individuals with lower scores may not be able to own pets, would be limited as to what schools their children could attend, or even be publicly labeled and shunned as "bad citizens" (Ma, 2018). This social credit system relies on massive surveillance, both personal and automated, including the use of facial recognition technologies and public cameras to document individuals' public behaviors and relying on both official monitors and peer reporting to identify more private behaviors (Figure 13.9).

FIGURE 13.9 China's social credit system uses computer systems to monitor individuals' behaviors and interactions. How would you feel if your every action was monitored and tallied and impacted your ability to get a job or rent a hotel room?
Bloomberg / Contributor

This increased surveillance may not be a concern to all, however. Studies of kids and young adults are revealing a decrease in concerns about surveillance and expectation of privacy (Desilver, 2013; Steeves, 2019). The change of expectation of personal privacy among younger individuals is the result of several convergent factors. First, younger individuals are already immersed in a surveillance culture and don't know otherwise. Cameras record our schools, public transit, and public spaces. Chicago, Illinois, has more than 32,000 video cameras (including police, ATM, traffic, and security cameras), averaging several cameras on every block, and leaving few places in Chicago out of the eye of some form of video surveillance (Glanton, 2019; Goldfine, 2018). Youths' activities in online games and learning modules have also always been monitored and tracked (Kumar et al., 2019), and every time they see someone swipe a credit card or scan another payment mechanism, they are aware it creates a digital record. From the perspective of younger individuals, "Big Brother" is and always has been watching. Second, younger individuals—digital natives—have grown up alongside social media and in a networked society. They are more used to sharing personal information with others online, whether that's their interests or class schedules. Third, younger users may be more comfortable than older users providing personal information or allowing their data to be tracked in exchange for goods and services (Desilver, 2013). More than 46% of those under the age of 29 view exchanging their personal information and/or user habits for the ability to stream music, use SNSs, receive discounts, and use digital tools to communicate as a fair trade (see Figure 13.10).

Finally, younger individuals may not be worried about their *personal* privacy being violated, considering their personal information is to be aggregated with other users. For example, individuals using fitness trackers did not report concern about the company using their unique user information, instead perceiving their personal information (e.g., run distances, routes, and health information) as useful to *them*, but aggregated user statistics (e.g., broad categories of people's fitness routines, frequency and intensity of workout, identifying popular running or biking

FIGURE 13.10 If you'd like an estimate of how much your personal information is worth to marketers, take a look at the *Financial Times*'s personal data calculator (https://ig.ft.com/how-much-is-your-personal-data-worth/). What demographics are worth more to marketers? Why do you think those personal characteristics make them more valuable or of interest to marketers?

routes within a city) was of interest and use to the fitness tracker company (Zimmer et al., 2020). In other words, because there is so much information about us online, our personal information may be lost amid the glut of others' information.

Even privacy-savvy individuals may not realize they are exchanging their personal information for a service or convenience. When you provide your local pizza parlor or sandwich shop your name, address, email, and mobile number in exchange for delivery, you are yielding some of your privacy by connecting your name, address, phone number, payment information, and purchasing behavior. The next time you order delivery, ask the company representative about their customer information privacy: many restaurants, particularly large franchises, sell customer information to marketing firms to supplement their revenue from food sales. Personal information is becoming a quantifiably valuable resource, and some individuals may simply not mind giving up some privacy or being surveilled for the benefits they receive. However, there's an important distinction between knowledgably trading privacy for services and simply being ambivalent or ignorant to such exchanges, as they are likely to increasingly occur as your personal information is either monetized or converted into social credit.

This discussion of online privacy comes with an important reminder that by now may border on trite. One last time: CMC tools are neither inherently good nor inherently bad. The same tools that may reduce our privacy and increase our surveillance can also be used to increase our privacy and circumvent monitoring. For example, the Telegram app was developed to facilitate encrypted messaging, and it has been used to facilitate and coordinate political protests, particularly in countries where such civic dissent could result in severe personal penalties from the government (Maréchal, 2018; Purbrick, 2019). Likewise, the WhatsApp messenger service has been used globally to securely transmit and receive group messages in order to coordinate and reinforce opposing political views and ideologies without fear of surveillance or repercussion (Cheeseman et al., 2020; Treré, 2020). The way communities and political dissidents have used emerging CMC tools to avoid surveillance hopefully can demonstrate that all digital tools are not innately means of surveillance, and that the social determinism perspective guides the conclusion that CMC tools can be used to surveil or avoid surveillance, based on how companies and users make use of digital communication technologies.

I Always Feel Like Somebody's Watchin' Me

What's your favorite color? Your oldest sibling's first name? What elementary school did you attend? These types of questions (and answers) are commonly found in two places: (1) social media profiles and (2) bank account recovery systems. Social media can be great tools to let us communicate with distant friends and strangers, but why may users disclose personal, private information? One theory often used to explore this tension is *privacy calculus*, a theory with roots in economics that suggests people will disclose or provide personal information when they think the benefits of providing that information will outweigh the potential harm of doing so. For example, you provide Facebook your name because it helps others identify and find you without opening you up to a lot of risk, but you don't tweet your bank account or email passwords because the bad things people could do with that are much more significant than the good you may get from doing so.

Dienlin and Metzger (2016) applied privacy calculus to understand how people provide personal information on SNSs. In a nationwide survey, they asked 1,156 Americans about the information they disclosed (or purposefully did not) on Facebook, the benefits they perceived from using Facebook, and their general privacy concerns. Exploring what factors led to self-disclosures on Facebook, the data revealed people were more likely to post personal information (both in profiles and in Facebook posts) if they perceived doing so as beneficial (including making new social and business connections, seeking social support, and feeling connected to others) and had lower general privacy concerns. Users were more likely to withhold information (including untagging themselves in photos or making their profile unsearchable) on Facebook if they had higher privacy concerns.

As more information about us is available online—not always authored by us—we'll need to make more decisions about what information about us we want others to have access to. How we decide appears to not be universal, as some individuals may be comfortable posting personal photos or vacation plans online while others are not. But it does seem that, at least for now, users do consider to some degree the risk-versus-reward of making their information available online. Just please don't post your bank account number.

Reference

Dienlin, T., & Metzger, M. J. (2016). An extended privacy calculus model for SNSs: Analyzing self-disclosure and self-withdrawal in a representative US sample. *Journal of Computer-Mediated Communication, 21*(5), 368–383. https://doi.org/10.1111/jcc4.12163

The Promises for Future CMC

Though the future of CMC comes with causes for concern, it also brings with it the potential for great personal and societal benefits. If we use CMC tools carefully and strategically, they can be used to make the human condition better. Current developments in CMC are already letting us become smarter, more culturally aware, and physically healthier, and developments will likely continue and enhance these benefits.

Increasing Access to Information and Education

One of the most evident—at least to us students and scholars—promises of CMC in the future is the opportunities CMC can present to enhance our knowledge and understanding. Distance education is not a new concept, as for decades individuals have enrolled in correspondence courses to take classes and engage in learning without having to attend a physical school or university (Garrison, Tang, & Schmeichel, 2016). Beginning as readings and assignments physically mailed between instructors and students, the introduction of home television and recording devices allowed correspondence courses to integrate video lectures mailed via videocassette. Students now can download readings from online and upload assignments to an online course management system, or watch a YouTube of a lecture and learn something new, whether it's a lesson on painting or on particle physics. In this respect, CMC tools have not fundamentally transformed the process of education and learning—information can still be exchanged and shared online, just as we have done in classrooms or via postal mail for decades. This may manifest simply as the global rush to quickly move offline classes online midsemester in March 2020 due to the COVID-19 pandemic. Even for the most determined and devoted of instructors, much of the adaptation to virtual learning during the pandemic was a patchwork of making the best of a terrible situation using the tools currently available. But when used deliberately, taking time to craft learning experiences *for* digital channels (rather than just to be conducted over digital channels), CMC tools have created new opportunities and directions for accessing, synthesizing, and generating knowledge.

Collaborative Knowledge Generation

One of the most apparent ways CMC tools have advanced education is through collaborative knowledge-generation tools like wikis, allowing multiple individuals to record and generate knowledge. With millions of entries generated by thousands of individuals on diverse subjects from aardvarks to the Zeeman effect, Wikipedia represents one of the largest repositories of knowledge ever gathered. Though Wikipedia is often derided as edited by amateurs, the sheer number of editors and automated programs constantly combing and updating its entries actually make Wikipedia a more accurate source than textbooks and encyclopedias (Kittur & Kraut, 2008). Consequently, this crowdsourced and collaborative database is typically a valid and reliable source of information, particularly as communities communicate in backchannel "Talk" pages to discuss and debate the truth and objectivity of entries before they are made public in the Wikipedia entry.

Similarly, YouTube has emerged as an effective way to collaboratively create and share knowledge. Numerous user-generated videos provide information, both theoretical and practical. Many individuals now can learn how to change their car's headlights by watching a user-generated tutorial, or find guidance to solve a math equation by looking at a math tutorial video's comments to see if other users have previously encountered and solved the same challenge. Crowdsourced videos can be particularly helpful for individuals to obtain procedural knowledge—understanding how to do something—because the verbal presentation of knowledge is supplemented and enhanced by a visual demonstration of the skill that viewers can then emulate and practice (Lee & Lehto, 2013).

The emergence of crowdsourced knowledge—whether in the form of wikis, databases, websites, or videos—that can be stored, tagged, indexed, and searched has given us the ability to create a global repository of knowledge. Through digital tools and distributed users, almost limitless information can be readily generated, archived, and retrieved. If you want to know about quantum theory, no longer must you enroll in a college-level course—numerous sources are freely available to let anyone interested access the desired information.

Massive Open Online Courses (MOOCs)

Speaking of courses, another change we see CMC driving related to education is the changing nature and form of classrooms. Beyond simply putting courses online, many schools and nonprofits have begun exploring the development and use of **massive open online courses** (MOOCs). Potentially serving hundreds of thousands of students in a single course, MOOCs are a way to scale education, providing it

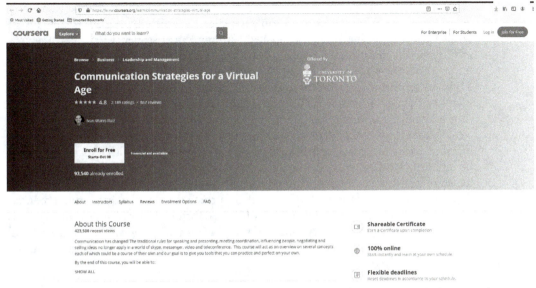

FIGURE 13.11 Even if you don't live near a prestigious university or don't want to pay a hefty course fee, MOOCs can help you access college-level knowledge, instruction, and co-learners. Here, the University of Toronto uses the Coursra MOOC platform to simultaneously provide a course on communication strategies to more than 93,000 learners.

often free or at low cost to those interested (Figure 13.11). Ever dreamed of taking a course on neoclassical art or nonlinear algebra? There's likely a course online, including from renowned higher education institutions and reputed faculty, you could take. Some MOOCs are accredited courses, meaning that upon successful completion of the course's requirements you can receive academic credit or make progress toward a degree.

Massive open online courses admittedly pose several problems in their present state. For example, the large class sizes of MOOCs limit interaction with peers and make individual engagement with instructors implausible (Khalil & Ebner, 2015). The challenge of communication with co-learners is particularly problematic, as engagement with peers strongly influences an individual's enjoyment of the course and learning outcomes (Carr, 2014; Carr et al., 2013), and strongly whether the individual will complete the course (Pursel et al., 2016). Additionally, though MOOCs have been touted for their ability to provide education to those without access to college and university education systems, at present many of those enrolled in MOOCs appear to be college students and college graduates (Christensen et al., 2013). This suggests the underserved populations MOOCs were promised to support are not yet being aided by these virtual classes. Finally, MOOC completion rates tend to be very low: around 5% of those who begin a MOOC may typically finish the course, and fewer still may pass the course with a satisfactory grade (Pursel et al., 2016). That few students who enroll or begin a MOOC complete the course may be influenced by several factors. As enrollment in MOOCs is often free (or inexpensive), there may be no perceived cost associated with not completing the course, unlike courses for which students must pay to enroll and therefore they view the course as an investment. A second reason for the low completion rate of MOOCs may be the lack of accountability or individual motivation to complete the course. Online learners may be less engaged or self-motivated, and without physical cues and reminders to complete readings or assignments, individuals logging onto an optional class may not have the same internal motivation or self-policing necessary to complete the MOOC as they would a traditional course (Zheng et al., 2015). A final reason completion rates may currently be so low is the lack of familiarity with online and/or self-paced learning. Most students have become very familiar with traditional classroom learning, including classroom practices, study habits, and ways to communicate with other students and instructors. Because MOOCs represent entirely new ways of being a student, learners may need time to adapt to the new interactions and processes of a MOOC, even beyond that which they may have encountered in an online class offered or taken through their school.

Although MOOCs currently have several limitations, the promise they offer to increase access to knowledge and reduce the costs associated with education cannot be ignored (Vardi, 2012). The form of MOOCs and their relationship with established educational institutions may continue to change, particularly as institutions determine how to adapt MOOCs and their structures. But the disruptive potential of MOOCs also presents opportunities for new forms of formal education systems beyond established secondary and tertiary educational systems. How students and instructors learn, interact, and share knowledge will continue to emerge alongside these MOOC systems.

Bridging Cultures

An additional way CMC may provide benefits in the future is by increasing our ability to access, interact with, and understand new cultures. As we discussed in Chapter 5, prejudice and distrust of broad groups is often a result of lack of exposure to and knowledge about the other groups (Allport, 1979). From this perspective, the contact hypothesis tells us that one way to make people more understanding and tolerant is to help them learn about and interact with different cultures. And yet physical travel can be challenging. A monetary cost is associated with traveling (both in the expense of taking time off of work and of the travel itself) and travelers face the potential discomfort of being immersed in a culture different from their own. Computer-mediated communication tools can and are beginning to help us bridge cultural divides. Not only can CMC tools make it easier to access new cultures, the affordances of CMC tools can be harnessed to make intercultural contact more comfortable.

Time and Space Compression

About 146 million Americans have a valid passport (US Department of State, 2019), indicating two-thirds of US citizens cannot go see and experience a different national culture even if they wanted to and had the money to do so. Even if you have a passport, traveling can be costly. The ability of CMC to cross time and space makes CMC tools particularly appealing to those interested in experiencing and understanding new cultures, but without having to buy a plane ticket.

Simply being able to access new cultures can help individuals become more aware of and feel more connected to other cultures and lifestyles. One of my favorite times to play *World of Warcraft* is December 24th (Don't judge my life choices.). The reason I enjoy playing the massive multiplayer online game on the 24th is that I play with other players from around the world, and as Christmas Eve wears on, you can watch players wish others a "Happy Christmas" as December 25th dawns in their time zone. It's a simple yearly reminder that the people I game with are from Australia, Germany, and California, made possible through the *Warcraft* game.

Similar experiences can be had by friending or following individuals from different cultures on social media, meeting and talking with others in chat rooms, and even setting up international pen pals with whom to exchange emails. For example, blogs and email can be used to practice second languages with native speakers (Hunter, Muilenburg, & Burnside, 2012), with the native speakers providing feedback and helping refine language skills. Similarly, pen pals have long been used in schools—now via email—to help students learn about other cultures, including daily experiences, language, family, and social structures (Barksdale, Watson, & Park, 2007). By compressing space and time, individuals will continue to use CMC tools to quickly access cultures that may be too remote—either geographically or socially—to access offline without incurring substantive costs. The ability to access other cultures and the subsequent exposure to other ideologies, lifestyles, and values can bridge cultural divides, and CMC is creating a space in which those different cultures can be brought together in a single global village (Ess & Sudweeks, 2001).

No Longer Constrained by Offline/Physical Traits
An additional mechanism by which CMC can bridge cultures and reduce cultural barriers is through the reduction of visual or nonverbal cues to cultural differences. Different accents, forms of dress, or physical appearances can quickly make others aware of cultural differences and activate group processes (Dovidio & Gluszek, 2012; Hargie, Dickson, & Nelson, 2003), which can lead to individuals seeking to distance themselves from the other culture. However, CMC can reduce these cues to intergroup differences. Text-based communication can make one's appearance or attire irrelevant, and it mostly hides accents and dialects. Increasingly, tools like Google Translate (translate.google.com) can even help overcome language barriers by helping non-native speakers quickly encode their thoughts into other languages and decode others' messages into their native tongue. Google Translate has already proven useful as a translation tool for medical providers to communicate with patients and family members in other languages, even with only about a 58% accuracy rate in translating medical phrases (Patil & Davies, 2014). By reducing physical and language cues that highlight and activate cultural differences, individuals can be more willing to interact with those from different cultures who may not look, act, or sound like them (Rochadiat, Tong, & Novak, 2018). Consistent with the contact hypothesis, these individual interactions can then lead to more favorable perceptions toward that individual's group and culture later (Walther et al., 2015). In other words, by hiding the cues that make our cultures apparently different behind the computer screen and interacting with other cultures interpersonally, we can be less aware of the cultural differences and foster cultural learning and understanding. Whether it's learning about or engaging with members of a different culture from the other side of your country or the other side of the globe, moving forward CMC's ability to selectively filter out cues can bridge cultural divides.

Telehealth and Medicine
One of the most exciting opportunities and applications of CMC moving forward will be in health contexts, as the ability to feel will help medical practitioners diagnose and treat patients. **Telehealth** (or *telemedicine*) refers to healthcare that "uses communications networks for delivery of … services and medical education from one geographical location to another" (Sood et al., 2007, 576). Telehealth can be as simple as physicians sending out automated digital reminders via text or email to patients to remind them to take medicine, follow a therapy regime, or otherwise engage in health management (Holtz & Lauckner, 2012). Other forms of telehealth may include more intensive communicative options like videoconferencing with a mental health professional (Cohen & Kerr, 1999) or online therapies and interventions (Rickwood, 2010). These telehealth options may be an especially desirable

option for individuals managing anxiety, challenged by social settings, who distrust formal medical settings, or who wish to receive in-home healthcare.

Telemedicine can also include remote surgery or virtual medical procedures (Ahn & Fox, 2017), where the medical staff is located in one location and operating on or physically manipulating a patient elsewhere via a robot or other agent controlled by the medical professional (see Figure 13.12). Computer-mediated communication will help provide access to medical services, including by allowing rural individuals to stay at home or go to a local clinic to be seen by a specialist or medical provider via a telehealth system. In rural communities, where access to medical practitioners may be limited and distances between medical services prohibitive, telehealth will help enable quality medical care for underserved populations (Smith et al., 2005). Particularly as haptic interfaces are integrated into doctors' offices and health clinics, doctors will remotely check on patients, including taking pulses and blood pressure and feeling for tenderness—all needing physical touch. Additionally, the ability to mindfully apply and perceive pressure digitally is critical to surgeons' practices, as surgeons need to carefully feel patients and surgical instruments remotely (Green et al., 1995). Emerging telesurgery tools have been literal lifesavers, allowing doctors to remotely treat patients in isolated or hostile areas, such as in disaster relief and combat zones, providing victims, responders, and soldiers immediate care (Ling, Rhee, & Ecklund, 2010). In such cases, haptic interfaces have allowed surgeons to feel injuries and physically assess patients, perceive tension on suture thread, and manipulate medical instruments to conduct surgeries, all without having to transport themselves or patients.

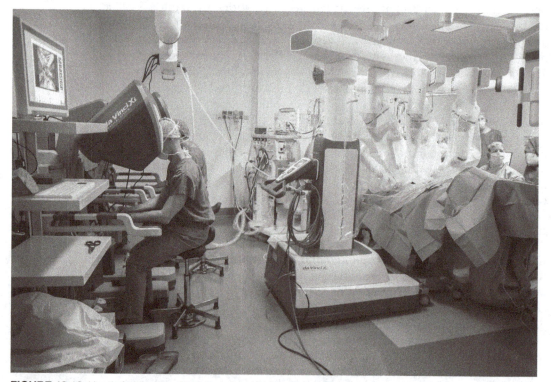

FIGURE 13.12 Haptic feedback in CMC tools would allow users to transmit touch. Haptic devices would allow you to digitally hug your mother. Such devices would also allow doctors to practice telemedicine, virtually conducting a surgery from afar but able to feel the tension and texture required for surgery.
THOMAS SAMSON / Contributor

Concluding the Future of CMC

The future of CMC has yet to be written. Some predictions of future computing have come remarkably close to true. One of the most notable of these predictions was Vannevar Bush's prediction in the 1930s of the *memex*—a device for storing, sorting, and cross-indexing the entirety of one's memory—which Wikipedia now closely resembles. Though some—including the memex, Dick Tracy's watch, and *Star Trek* tricorders—have come true, our past is littered with even more flawed predictions of computing and humans' use of technologies to interact. It can be hard to predict not only what technologies will come to pass, but how individuals will use them, both for their intended purposes and in ways designers and inventors may never have imagined. Importantly, though, the future of CMC will involve the same fundamental human communicative processes that have sustained us and our society for millennia. The same basic needs—for relationships, organizing, and understanding ourselves—will still drive us and be met communicatively. We'll just increasingly meet these needs through technical channels. It is my hope that these technologies will allow us to communicate from greater geographic distance just as they bring us closer together socially. But ultimately, how we use CMC and the impacts it has on our lives are decided by us.

Key Terms

haptic, 267

gustatory, 269

olfactory, 269

Internet of Things, 269

first-level digital divide, 271

second-level digital divide, 272

self-efficacy, 273

third-level digital divide, 274

massive open online courses (MOOCs), 280

telehealth, 283

Review Questions

1. How will adding smell, touch, and taste cues change the way we can communicate in the near future from our present emphasis on text and visual cues?

2. Even as CMC technologies are becoming more widespread, the differences in users' abilities are also continuing to diverge. What computer and/or communication skills are necessary to overcome the second- and third-level digital divides?

3. How can CMC help individuals access and understand cultures beyond their immediate environment?

4. What possibilities could augmented and/or virtual reality offer to enhance educational processes and practices?

5. How will the ability for haptic communication enable telemedicine to provide services to remote or underserved areas and populations?

absent ties Relationships that do not exist, although relational structures are in place such that there is the potential for future relational development

actual self The physical, personal, social, and psychological attributes representing an individual that the individual believes they actually possess

advergames [Video] Games developed primarily to advertise a product or service

advertising The nonpersonal communication of information, usually paid for and usually persuasive in nature about products, services, or ideas by identified sponsors through the various media

affordances Potential opportunities enabled by the properties of a given channel; things the medium may allow you to do

agent A visual online representation of a computer program intended to look and interact like a human

ambient awareness Knowledge or understanding of social others as a result of frequent and repeated exposure to fragmented personal information, particularly via social media

anthromorphize Thinking of or treating an inanimate object as something person-like, with humanistic qualities or agency

astroturfing The use of messages perceived by members of the public at large or an engaged audience as in support of a position, but are actually authored or influenced by the target the messages address

augmented reality The use of digital tools to influence or affect how we see and interact with the corporeal world around us by augmenting the physical space with computer-generated information that appears to coexist with the "real world"

avatar A visual online representation of a human user

behavioral confirmation A process by which an individual's expectations of a partner affect response patterns, so that the partner's attitudes and behaviors conform to the individual's preconceived expectations

brand community A specialized, non-geographically bound group of individuals associated based on a structured set of social relations among admirers of a brand

catfishing The process of presenting alternate personal characteristics or pretending to be someone else entirely in an effort to establish a—often romantic—relationship

channel entrainment A sender's ability to synchronize a message with a strategic self-presentation as facilitated by a particular communication channel

chronemics The use of time to communicate

common-bond group An association of individuals brought together and who maintain relationships due to close interpersonal ties

common-identity group An association of individuals brought together and who maintain relationships due to their affinity toward the group's shared interest or the group as a whole

compliance Giving into or going along with an implicit or explicit request

computer-mediated communication (CMC) The process of sending and receiving messages to exchange meaning among two or more humans through digital, networked channels

conformity The process by which real or perceived pressure from others causes an individual to act differently from how they would act if alone

contact hypothesis The prediction that if individuals first have an opportunity to communicate with each other at an interpersonal level or based on a common group identity, the interpersonal relationship may minimize subsequent interactions at an intergroup level

cookies Small files created by websites that track users and store preferences for subsequent visits to that site

crowdfunding Seeking small financial contributions from a larger number of individuals to provide financial support

crowdturfing The creation or purchase of fake accounts or reviews to influence consumer behavior

cues filtered out (CFO) perspective An early approach to CMC that suggested online channels could not facilitate the message signals (particularly nonverbal messages) necessary for interpersonal communication

cyberbullying Any behavior performed through electronic or digital media by individuals or groups that repeatedly communicates hostile or aggressive messages intended to inflict harm or discomfort on others

decentralized network An infrastructure in which a few key locations link all other local and distant points

deindividuation The loss of awareness of one's self as an individual, often while in a group

dialogic A communicative approach that emphasizes communication as a tool to cocreate and negotiate (rather than manage) relationships with publics and stakeholders, often used in public relations

digital divide The division in societies between those who have meaningful access to digital technologies and those who do not

digital immigrants Those individuals who for whom digital communication technologies were introduced and popularized after their childhood, and have had to learn and adapt to the role of email, video games, and other CMC channels in their lives

digital natives Those individuals who have grown up with digital communication technologies, being immersed their entire lives in email, video games, and other CMC channels

discursive anonymity A state in which specific comments or statements cannot be linked to a particular sender

disentrained Channels and messages that allow users an opportunity for some degree of selective self-presentation, either through asynchronous interaction or by augmenting synchronous interactions

disinhibition effect When an individual's concern about the outcomes of their actions is reduced, and that individual is therefore able to act more freely and without regard for consequences

distributed network An infrastructure in which information can be routed through many different network connections, so that if one connection fails the information may be routed through an alternate node

dyadic interaction Interaction via direct one-to-one communication among two relational partners

echo chambers Closed networks in which individuals with similar ideologies share homophilous information, perspectives, and attitudes that reinforce their existing views

email overload The perception of being overwhelmed by a constant flow of messages into one's inbox and the inability to effectively manage or keep up with the high volume of communication

embodiment Actions taken and experienced through digital representations of a physical self, allowing the user to take on the perception and feeling of the avatar being controlled

emojis Pictographic representations of faces, objects, and other subjects entered as characters into an otherwise text-based exchange

emoticon Configurations of characters typed from a computer keyboard meant to represent facial expressions or physical actions

engagement User-initiated action leading to the cocreation of value, often with behavioral, cognitive, and emotional dimensions

enterprise social media Proprietary social network sites created for exclusive use by an organization that help the members of that organization identify themselves, their knowledge and competencies, and their network connections to be accessed and used by other organizational members

equalization effects The leveling of status roles and subsequent communicative behaviors because of the obfuscation of communicators' roles or hierarchical differences

equivocality With respect to media richness, individuals' needs to identify one option among several, all of which may be vague or nebulous, which can be reduced through iterative exchanges of messages and cocreation of understanding

extractive information Information about a target individual obtained from online tools

extractive uncertainty reduction Learning about others through observing the traces and records of their online self

face-to-face communication (FtF) The process of sending and receiving messages to exchange meaning among two or more humans who are physically colocated, without the use of a communication medium

faithful adoption Using a channel consistent with its intended purpose, function, and design

first-level divide A societal separation between individuals who have access to digital communication tools and those who do not

flaming Hostile intentions that inflict harm to a person or an organization resulting from uninhibited behavior, characterized by words of profanity, obscenity, and insults

gatekeepers The individuals or organizations who control the flow of information through media channels by editorial intervention and validation of messages

geolocated services Tools that connect users' information and communication to their physical location

ghosting Withdrawing access to a communication partner via one or multiple communication channels, unilaterally ceasing communication to reduce or end a relationship

group communication The semi-frequent inter-action among at least two or more individuals who are associated through some shared attribute or social identity

guilds Strong, formal, persistent associations among large groups of players of a video game, used to collaboratively complete end-game events

gustatory The use of smell for communication

haptic The use of touch for communication

hyperpersonal relationship A connection between two communicators online in which the receiver's impressions of the sender are stronger and more salient than similar impressions that may be developed face-to-face, because they surpass the lev-els of perceptions and attributions typically occur-ring in an interpersonal relationship through taking advantage of the digital channel's characteristics

hypertext The content on a webpage serving as logical links between elements of that webpage or among other websites

hypertext markup language (HTML) The code that runs the World Wide Web, describing and structuring the contents of a web page

ideal self The physical, personal, social, and psy-chological attributes representing an individual that the individual desires to possess

identifiability A state in which the source of a message is distinguishable and persistent

identity shift A process of self-transformation occurring when an individual makes deliberate mediated self-presentations

identity trial A process in which an individ-ual can try out parts of their identity or facets of their self while online to determine whether the portrayed self is a comfortable self or attribute to express, without fear of those messages or displays being connected back with them offline

immersive The experience of digital devices sim-ulating multiple senses so that users can experience sight, sound, touch, taste, and/or smells in the vir-tual environment around them

impersonal communication The exchanges between participants facilitate a message but do not allow the participants to form a meaningful relationship based on their individual selves, traits, and personalities

ingroup An association of individuals who share a social identity or category, typically associated with favoritism and similar attitudes and behaviors with respect to the shared identity

interactivity The extent to which messages in a sequence are connected to and reference each other, particularly the degree to which later messages are related to earlier messages

Internet The worldwide public network of net-worked computers used to transmit information

Internet of Things The interlinking of everyday objects to the Internet in a way that makes those objects devices for communication, interaction, and task facilitation

interpersonal communication The exchanges between at least two interactants that allow the participants to form a meaningful understand-ing and/or relationship based on their individual selves, traits, and personalities

intrapersonal communication The communica-tion an individual has with and within their self

ironic adoption Using a channel in a way unin-tended by designers, inconsistent with its intended purpose, function, and design

last-mile problem The challenge of connecting many individuals to the Internet due to infrastruc-ture expenses related to the final short distance between the Internet backbone or trunk and an individual home or business

latent ties One's social connections that are not currently active

lean media Those channels that are effective at transmitting raw data, but not a sense of the per-sonality or social presence of the message sender

least common denominator self The presenta-tion of one's attributes which are perceived as the least objectionable to the widest or most critical members of the groups in an individual's social network

long tail Distribution of content creation or use among users reflecting an exponential relationship curve, whereby either (a) a large number of indi-viduals either use or generate a relative use of a minority of content or (b) a small number of indi-viduals either use or generate a majority of content

massive open online course (MOOC) A class offered online, structured to be delivered at an almost unlimited scale of enrollment

masspersonal communication A broadly acces-sible message intended or targeted to a narrow or specific receiver, transmitted via an intermediary channel

media multiplexity The tendency of individuals who are tightly connected relationally to use more available channels for communication

memes Units of culture that spread from person to person by means of copying or imitation

minimal group paradigm The manner in which simply being assigned or categorized to an arbitrary group is enough to activate intergroup processes

modality switch Moving interactions from one channel to another

news finds me A perspective of current events consumption that suggests individuals who do not actively attend to professional news outlets will remain politically informed through peer networks via social media

olfactory The use of smell for communication

online surveillance Conscious, furtive, and deliberate monitoring via digital online channels to increase awareness of another's offline and online behaviors

organizational communication The flow of messages within a network of interdependent relationships, often guided by hierarchical structures or power differences among network members

other-anonymity When a receiver is unable to identify or distinguish the sender

ought self The physical, personal, social, and psychological attributes representing an individual that others desire the individual to possess

outgroup Associations of individuals who share a different social identity or category than the relevant social group

paralinguistic digital affordances (PDAs) Simple, usually one-click or hover-click cues that are easy to transmit and whose receipt is often displayed in the aggregate

paralinguistics The parts of verbal communication beyond the actual words spoken, including vocalic features like tone and volume of voice, speed of delivery or pauses, and pronunciation and enunciation

parasocial A one-way communication in which a mass media message from a personality results in perceptions of closeness or attraction from an individual receiver

parasocial interaction When an audience responds to a mediated personality as if the communication was unmediated

passive uncertainty reduction Learning about others through discreet observation of their behaviors

persuasion An intentional effort at influencing another's mental state through communication in a circumstance in which the persuadee has some measure of freedom

phatic communication Messages that do not necessarily convey a specific meaning, but do establish a mood, share acknowledgment, or demonstrate sociability

physical anonymity A state in which one cannot determine or sense the presence of a corporeal or visual message of a source

political campaign Organized efforts to influence public sentiment favorably toward a political candidate or issue in an upcoming election

propinquity The perception of relational or psychological closeness felt toward another person

Proteus effect Changes in self-behavior and self-perception based on changes in physical representation of the self

pseudonymous A state in which the source of the message is distinguishable and persistent, but is not faithful to the offline sender

public relations A strategic communication process that builds mutually beneficial relationships between organizations and their publics

requisite variety The principle that the diversity within the system needs to match the diversity of the broader system or organization in which the group operates for the system to be functional

rich media Those channels that are effective at transmitting many social cues and a sense of the message sender's personality and social presence

scrape A technical process wherein the contents of the comments of an online space are automatically obtained and saved in a remote location

second-level digital divide A societal separation between individuals who use the digital communication tools they have available and those who do not

self-anonymity When a sender perceives herself/himself to be unidentifiable to the receiver

self-efficacy An individual's belief they can successfully use a tool or skill

slacktivism Political activities that are easily performed, but may be more effective in making the participant feel good about themself than achieving the stated political goals

social categories The salient demographics or psychographics shared by a group of individuals that guide a sense of social collectivity

social cues Information that may signal one's social identity or category, and can guide group (rather than interpersonal) processes

social determinism An approach to digital tools taking the position that society guides how a technology is adopted and used

social identity The characteristics and traits associated with a group of individuals

social influence Symbolic efforts designed (a) to preserve or change the behavior of another individual or (b) to maintain or modify aspects of another individual that are proximal to behavior, such as cognitions, emotions, and identities

social media Internet-based, persistent, and disentrained channels of masspersonal communication facilitating perceptions of interactions among users, deriving value primarily from user-generated content

social network site (SNS) Web-based services that allow individuals to (1) construct a public or semipublic profile within a bounded system, (2) articulate a list of other users with whom they share a connection, and (3) view and traverse their list of connections and those made by others within the system

social norms The rules and standards understood by members of a group that guide and/or constrain social behavior without the force of laws

social support Verbal and nonverbal communication between recipients and providers that enhances a perception of personal control in one's experience by reducing uncertainty about a situation, the self, the other, or the relationship

socioemotional cues The nonverbal and verbal indicators we have of a communicative partner's social presence, emotion, and interpersonal closeness for interaction

static website A webpage with fixed, nondynamic content, displaying the same information to each viewer

strong ties Close interpersonal relationships between two individuals, signified by significant time spent together, mutual emotional and social support, reciprocal communication, and relational closeness

technodeterminism An approach to digital tools taking the position that the mere presence of a technology changes society

technology Any application of knowledge for practical purposes that enhance work processes or flows

telehealth Healthcare that uses communication networks for delivery of services and medical education from one geographical location to another

third-level digital divide A societal separation between individuals who have the confidence and belief in their skills and abilities to use via the digital communication tools they have available and those who do not

third-spaces Locations people socially congregate beyond home and work

uncertainty With respect to media richness, individuals' needs to access specific, objective information to make decisions that can be reduced through the exchange of concrete information or data

viral Online content that yields a large number of views or iterative instances within a short period of time due to sharing

virtual reality A digitally created space humans can access and immerse themselves in by wearing computer equipment providing sensory input

voice The language styles a company uses to express a distinctive personality or set of values that will differentiate its brands from those of competitors

warranting The connection between an online claim and the perceiver's belief the online target truly possesses the espoused characteristics offline

warranting value The strength of the connection online information presents to connect the offline individual to the online persona

weak ties Distal interpersonal relationships between two individuals, signified by minimal time spent together, limited mutual emotional and social support, low reciprocity of communication, and relational distance

wikis Online information repositories whose content and structure can be edited collaboratively

World Wide Web An information system for sharing digital content interconnected by hyperlinks, accessible via the Internet

References

99 Firms. (2019). *How many email users are there?* https://99firms.com/blog/how-many-email-users-are-there/

Adamic, L. A., & Glance, N. (2005, August 21–25). *The political blogosphere and the 2004 US election: Divided they blog* [Paper presentation]. 3rd International Workshop on Link Discovery, Chicago, IL.

Agosto, D. E. (2005). The digital divide & public libraries: A first-hand view. *Progressive Librarian* (25), 23–27.

Ahn, S. J., & Fox, J. (2017). Immersive virtual environments, avatars, and agents for Health. In R. Parrott (Ed.), *Encyclopedia of health and risk message design and processing*. Oxford University Press.

Ahn, S. J., Le, A. M. T., & Bailenson, J. (2013). The effect of embodied experiences on self-other merging, attitude, and helping behavior. *Media Psychology, 16*(1), 7–38. https://doi.org/10.1080/15213269.2012.755877

Aichner, T., Grünfelder, M., Maurer, O., & Jegeni, D. (in press). Twenty-five years of social media: A review of social media applications and definitions from 1994 to 2019. *Cyberpsychology, Behavior, and Social Networking.* https://doi.org/10.1089/cyber.2020.0134

Albrecht, T. L., & Adelman, M. B. (1987). Communication networks as structures of social support. In T. L. Albrecht & M. B. Adelman (Eds.), *Communicating social support* (pp. 40–63). Sage.

Alhabash, S., & Ma, M. (2017). A tale of four platforms: Motivations and uses of Facebook, Twitter, Instagram, and Snapchat among college students? *Social Media + Society, 3*(1). https://doi.org/10.1177/2056305117691544

Allport, G. W. (1979). *The nature of prejudice.* Basic Books.

Al-Rawi, A., & Rahman, A. (2020). Manufacturing rage: The Russian Internet Research Agency's political astroturfing on social media. *First Monday, 25*(9). https://doi.org/10.5210/fm.v25i9.10801

Altman, I., & Taylor, D. A. (1973). *Social penetration: The development of interpersonal relationships.* Holt, Rinehart and Winston.

Amichai-Hamburger, Y., & McKenna, K. Y. A. (2006). The contact hypothesis revisited: Interacting via the Internet. *Journal of Computer-Mediated Communication, 11*(3), 825–843. https://doi.org/10.1111/j.1083-6101.2006.00037.x

Amir, Y. (1969). Contact hypothesis in ethnic relations. *Psychological Bulletin, 71*(5), 319–342. https://doi.org/10.1037/h0027352

An, S., & Kang, H. (2014). Advertising or games? Advergames on the Internet gaming sites targeting children. *International Journal of Advertising, 33*(3), 509–532. https://doi.org/10.2501/IJA-33-3-509-532

Anderson, C. (2004). The long tail. *Wired, 12*(10). http://www.wired.com/wired/archive/12.10/tail.html

Anderson, M. (2015, August 20). *How having smartphones (or not) shapes the way teens communicate.* Pew Research Center. https://www.pewresearch.org/fact-tank/2015/08/20/how-having-smartphones-or-not-shapes-the-way-teens-communicate/

Angwin, J., & Stecklow, S. (2010, October 11). "Scrapers" dig deep for data on Web. *Wall Street Journal.* http://online.wsj.com/article/SB10001424052748703358504575544381288117888.html

Anonymous. (1998). To reveal or not to reveal: A theoretical model of anonymous communication. *Communication Theory, 8*(4), 381–407. https://doi.org/10.1111/j.1468-2885.1998.tb00226.x

Antheunis, M. L., Schouten, A. P., & Walther, J. B. (2012). The hyperpersonal effect in online dating: Effects of text-based CMC vs. videoconferencing before meeting face-to-face. *Media Psychology, 23*(6), 820 –839. https://doi.org/10.1080/15213269.2019.1648217

Antorini, Y. M., Muñiz Jr., A. M., & Askildsen, T. (2012). Collaborating with customer communities: Lessons from the LEGO Group. *MIT Sloan Management Review, 53*(3), 73–95.

Armstrong, M. (2019). *MySpace isn't dead.* Statista. https://www.statista.com/chart/17392/myspace-global-traffic/

Ashby, W. R. (1958). Requisite variety and its implications for the control of complex systems. *Cybernetica, 2*(3), 83–99.

Augenbraun, E. (2011, September 29). Occupy Wall Street and the limits of spontaneous street protest. *The Guardian.* https://www.theguardian.com/commentisfree/cifamerica/2011/sep/29/occupy-wall-street-protest

Auxier, B., & Anderson, M. (2020). *As schools close due to the coronavirus, some U.S. students face a digital 'homework gap'.* Pew Research Center. https://www.pewresearch.org/

fact-tank/2020/03/16/as-schools-close-due-to-the-coronavirus-some-u-s-students-face-a-digital-homework-gap/

Azuma, R., Baillot, Y., Behringer, R., Feiner, S., Julier, S., & MacIntyre, B. (2001). Recent advances in augmented reality. *IEEE Computer Graphics and Applications*, 21(6), 34–47. https://doi.org/10.1109/38.963459

Backstrom, L., & Kleinberg, J. (2014, February 15–19). *Romantic partnerships and the dispersion of social ties: A network analysis of relationship status on Facebook* [Paper presentation]. 17th ACM conference on Computer Supported Cooperative Work & Social Computing (CSCW'14), Baltimore, MD.

Bakshy, E., Messing, S., & Adamic, L. A. (2015). Exposure to ideologically diverse news and opinion on Facebook. *Science*, 348(6239), 1130–1132. https://doi.org/10.1126/science.aaa1160

Baltes, B. B., Dickson, M. W., Sherman, M. P., Bauer, C. C., & LaGanke, J. S. (2002). Computer-mediated communication and group decision making: A meta-analysis. *Organizational Behavior and Human Decision Processes*, 87(1), 156–179. https://doi.org/10.1006/obhd.2001.2961

Banks, J., & Carr, C. T. (2019). Toward a relational matrix model of avatar-mediated interactions. *Psychology of Popular Media Culture*, 8(3), 287–295. https://doi.org/10.1037/ppm0000180

Barber, A. E. (1998). *Recruiting employees: Individual and organizational perspectives*. Sage.

Barberá, P., Jost, J. T., Nagler, J., Tucker, J. A., & Bonneau, R. (2015). Tweeting from left to right: Is online political communication more than an echo chamber? *Psychological Science*, 26(10), 1531–1542. https://doi.org/10.1177/0956797615594620

Barbieri, F., Kruszewski, G., Ronzano, F., & Saggion, H. (2016). *How cosmopolitan are emojis? Exploring emojis usage and meaning over different languages with distributional semantics* [Paper presentation]. 24th ACM international conference on Multimedia (MM'16), Amsterdam, The Netherlands.

Bargh, J. A., McKenna, K. Y. A., & Fitzsimons, G. M. (2002). Can you see the real me? Activation and expression of the "True Self" on the Internet. *Journal of Social Issues*, 58, 33–48. https://doi.org/10.1111/1540-4560.00247

Barksdale, M. A., Watson, C., & Park, E. S. (2007). Pen pal letter exchanges: Taking first steps toward developing cultural understandings. *Reading Teacher*, 61(1), 58–68. https://doi.org/10.1598/RT.61.1.6

Barnes, N. G., Mazzola, A., & Killeen, M. (2020, February 11). *Oversaturation & disengagement: The 2019 Fortune 500 social media dance*. UMass – Dartmouth: Center for Marketing Research. https://www.umassd.edu/cmr/research/2019-fortune-500.html

Barnidge, M. (2017). Exposure to political disagreement in social media versus face-to-face and anonymous online settings. *Political Communication*, 34(2), 302–321. https://doi.org/10.1080/10584609.2016.1235639

Baum, M. A., & Groeling, T. (2008). New media and the polarization of American political discourse. *Political Communication*, 25(4), 345–365. https://doi.org/10.1080/10584600802426965

Baym, N. K. (2000). *Tune in, log on: Soaps, fandom, and online community* (Vol. 3). Sage.

Baym, N. K., & Ledbetter, A. (2009). Tunes that bind? Predicting friendship strength in a music-based social network. *Information, Communication & Society*, 12(3), 408–427. https://doi.org/10.1080/13691180802635430

Bekafigo, M. A., & McBride, A. (2013). Who tweets about politics? Political participation of Twitter users during the 2011 gubernatorial elections. *Social Science Computer Review*, 31(5), 625–643. https://doi.org/10.1177/0894439313490405

Bem, D. J. (1972). Self-perception theory. In L. Berkowitz (Ed.), *Advances in experimental social psychology* (Vol. 6, pp. 1–62). Elsevier.

Bennett, W. L. (2008). Changing citizenship in the digital age. In W. L. Bennett (Ed.), *Civic life online: Learning how digital media can engage youth* (pp. 1–24). MIT Press. https://doi.org/10.1162/dmal.9780262524827.001

Bennett, W. L., & Livingston, S. (2018). The disinformation order: Disruptive communication and the decline of democratic institutions. *European Journal of Communication*, 33(2), 122–139. https://doi.org/10.1177/0267323118760317

Berger, C. R., & Calabrese, R. J. (1975). Some explorations in initial interaction and beyond: Toward a developmental theory of interpersonal communication. *Human Communication Research*, 1(2), 99–112. https://doi.org/10.1111/j.1468-2958.1975.tb00258.x

Berger, J., & Milkman, K. L. (2013). Emotion and virality: What makes online content go viral? *Marketing Intelligence Review*, 5(1), 18–23. https://doi.org/10.2478/gfkmir-2014-0022

Berlo, D. K. (1960). *The process of communication*. Rinehart & Winston.

Bernardin, H. J. (1986). Subordinate appraisal: A valuable source of information about managers. *Human Resource Management*, 25(3), 421–439. https://doi.org/10.1002/hrm.3930250307

Berscheid, E., & Walster, E. H. (1969). *Interpersonal attraction*. Wesley.

Bessière, K., Seay, A. F., & Kiesler, S. (2007). The ideal elf: Identity exploration in *World* of *Warcraft. CyberPsychology & Behavior, 10*(4), 530–535. https://doi.org/10.1089/cpb.2007.9994

Bevan, J. L. (2013). *The communication of jealousy*. Peter Lang.

Bevan, J. L. (2017). Liking, creeping, and password sharing. In N. M. Punyanunt-Carter & J. S. Wrench (Eds.), *The impact of social media in modern romantic relationships* (pp. 165–180). Lexington Books.

Billedo, C. J., Kerkhof, P., & Finkenauer, C. (2015). The use of social networking sites for relationship maintenance in long-distance and geographically close romantic relationships. *Cyberpsychology, Behavior, and Social Networking, 18*(3), 152–157. https://doi.org/10.1089/cyber.2014.0469

Billig, M., & Tajfel, H. (1973). Social categorization and similarity in intergroup behaviour. *European Journal of Social Psychology, 3*(1), 27–52. https://doi.org/10.1002/ejsp.2420030103

Bisel, R. S., & Arterburn, E. N. (2012). Making sense of organizational members' silence: A sensemaking-resource model. *Communication Research Reports, 29*(3), 217–226. https://doi.org/10.1080/08824096.2012.684985

Blascovich, J., & Bailenson, J. (2011). *Infinite reality: Avatars, eternal life, new worlds, and the dawn of the virtual revolution*. William Morrow.

Blum, A. (2013). *Tubes: A journey to the center of the Internet*. HarperCollins.

Bochner, A. P. (1989). Interpersonal communication. In E. Barnouw, G. Gerbner, W. Schramm, T. L. Worth, & L. Gross (Eds.), *International encyclopedia of communications* (pp. 336–340). Oxford University Press.

Bogomoletc, E., & Lee, N. M. (2021). Frozen meat against COVID-19 misinformation: An analysis of Steak-Umm and positive expectancy violations. *Journal of Business and Technical Communication, 35*(1), 118-125. https://doi.org/10.1177/1050651920959187

Bohnert, D., & Ross, W. H. (2010). The influence of social networking web sites on the evaluation of job candidates. *Cyberpsychology, Behavior, and Social Networking, 13*(3), 341–347. https://doi.org/10.1089/cyber.2009.0193

Bor, S. E. (2014). Using social network sites to improve communication between political campaigns and citizens in the 2012 election. *American Behavioral Scientist, 58*(9), 1195–1213. https://doi.org/10.1177/0002764213490698

Booth, P. (2010). *Digital fandom: New media studies*. Peter Lang.

Bovee, C. L., & Arens, W. F. (1986). *Contemporary advertising* (2nd ed.). Irwin.

boyd, d. m., & Ellison, N. B. (2007). Social network sites: Definition, history, and scholarship. *Journal of Computer-Mediated Communication, 13*(1), 210–230. https://doi.org/10.1111/j.1083-6101.2007.00393.x

Bruns, A., & Highfield, T. (2013). Political networks on Twitter: Tweeting the Queensland state election. *Information, Communication & Society, 16*(5), 667–691. https://doi.org/10.1080/1369118X.2013.782328

Brady, W. J., Gantman, A. P., & Van Bavel, J. J. (2020). Attentional capture helps explain why moral and emotional content go viral. *Journal of Experimental Psychology: General, 149*(4), 746–756. https://doi.org/10.1037/xge0000673

Braithwaite, D. O., Waldron, V. R., & Finn, J. (1999). Communication of social support in computer-mediated groups for people with disabilities. *Health Communication, 11*(2), 123–151. https://doi.org/10.1207/s15327027hc1102_2

Braun, S., Hernandez Bark, A., Kirchner, A., Stegmann, S., & Van Dick, R. (2019). Emails from the boss: Curse or blessing? Relations between communication channels, leader evaluation, and employees' attitudes. *International Journal of Business Communication, 56*(1), 50–81. https://doi.org/10.1177/2329488415597516

Brewer, M. B. (1999). The psychology of prejudice: Ingroup love or outgroup hate? *Journal of Social Issues, 55*(3), 429–444. https://doi.org/10.1111/0022-4537.00126

Brody, N., LeFebvre, L. E., & Blackburn, K. G. (2017). Post-dissolution surveillance on social networking sites. In N. Punyanunt-Carter & J. S. Wrench (Eds.), *Swipe right for love: The impact of social media in modern romantic relationships* (pp. 237–258). Rowman & Littlefield.

Burgess, J., & Baym, N. K. (2020). *Twitter: A biography*. New York University Press.

Burgoon, J. K., Bonito, J. A., Bengtsson, B., Ramirez Jr., A., Dunbar, N. E., & Miczo, N. (1999). Testing the interactivity model: Communication processes, partner assessments, and the quality of collaborative work. *Journal of Management Information Systems, 16*(3), 33–56. https://doi.org/10.1080/07421222.1999.11518255

Burgoon, J. K., Bonito, J. A., Ramirez, A., Jr., Dunbar, N. E., Kam, K., & Fischer, J. (2002). Testing the interactivity principle: Effects of mediation, propinquity, and verbal and nonverbal modalities in interpersonal interaction. *Journal of Communication, 52*(3), 657–677. https://doi.org/10.1111/j.1460-2466.2002.tb02567.x

Burgoon, J., & Hale, J. L. (1987). Validation and measurement of the fundamental themes of relational communication. *Communication*

Monographs, 54(1), 19–41. https://doi.org/10.1080/03637758709390214

Burke, K. (2018, November). *How many texts do people send every day (2018)?* Text Request. https://www.textrequest.com/blog/how-many-texts-people-send-per-day/

Burke, M., & Kraut, R. E. (2014, April 26–May 1). *Growing closer on Facebook: Changes in tie strength through social network site use* [Paper presentation]. SIGCHI conference on human factors in computing systems (CHI'14), Toronto, ON.

Bush, V. (1945, July). As we may think. *The Atlantic*, 101–108. https://www.theatlantic.com/magazine/archive/1945/07/as-we-may-think/303881/

Bustamante, J. (2019). *IoT statistics*. iProperty Management. https://ipropertymanagement.com/iot-statistics

Buys, P., Dasgupta, S., Thomas, T. S., & Wheeler, D. (2009). Determinants of a digital divide in sub-Saharan Africa: A spatial econometric analysis of cell phone coverage. *World Development, 37*(9), 1494–1505. https://doi.org/10.1016/j.worlddev.2009.01.011

Cabalquinto, E. C. (2018). Home on the move: Negotiating differential domesticity in family life at a distance. *Media, Culture & Society, 40*(6), 795–816. https://doi.org/10.1177/0163443717737611

Cao, B., & Lin, W.-Y. (2017). Revisiting the contact hypothesis: Effects of different modes of computer-mediated communication on intergroup relationships. *International Journal of Intercultural Relations, 58*, 23–30. https://doi.org/10.1016/j.ijintrel.2017.03.003

Carey, J. (1980, June 19–22). *Paralanguage in computer mediated communication* [Paper presentation]. 18th annual meeting on Association for Computational Linguistics, Philadelphia, PA.

Carney, D. R., Cuddy, A. J. C., & Yap, A. J. (2010). Power posing: Brief nonverbal displays affect neuroendocrine levels and risk tolerance. *Psychological Science, 21*(10), 1363–1368. https://doi.org/10.1177/0956797610383437

Carr, C. T. (2014). Applying a model of communicative influence in education in closed online and offline courses. *Journal of Asynchronous Learning Networks, 18*(1), 115–129.

Carr, C. T. (2017). A social identification approach to the effects of religious disclosures in business communication. *Journal of Social Psychology, 157*, 571–587. https://doi.org/10.1080/00224545.2016.1248810

Carr, C. T. (2016). An uncertainty reduction approach to applicant information-seeking in social media: Effects on attributions and hiring. In R. N. Landers & G. B. Schmidt (Eds.), *Using social media in employee selection: Theory, practice, and future research* (pp. 59–78). Springer.

Carr, C. T. (2019). Have you heard? Testing the warranting value of third-party employer reviews. *Communication Research Reports, 36*(5), 371–382. https://doi.org/10.1080/08824096.2019.1683529

Carr, C. T. (2020a). CMC is dead, long live CMC! Situating computer-mediated communication scholarship beyond the digital age. *Journal of Computer-Mediated Communication, 25*(1), 9–22. https://doi.org/10.1093/jcmc/zmz018

Carr, C. T. (2020b). The delocalization of the local election. *Social Media+ Society, 6*(2), 1–4. https://doi.org/10.1177/2056305120924772

Carr, C. T. (in press). Identity shift effects of personalization of self-presentation on extraversion. *Media Psychology*. https://doi.org/10.1080/15213269.2020.1753540

Carr, C. T., & Hayes, R. A. (2015). Social media: Defining, developing, and divining. *Atlantic Journal of Communication, 23*(1), 46–65. https://doi.org/10.1080/15456870.2015.972282

Carr, C. T., & Hayes, R. A. (2019). Identity shift effects of self-presentation and confirmatory & disconfirmatory feedback on self-perceptions of brand identification. *Media Psychology, 22*(3), 418–444. https://doi.org/10.1080/15213269.2017.1396228

Carr, C. T., Hayes, R. A., Smock, A., & Zube, P. (2016). Facebook in presidential elections: Status of effects. In G. W. Richardson (Ed.), *Social media and politics: A new way to participate in the political process* (pp. 41–70). Praeger.

Carr, C. T., Schrock, D. B., & Dauterman, P. (2012). Speech acts within Facebook status messages. *Journal of Language and Social Psychology, 31*(2), 176–196. https://doi.org/10.1177/0261927X12438535

Carr, C. T., & Stefaniak, C. (2012). Sent from my iPhone: The medium and message as cues of sender professionalism in mobile telephony. *Journal of Applied Communication Research, 40*(4), 403–424. https://doi.org/10.1080/00909882.2012.712707

Carr, C. T., & Van der Heide, B. (2009, September). *Communication technologies facilitating social and task dynamics: Examining intra-group relational satisfaction in* World of Warcraft [Paper presentation]. 6th Conference for Media Psychology, Duisburg, Germany.

Carr, C. T., Varney, E. J., & Blesse, J. R. (2016). Social media and intergroup communication: Collapsing and expanding group contexts. In H. Giles & A. Maass (Eds.), *Advances in and prospects for intergroup communication* (pp. 155–173). Peter Lang.

Carr, C. T., Wohn, D. Y., & Hayes, R. A. (2016). 👍 as social support: Relational closeness, automaticity, and interpreting social support from paralinguistic digital affordances in social media. *Computers in Human Behavior, 62*, 385–393. https://doi.org/10.1016/j.chb.2016.03.087

Carr, C. T., Zube, P., Dickens, E., Hayter, C. A., & Barterian, J. A. (2013). Toward a model of sources of influence in online education: Cognitive learning and the effects of Web 2.0. *Communication Education, 62*(1), 61–85. https://doi.org/10.1080/03634523.2012.724535

Carroll, L. (1865). *Alice's adventures in Wonderland*. Macmillan.

Cassino, D., Woolley, P., & Jenkins, K. (2012). *What you know depends on what you watch: Current events knowledge across popular news sources*. Fairleigh Dickinson University's Public Mind Poll. http://publicmind.fdu.edu/2012/confirmed/

Cathcart, R., & Gumpert, G. (1985). The person-computer interaction: A unique source. In B. D. Ruben (Ed.), *Information and behavior* (*Vol. 1*, pp. 113–124). Transaction Books.

Center for Responsive Politics. (2019). *Top in-state vs. out-of-state*. OpenSecrets.org. https://www.opensecrets.org/overview/instvsout.php?cycle=2018&display=

Cerf, V. G. (2013). *The open Internet and the Web*. CERN: Accelerating science. https://home.cern/news/opinion/computing/open-internet-and-web

Chae, B. K., McHaney, R., & Sheu, C. (2020). Exploring social media use in B2B supply chain operations. *Business Horizons, 63*(1), 73–84. https://doi.org/10.1016/j.bushor.2019.09.008

Chang, A., O'Modhrain, S., Jacob, R., Gunther, E., & Ishii, H. (2002, June). *ComTouch: Design of a vibrotactile communication device* [Paper presentation]. 4th conference on Designing interactive systems: Processes, practices, methods, and techniques (DIS'02), London, England.

Chau, T., & Maurer, F. (2005, October 2–5). *A case study of wiki-based experience repository at a medium-sized software company* [Paper presentation.]. 3rd international conference on knowledge capture (K-CAP'05), Banff, Alberta, Canada.

Cheeseman, N., Fisher, J., Hassan, I., & Hitchen, J. (2020). Social media disruption: Nigeria's WhatsApp politics. *Journal of Democracy, 31*(3), 145–159. https://doi.org/10.1353/jod.2020.0037

Chester, A., & Gwynne, G. (1998). Online teaching: Encouraging collaboration through anonymity. *Journal of Computer-Mediated Communication, 4*(2). https://doi.org/10.1111/j.1083-6101.1998.tb00096.x

Christensen, G., Steinmetz, A., Alcorn, B., Bennett, A., Woods, D., & Emanuel, E. (2013). *The MOOC phenomenon: Who takes massive open online courses and why?* SSRN. https://pdfs.semanticscholar.org/43f5/ab1690524d291ae21946683ae63b221564e4.pdf

Christensen, H. S. (2011). Political activities on the Internet: Slacktivism or political participation by other means? *First Monday, 16*(2). https://doi.org/10.5210/fm.v16i2.3336

Cho, C. H., Martens, M. L., Kim, H., & Rodrigue, M. (2011). Astroturfing global warming: It isn't always greener on the other side of the fence. *Journal of Business Ethics, 104*(4), 571–587. https://doi.org/10.1007/s10551-011-0950-6

Choney, S. (2009). *Obama gets to keep his BlackBerry*. NBC News. http://www.nbcnews.com/id/28780205/ns/technology_and_science-tech_and_gadgets/t/obama-gets-keep-his-blackberry/#.Xeh939VMGUk

Cialdini, R. B. (2009). *Influence: Science and practices*. William Morrow.

Cialdini, R. B., & Trost, M. R. (1998). Social influence: Social norms, conformity and compliance. In D. T. Gilbert, S. T. Fiske, & G. Lindzey (Eds.), *The handbook of social psychology* (pp. 151–192). McGraw-Hill.

Clement, J. (2020, March 25). *Number of e-mail users worldwide 2017–2024*. Statista. https://www.statista.com/statistics/255080/number-of-e-mail-users-worldwide/

Cohen, G. E., & Kerr, B. A. (1999). Computer-mediated counseling: An empirical study of a new mental health treatment. *Computers in Human Services, 15*(4), 13–26. https://doi.org/10.1300/J407v15n04_02

Computer History Museum. (2020). *The Babbage engine*. http://www.computerhistory.org/babbage/

Conover, M. D., Ratkiewicz, J., Francisco, M., Gonçalves, B., Menczer, F., & Flammini, A. (2011, July 17–21). *Political polarization on Twitter* [Paper presentation]. 5th International AAAI Conference on Weblogs and Social Media, Barcelona, Spain.

Conroy, M., Feezell, J. T., & Guerrero, M. (2012). Facebook and political engagement: A study of online political group membership and offline political engagement. *Computers in Human Behavior, 28*(5), 1535–1546. https://doi.org/10.1016/j.chb.2012.03.012

Cookson, C. (2013, May 10). Time is money when it comes to microwaves. *Financial Times*. https://www.ft.com/content/2bf37898-b775-11e2-841e-00144feabdc0

Cotten, S. R., Ford, G., Ford, S., & Hale, T. M. (2014). Internet use and depression among retired older adults in the United States: A longitudinal analysis. *Journals of Gerontology B*, 69(5), 763–771. https://doi.org/10.1093/geronb/gbu018

Coulson, N. S. (2005). Receiving social support online: An analysis of a computer-mediated support group for individuals living with irritable bowel syndrome. *CyberPsychology & Behavior*, 8(6), 580–584. https://doi.org/10.1089/cpb.2005.8.580

Coulson, N. S., Buchanan, H., & Aubeeluck, A. (2007). Social support in cyberspace: A content analysis of communication within a Huntington's disease online support group. *Patient Education and Counseling*, 68(2), 173–178. https://doi.org/10.1016/j.pec.2007.06.002

Cover, R. (2012). Performing and undoing identity online: Social networking, identity theories and the incompatibility of online profiles and friendship regimes. *Convergence*, 18(2), 177–193. https://doi.org/10.1177/1354856511433684

Coyne, S. M., Stockdale, L., Busby, D., Iverson, B., & Grant, D. M. (2011). "I luv u :)!" A descriptive study of the media use of individuals in romantic relationships. *Family Relations*, 5, 150–162. https://doi.org/10.1111/j.1741-3729.2010.00639.x

Crook, B., Glowacki, E. M., Suran, M., K. Harris, J., & Bernhardt, J. M. (2016). Content analysis of a live CDC Twitter chat during the 2014 Ebola outbreak. *Communication Research Reports*, 33(4), 349–355. https://doi.org/10.1080/08824096.2016.1224171

Crusco, A. H., & Wetzel, C. G. (1984). The Midas touch: The effects of interpersonal touch on restaurant tipping. *Personality and Social Psychology Bulletin*, 10(4), 512–517. https://doi.org/10.1177/0146167284104003

Cui, R., Gallino, S., Moreno, A., & Zhang, D. J. (2018). The operational value of social media information. *Production and Operations Management*, 27(10), 1749–1769. https://doi.org/10.1111/poms.12707

Culnan, M. J., & Markus, M. L. (1987). Information technologies. In F. M. Jablin, L. L. Putnam, K. H. Roberts, & L. W. Porter (Eds.), *Handbook of organizational communication: An interdisciplinary perspective* (pp. 421–443). Sage.

Curry, C., James, M. S., & Harris, D. (2013, January 16). *Notre Dame: Football star Manti Te'o was "Catfished" in girlfriend hoax*. ABC News. https://abcnews.go.com/US/notre-dame-football-star-manti-teo-dead-girlfriend/story?id=18232374

Cutrona, C. E., & Suhr, J. A. (1992). Controllability of stressful events and satisfaction with spouse support behaviors. *Communication Research*, 19(2), 154–174. https://doi.org/10.1177/009365092019002002

Daft, R. L., & Lengel, R. H. (1986). Organizational information requirements, media richness and structural design. *Management Science*, 32, 554–571. https://doi.org/10.1287/mnsc.32.5.554

Danet, B., Ruedenberg-Wright, L., & Rosenbaum-Tamari, Y. (1997). Hmmm... where's that smoke coming from? Writing, play and performance on Internet relay chat, Network and Netplay: Virtual groups on the Internet. *Journal of Computer-Mediated Communication*, 2(4). https://doi.org/10.1111/j.1083-6101.1997.tb00195.x

Davis, C. B., Glantz, M., & Novak, D. R. (2016). "You Can't Run Your SUV on Cute. Let's Go!": Internet memes as delegitimizing discourse. *Environmental Communication*, 10(1), 62–83. https://doi.org/10.1080/17524032.2014.991411

Dawkins, R. (1976). *The selfish gene*. Oxford University Press.

DeAndrea, D. C., & Carpenter, C. J. (2018). Measuring the construct of warranting value and testing warranting theory. *Communication Research*, 45(8), 1193–1215. https://doi.org/10.1177/0093650216644022

DeAndrea, D. C., Tong, S. T., & Lim, Y.-S. (2018). What causes more mistrust: Profile owners deleting user-generated content or website contributors masking their identities? *Information, Communication & Society*, 21(8), 1068–1080. https://doi.org/10.1080/1369118X.2017.1301523

DeAndrea, D. C., Van der Heide, B., Vendemia, M. A., & Vang, M. H. (2018). How people evaluate online reviews. *Communication Research*, 45(5), 719–736. https://doi.org/10.1177/0093650215573862

Das Swain, V., Saha, K., Reddy, M. D., Rajvanshy, H., Abowd, G. D., & De Choudhury, M. (2020, April 25–30). *Modeling organizational culture with workplace experiences shared on GlassGoor* [Paper presentation]. 2020 CHI Conference on Human Factors in Computing Systems, Honolulu, HI.

Dede, C. (2009). Immersive interfaces for engagement and learning. *Science*, 323(5910), 66–69. https://doi.org/10.1126/science.1167311

Delin, J. (2005). Brand tone of voice. *Journal of Applied Linguistics and Professional Practice*, 2(1), 1–44. https://doi.org/10.1558/japl.v2.i1.1

Denisova, A. (2019). *Internet memes and society: Social, cultural, and political contexts*. Routledge.

Desilver, D. (2013). *Young Americans and privacy: "It's complicated."* Pew Research Center. https://

www.pewresearch.org/fact-tank/2013/06/20/young-americans-and-privacy-its-complicated/

Derlega, V. J., & Chaikin, A. L. (1977). Privacy and self-disclosure in social relationships. *Journal of Social Issues, 33*(3), 102–115. https://doi.org/10.1111/j.1540-4560.1977.tb01885.x

Desanctis, G., & Gallupe, R. B. (1987). A foundation for the study of group decision support systems. *Management Science, 33*(5), 589–609. https://doi.org/10.1287/mnsc.33.5.589

DeSanctis, G., & Poole, M. S. (1994). Capturing the complexity in advanced technology use: Adaptive structuration theory. *Organization Science, 5*(2), 121–147. https://doi.org/10.1287/orsc.5.2.121

de la Peña, N., Weil, P., Llobera, J., Giannopoulos, E., Pomés, A., Spanlang, B., Friedman, D., Sanchez-Vives, M. V., & Slater, M. (2010). Immersive journalism: Immersive virtual reality for the first-person experience of news. *Presence: Teleoperators and Virtual Environments, 19*(4), 291–301. https://doi.org/10.1162/PRES_a_00005

Di Gennaro, C., & Dutton, W. (2006). The Internet and the public: Online and offline political participation in the United Kingdom. *Parliamentary Affairs, 59*(2), 299–313. https://doi.org/10.1093/pa/gsl004

Dibbell, J. (1993, December 23). A rape in cyberspace or how an evil clown, a Haitian trickster spirit, two wizards, and a cast of dozens turned a database into a society. *Village Voice.* http://www.juliandibbell.com/texts/bungle_vv.html

Dibble, J. L., Hartmann, T., & Rosaen, S. F. (2016). Parasocial interaction and parasocial relationship: Conceptual clarification and a critical assessment of measures. *Human Communication Research, 42*(1), 21–44. https://doi.org/10.1111/hcre.12063

Dienlin, T., & Metzger, M. J. (2016). An extended privacy calculus model for SNSs: Analyzing self-disclosure and self-withdrawal in a representative US sample. *Journal of Computer-Mediated Communication, 21*(5), 368–383. https://doi.org/10.1111/jcc4.12163

Digital Information World. (2019). *Revealed: The social media platforms that make the most revenue off their users.* https://www.digitalinformationworld.com/2019/12/revenue-per-social-media-user.html

DiGiuseppe, N., & Nardi, B. (2007). Real genders choose fantasy characters: Class choice in *World of Warcraft. First Monday, 12*(5). https://doi.org/10.5210/fm.v12i5.1831

Dillard, J. P., Anderson, J. W., & Knobloch, L. K. (2002). Interpersonal influence. In M. L. Knapp & J. A. Daly (Eds.), *Handbook of interpersonal communication* (3rd ed., pp. 425–474). Sage.

Dindia, K., & Canary, D. J. (1993). Definitions and theoretical perspectives on maintaining relationships. *Journal of Social and Personal Relationships, 10*(2), 163–173. https://doi.org/10.1177/026540759301000201

Dolcos, S., Sung, K., Argo, J. J., Flor-Henry, S., & Dolcos, F. (2012). The power of a handshake: Neural correlates of evaluative judgments in observed social interactions. *Journal of Cognitive Neuroscience, 24*(12), 2292–2305. https://doi.org/10.1162/jocn_a_00295

Donner, J. (2007). The rules of beeping: Exchanging messages via intentional "missed calls" on mobile phones. *Journal of Computer-Mediated Communication, 13*(1), 1–22. https://doi.org/10.1111/j.1083-6101.2007.00383.x

Donovan, J. (2019). How memes got weaponized: A short history. *MIT Technology Review.* https://www.technologyreview.com/s/614572/political-war-memes-disinformation/

Doubek, J. (2015). *Political campaigns go social, but email is still king.* NPR. https://www.npr.org/sections/itsallpolitics/2015/07/28/426022093/as-political-campaigns-go-digital-and-social-email-is-still-king

Dougiamas, M., & Taylor, P. (2003, June 22). *Moodle: Using learning communities to create an open source course management system* [Paper presentation]. EdMedia+ Innovate Learning, Honolulu, HI.

Dovidio, J. F., & Gluszek, A. (2012). Accents, nonverbal behavior, and intergroup bias. In H. Giles (Ed.), *The handbook of intergroup communication* (pp. 109–121). Routledge.

Druckman, J. N. (2003). The power of television images: The first Kennedy-Nixon debate revisited. *Journal of Politics, 65*(2), 559–571. https://doi.org/10.1111/1468-2508.t01-1-00015

Druckman, J. N., Kifer, M. J., & Parkin, M. (2007). The technological development of congressional candidate web sites: How and why candidates use web innovations. *Social Science Computer Review, 25*(4), 425–442. https://doi.org/10.1177/0894439307305623

Duggan, M., & Smith, A. (2016). *The tone of social media discussions around politics.* Pew Research Center – Internet & Technology. https://www.pewresearch.org/internet/2016/10/25/the-tone-of-social-media-discussions-around-politics/

Dunbar, R. I. M. (1992). Neocortex size as a constraint on group size in primates. *Journal of Human Evolution, 22*(6), 469–493. https://doi.org/10.1016/0047-2484(92)90081-J

Dunbar, R. I. M. (2016). Do online social media cut through the constraints that limit the size of offline social networks? *Royal Society Open Science, 3*(1). https://doi.org/10.1098/rsos.150292

Eastin, M. S., & LaRose, R. (2005). Alt. support: Modeling social support online. *Computers in Human Behavior, 21*(6), 977–992. https://doi.org/10.1016/j.chb.2004.02.024

The Economist. (2018). Half the world will be online in 2019. https://www.economist.com/the-world-in/2018/12/31/half-the-world-will-be-online-in-2019

Edwards, A., Edwards, C., Shaver, C., & Oaks, M. (2009). Computer-mediated word-of-mouth communication on RateMyProfessors.com: Expectancy effects on student cognitive and behavioral learning. *Journal of Computer-Mediated Communication, 14*(2), 368–392. https://doi.org/10.1111/j.1083-6101.2009.01445.x

Edwards, C., Edwards, A., Qing, Q., & Qahl, S. (2007). The influence of computer-mediated word-of-mouth communication on student perceptions of instructors and attitudes toward learning course content. *Communication Education, 55*(3), 255–277. https://doi.org/10.1080/03634520701236866

Eklund, L. (2011). Doing gender in cyberspace: The performance of gender by female *World of Warcraft* players. *Convergence, 17*(3), 323–342. https://doi.org/10.1177/1354856511406472

El Ouirdi, M., Segers, J., El Ouirdi, A., & Pais, I. (2015). Predictors of job seekers' self-disclosure on social media. *Computers in Human Behavior, 53*, 1–12. https://doi.org/10.1016/j.chb.2015.06.039

Ellison, N. B., & boyd, d. (2013). Sociability through social network sites. In W. H. Dutton (Ed.), *The Oxford handbook of Internet studies* (pp. 151–172). Oxford University Press.

Ellison, N., Heino, R., & Gibbs, J. (2006). Managing impressions online: Self-presentation processes in the online dating environment. *Journal of Computer-Mediated Communication, 11*(2), 415–441. https://doi.org/10.1111/j.1083-6101.2006.00020.x

Ellison, N. B., Steinfield, C., & Lampe, C. (2007). The benefits of Facebook "friends": Social capital and college students' use of online social network sites. *Journal of Computer-Mediated Communication, 12*(4), 1143–1168. https://doi.org/10.1111/j.1083-6101.2007.00367.x

Elphinston, R. A., & Noller, P. (2011). Time to face it! Facebook intrusion and the implications for romantic jealousy and relationship satisfaction. *Cyberpsychology, Behavior, and Social Networking, 14*(11), 631–635. https://doi.org/10.1089/cyber.2010.0318

Engstrom, M. E., & Jewett, D. (2005). Collaborative learning the wiki way. *TechTrends, 49*(6), 12–15. http://educ116eff11.pbworks.com/w/file/fetch/44828198/engstromwiki.pdf

Erdody, L. (2018, September 19). Political campaigns boost investment in social media ads. *Indianapolis Business Journal.* https://www.ibj.com/articles/70545-political-campaigns-boost-investment-in-social-media-ads

Ess, C., & Sudweeks, F. (Eds.). (2001). *Culture, technology, communication: Towards an intercultural global village.* State University of New York Press.

Faber, A. J., Willerton, E., Clymer, S. R., MacDermid, S. M., & Weiss, H. M. (2008). Ambiguous absence, ambiguous presence: A qualitative study of military reserve families in wartime. *Journal of Family Psychology, 22*(2), 222–230. https://doi.org/10.1037/0893-3200.22.2.222

Facebook. (2019). *Stats.* Facebook Newsroom. https://newsroom.fb.com/company-info/

Farris, P., Raggio, R. D., DesMarteau, P., Mazakov, A., & Murphy, L. (2014). *Pitching J. Crew maternity apparel to Mickey Drexler* (Darden Case No. UVA-M-0854.) [Case Study]. https://papers.ssrn.com/sol3/papers.cfm?abstract_id=2974719

Farrow, H., & Yuan, Y. C. (2011). Building stronger ties with alumni through Facebook to increase volunteerism and charitable giving. *Journal of Computer-Mediated Communication, 16*(3), 445–464. https://doi.org/10.1111/j.1083-6101.2011.01550.x

Felmlee, D. H. (2001). No couple is an island: A social network perspective on dyadic stability. *Social Forces, 79*(4), 1259–1287. https://doi.org/10.1353/sof.2001.0039

Fenton, N., & Barassi, V. (2011). Alternative media and social networking sites: The politics of individuation and political participation. *Communication Review, 14*(3), 179–196. https://doi.org/10.1080/10714421.2011.597245

File, T., & Ryan, C. (2014). *Computer and Internet use in the United States: 2013.* American community survey reports, Issue. https://www.census.gov/history/pdf/acs-internet2013.pdf

Fleischer, D. (Director). (1933). *Popeye the Sailor* [Film]. Paramount Pictures.

Flynn, N. (2012). *The social media handbook: Rules, policies, and best practices to successfully manage your organization's social media presence, posts, and potential.* Pfeiffer.

Fowler, G. A. (2012, October 13). When the most personal secrets get outed on Facebook. *Wall Street Journal.* http://online.wsj.com/news/articles/SB10001424052970204577800452032240

Fox, J., Arena, D., & Bailenson, J. N. (2009). Virtual reality: A survival guide for the social scientist. *Journal of Media Psychology, 21*(3), 95–113. https://doi.org/10.1027/1864-1105.21.3.95

Fox, J., & Tokunaga, R. S. (2015). Romantic partner monitoring after breakups: Attachment, dependence, distress, and post-dissolution online surveillance via social networking sites. *Cyberpsychology, Behavior, and Social Networking, 18*(9), 491–498. https://doi.org/10.1089/cyber.2015.0123

Fox, J., & Vendemia, M. A. (2016). Selective self-presentation and social comparison through photographs on social networking sites. *Cyberpsychology, Behavior, and Social Networking, 19*(10), 593–600. https://doi.org/10.1089/cyber.2016.0248

Fox, J., & Warber, K. M. (2014). Social networking sites in romantic relationships: Attachment, uncertainty, and partner surveillance on Facebook. *Cyberpsychology, Behavior, and Social Networking, 17*(1), 3–7. https://doi.org/10.1089/cyber.2012.0667

Franceschi-Bicchierai, L. (2013). *Facebook hacker breaks into Zuckerberg's timeline to report bug.* Mashable. https://mashable.com/2013/08/18/facebook-hacker-zuckerberg-timeline/

Freedom House. (2013). *North Korea.* Freedom House. https://freedomhouse.org/report/freedom-press/2013/north-korea

Freelon, D. G. (2010). Analyzing online political discussion using three models of democratic communication. *New Media & Society, 12*(7), 1172–1190. https://doi.org/10.1177/1461444809357927

Freelon, D., & Wells, C. (2020). Disinformation as political communication. *Political Communication, 37*(2), 145–156. https://doi.org/10.1080/10584609.2020.1723755

Frey, L. R. (1999). Introduction. In L. R. Frey, D. S. Gouran & M. S. Poole (Eds.), *The handbook of group communication theory & research* (pp. ix–xxi). Sage.

Friggeri, A. (2014, February 15). *When love goes awry* [Note]. Facebook. https://www.facebook.com/notes/facebook-data-science/when-love-goes-awry/10152066701893859

Fritz, N., & Gonzales, A. (2018). Not the normal trans story: Negotiating trans narratives while crowdfunding at the margins. *International Journal of Communication, 12,* 1189–1208.

Fulk, J., Steinfield, C. W., Schmitz, J., & Power, J. G. (1987). A social information processing model of media use in organizations. *Communication Research, 14*(5), 529–552. https://doi.org/10.1177/009365087014005005

Gainous, J., & Wagner, K. M. (2013). *Tweeting to power: The social media revolution in American politics.* Oxford University Press.

Garimella, V. R. K., Weber, I., & Dal Cin, S. (2014, November). *From "I love you babe" to "leave me alone": Romantic relationship breakups on Twitter* [Paper presentation]. 6th International Conference on Social Informatics (SocInfo), Barcelona, Spain.

Garrison, K. E., Tang, D., & Schmeichel, B. J. (2016). Embodying power: A preregistered replication and extension of the power pose effect. *Social Psychological and Personality Science, 7*(7), 623–630. https://doi.org/10.1177/1948550616652209

Gilbert, E., Bergstrom, T., & Karahalios, K. (2009, January 5–8). *Blogs are echo chambers: Blogs are echo chambers* [Paper presentation]. 42nd Hawaii International Conference on System Sciences (HICSS'09), Big Island, HI.

Gil de Zúñiga, H., & Diehl, T. (2019). News finds me perception and democracy: Effects on political knowledge, political interest, and voting. *New Media & Society, 21*(6), 1253–1271. https://doi.org/10.1177/1461444818817548

Gil de Zúñiga, H., Puig-I-Abril, E., & Rojas, H. (2009). Weblogs, traditional sources online and political participation: An assessment of how the Internet is changing the political environment. *New Media & Society, 11*(4), 553–574. https://doi.org/10.1177/1461444809102960

Gil de Zúñiga, H., Weeks, B., & Ardèvol-Abreu, A. (2017). Effects of the news-finds-me perception in communication: Social media use implications for news seeking and learning about politics. *Journal of Computer-Mediated Communication, 22*(3), 105–123. https://doi.org/10.1111/jcc4.12185

Gladstein, D. L. (1984). Groups in context: A model of task group effectiveness. *Administrative Science Quarterly, 29*(4), 499–517. https://doi.org/10.2307/2392936

Glanton, D. (2019). In a city with tens of thousands of surveillance cameras, who's watching whom? *Chicago Tribune.* https://www.chicagotribune.com/columns/dahleen-glanton/ct-met-dahleen-glanton-video-cameras-chicago-20190225-story.html

Goldfine, S. (2018). *32K surveillance cameras aim to keep Chicago safe.* Campus Safety. https://www.campussafetymagazine.com/technology/surveillance-cameras-keeping-chicago-safe/

Goldhaber, G. M. (1974). *Organizational communication.* William C. Brown.

Goldie, D., Linick, M., Jabbar, H., & Lubienski, C. (2014). Using bibliometric and social media analyses to explore the "echo chamber" hypothesis. *Educational Policy, 28*(2), 281–305. https://doi.org/10.1177/0895904813515330

Goldstein, S. B., & Kim, R. I. (2006). Predictors of US college students' participation in study abroad programs: A longitudinal study. *International Journal of Intercultural Relations, 30*(4), 507–521. https://doi.org/10.1016/j.ijintrel.2005.10.001

Gomez-Uribe, C. A., & Hunt, N. (2015). The Netflix recommender system: Algorithms, business value, and innovation. *ACM Transactions on Management Information Systems (TMIS)*, 6(4), 1–19. https://doi.org/10.1145/2843948

Gonzales, A. L., & Hancock, J. T. (2008). Identity shift in computer-mediated environments. *Media Psychology*, 11(2), 167–185. https://doi.org/10.1080/15213260802023433

Gossett, L. M., & Kilker, J. (2006). My job sucks: Examining counterinstitutional web sites as locations for organizational member voice, dissent, and resistance. *Management Communication Quarterly*, 20(1), 63–90. https://doi.org/10.1177/0893318906291729

Gouran, D. S. (2016). Group communication. In K. B. Jensen & R. T. Craig (Eds.), *The international encyclopedia of communication theory and philosophy* (pp. 1–12). Wiley.

Granovetter, M. S. (1973). The strength of weak ties. *American Journal of Sociology*, 78, 1360–1380. https://doi.org/10.1086/225469

Gray, J., Sandvoss, C., & Harrington, C. L. (Eds.). (2017). *Fandom: Identities and communities in a mediated world*. New York University Press.

Gray, R., Ellison, N. B., Vitak, J., & Lampe, C. (2013, February). *Who wants to know? Question-asking and answering practices among Facebook users* [Paper presentation]. Computer Supported Cooperative Work (CSCW'13), San Antonio, TX.

Green, P. S., Hill, J. W., Jensen, J. F., & Shah, A. (1995). Telepresence surgery. *IEEE Engineering in Medicine and Biology Magazine*, 14(3), 324–329. https://doi.org/10.1109/51.391769

Griffin, E. M., & Fingerman, K. L. (2018). Online dating profile content of older adults seeking same-and cross-sex relationships. *Journal of GLBT Family Studies*, 14(5), 446–466. https://doi.org/10.1080/1550428X.2017.1393362

Grigsby Smith, D. (2019). Centennial Airport: A case study for small airport social media strategy. *Journal of Airport Management*, 13(3), 225–237.

Gonzales, A. L., Kwon, E. Y., Lynch, T., & Fritz, N. (2018). "Better everyone should know our business than we lose our house": Costs and benefits of medical crowdfunding for support, privacy, and identity. *New Media & Society*, 20(2), 641–658. https://doi.org/10.1177/1461444816667723

Guadagno, R. E., Muscanell, N. L., Rice, L. M., & Roberts, N. (2013). Social influence online: The impact of social validation and likability on compliance. *Psychology of Popular Media Culture*, 2(1), 51–60. https://doi.org/10.1037/a0030592

Guadagno, R. E., Rempala, D. M., Murphy, S., & Okdie, B. M. (2013). What makes a video go viral? An analysis of emotional contagion and Internet memes. *Computers in Human Behavior*, 29(6), 2312–2319. https://doi.org/10.1016/j.chb.2013.04.016

Guadagno, R. E., Swinth, K. R., & Blascovich, J. (2011). Social evaluations of embodied agents and avatars. *Computers in Human Behavior*, 27(6), 2380–2385. https://doi.org/10.1016/j.chb.2011.07.017

Guzman, A. L. (2018). What is human-machine communication, anyway? In A. L. Guzman (Ed.), *Human-machine communication: Rethinking communication, technology, and ourselves* (pp. 1–28). Peter Lang.

Guzman, A. L. (2019). Voices in and of the machine: Source orientation toward mobile virtual assistants. *Computers in Human Behavior*, 90, 343–350. https://doi.org/10.1016/j.chb.2018.08.009

Haans, A., & IJsselsteijn, W. (2006). Mediated social touch: A review of current research and future directions. *Virtual Reality*, 9(2–3), 149–159. https://doi.org/10.1007/s10055-005-0014-2

Hague, B. N., & Loader, B. D. (Eds.). (1999). *Digital democracy: Discourse and decision making in the information age*. Routledge.

Halupka, M. (2014). Clicktivism: A systematic heuristic. *Policy & Internet*, 6(2), 115–132. https://doi.org/10.1002/1944-2866.POI355

Hancock, J. T., Thom-Santelli, J., & Ritchie, T. (2004, 24 April). *Deception and design: The impact of communication technology on lying behavior* [Paper presentation]. SIGCHI conference on Human Factors in Computing Systems (CHI 2004), Vienna, Austria.

Hancock, J. T., & Toma, C. L. (2009). Putting your best face forward: The accuracy of online dating photographs. *Journal of Communication*, 59(2), 367–386. https://doi.org/10.1111/j.1460-2466.2009.01420.x

Haritaipan, L., Hayashi, M., & Mougenot, C. (2018). Design of a massage-inspired haptic device for interpersonal connection in long-distance communication. *Advances in Human-Computer Interaction*, Article ID 5853474. https://doi.org/10.1155/2018/5853474

Hargie, O., Dickson, D., & Nelson, S. (2003). Working together in a divided society: A study of intergroup communication in the Northern Ireland workplace. *Journal of Business and Technical Communication*, 17(3), 285–318. https://doi.org/10.1177/1050651903017003002

Hargittai, E. (2002). Second-level digital divide: Differences in people's online skills. *First Monday*, 7(4). https://doi.org/10.5210/fm.v7i4.942

Hargittai, E., Gallo, J., & Kane, M. (2008). Cross-ideological discussions among conservative and liberal bloggers. *Public Choice*, *134*(1–2), 67–86. https://doi.org/10.1007/s11127-007-9201-x

Harris, L., & Dennis, C. (2011). Engaging customers on Facebook: Challenges for e-retailers. *Journal of Consumer Behaviour*, *10*(6), 338–346. https://doi.org/10.1002/cb.375

Hayes, R., & Carr, C. T. (2015). Does being social matter? The effect of enabled comments on credibility and brand attitude in social media. *Journal of Promotion Management*, *21*(3), 371–390. https://doi.org/10.1080/10496491.2015.1039178

Hayes, R. A., & Carr, C. T. (2021). Getting called out: Effects of feedback to social media corporate responsibility statements. *Public Relations Review*, *47*(1), article 101962. https://doi.org/10.1016/j.pubrev.2020.101962

Hayes, R. A., Carr, C. T., & Wohn, D. Y. (2016). One click, many meanings: Interpreting paralinguistic digital affordances in social media. *Journal of Broadcasting & Electronic Media*, *60*(1), 171–187. https://doi.org/10.1080/08838151.2015.1127248

Hayes, R. A., Smock, A., & Carr, C. T. (2015). Face[book] management: Self-presentation of political views on social media. *Communication Studies*, *66*(5), 549–568. https://doi.org/10.1080/10510974.2015.1018447

Hayes, R. A., Waddell, J. C., & Smudde, P. M. (2017). Our thoughts and prayers are with the victims: Explicating the public tragedy as a public relations challenge. *Public Relations Inquiry*, *6*(3), 253–274. https://doi.org/10.1177/2046147X16682987

Hayes, R. A., Wesselmann, E. D., & Carr, C. T. (2018). When nobody "Likes" you: Perceived ostracism through paralinguistic digital affordances within social media. *Social Media + Society*, *4*(3). https://doi.org/10.1177/2056305118800309

Haythornthwaite, C. (2000). Online personal networks: Size, composition and media use among distance learners. *New Media & Society*, *2*(2), 195–226. https://doi.org/10.1177/14614440022225779

Haythornthwaite, C. (2002). Strong, weak, and latent ties and the impact of new media. *Information Society*, *18*(5), 385–401. https://doi.org/10.1080/01972240290108195

Haythornthwaite, C. (2005). Social networks and Internet connectivity effects. *Information, Community & Society*, *8*(2), 125–147. https://doi.org/10.1080/13691180500146185

Heisler, Y. (2016). *Mobile Internet usage surpasses desktop usage for the first time in history*. BGR. https://bgr.com/2016/11/02/internet-usage-desktop-vs-mobile/

Heisler, J. M., & Crabill, S. L. (2006). Who are "stinkybug" and "Packerfan4"? Email pseudonyms and participants' perceptions of demography, productivity, and personality. *Journal of Computer-Mediated Communication*, *12*, 114–135. https://doi.org/10.1111/j.1083-6101.2006.00317.x

Helme-Guizon, A., & Magnoni, F. (2019). Consumer brand engagement and its social side on brand-hosted social media: How do they contribute to brand loyalty? *Journal of Marketing Management*, *35*(7–8), 716–741. https://doi.org/10.1080/0267257X.2019.1599990

Hershfield, H. E., Goldstein, D. G., Sharpe, W. F., Fox, J., Yeykelis, L., Carstensen, L. L., & Bailenson, J. N. (2011). Increasing saving behavior through age-progressed renderings of the future self. *Journal of Marketing Research*, *48*(SPL), S23–S37. https://doi.org/10.1509/jmkr.48.SPL.S23

Higgins, E. T. (1987). Self-discrepancy theory: A theory relating self and affect. *Psychological Review*, *94*(3), 1120–1134.

High, A. C., & Solomon, D. H. (2014). Communication channel, sex, and the immediate and longitudinal outcomes of verbal person-centered support. *Communication Monographs*, *81*(4), 439–468. https://doi.org/10.1080/03637751.2014.933245

Hill, K. (2012). How Target figured out a teen girl was pregnant before her father did. *Forbes*. https://www.forbes.com/sites/kashmirhill/2012/02/16/how-target-figured-out-a-teen-girl-was-pregnant-before-her-father-did/#260996946668

Hill, K. (2012b). The Facebook privacy setting that tripped up Randi Zuckerberg. *Forbes*. https://www.forbes.com/sites/kashmirhill/2012/12/26/the-facebook-privacy-setting-that-tripped-up-randi-zuckerberg/#52099281d0bc

Hiltz, S. R., Johnson, K., & Turoff, M. (1986). Experiments in group decision making: Communication processes and outcome in face-to-face versus computerized conferences. *Human Communication Research*, *13*, 225–252. https://doi.org/10.1111/j.1468-2958.1986.tb00104.x

Himelboim, I., McCreery, S., & Smith, M. (2013). Birds of a feather tweet together: Integrating network and content analyses to examine cross-ideology exposure on Twitter. *Journal of Computer-Mediated Communication*, *18*(2), 154–174. https://doi.org/10.1111/jcc4.12001

Hinck, A. (2012). Theorizing a public engagement keystone: Seeing fandom's integral connection to civic engagement through the case of the Harry Potter Alliance. *Transformative Works and Cultures, 10.* https://doi.org/10.3983/twc.2012.0311

Hinck, A. S., & Carr, C. T. (in press). Advancing a dual-process model to explain interpersonal versus intergroup communication in social media. *Communication Theory.* https://doi.org/10.1093/ct/qtaa012

History.com. (2009, June 6). *Morse code & the telegraph.* History Channel. https://www.history.com/topics/inventions/telegraph

Hmielowski, J. D., Hutchens, M. J., & Cicchirillo, V. J. (2014). Living in an age of online incivility: Examining the conditional indirect effects of online discussion on political flaming. *Information, Communication & Society, 17*(10), 1196–1211. https://doi.org/10.1080/1369118X.2014.899609

Hoffman, T. K., & DeGroot, J. M. (2014). Communicative responses to perceived Facebook jealousy. *Journal of the Communication, Speech & Theatre Association of North Dakota, 27,* 15–22.

Hogan, B. (2010). The presentation of self in the age of social media: Distinguishing performances and exhibitions online. *Bulletin of Science, Technology & Society, 30*(6), 377–386. https://doi.org/10.1177/0270467610385893

Holtz, B., & Lauckner, C. (2012). Diabetes management via mobile phones: A systematic review. *Telemedicine and e-Health, 18*(3), 175–184. https://doi.org/10.1089/tmj.2011.0119

Honan, M. (2012, August 6). How Apple and Amazon security flaws led to my epic hacking. *Wired.* https://www.wired.com/2012/08/apple-amazon-mat-honan-hacking/

Honeycutt, J. M., & Hatcher, L. C. (2016). Imagined interactions. In C. R. Berger & M. E. Roloff (Eds.), *The International encyclopedia of interpersonal communication.* Wiley.

Honeycutt, J. M., Zagacki, K. S., & Edwards, R. (1990). Imagined interaction and interpersonal communication. *Communication Reports, 3*(1), 1–8. https://doi.org/10.1080/08934219009367494

Hornyak, T. (2019). Defining the Heisei Era: Part 8—Communication. *Japan Times.* https://features.japantimes.co.jp/heisei-moments-part-8-communication/

Höpli, K., & Cuervo-Alvarez, R. (2016). *Virtual buzz: Simulating visual influences of alcohol in an augmented-reality app* [Unpublished term project]. HSR Hochschule für Technik Rapperswil

Horton, D., & Wohl, R. R. (1956). Mass communication and para-social interaction: Observations on intimacy at a distance. *Psychiatry, 19*(3), 215–229. https://doi.org/10.1521/00332747.1956.11023049

Houston, J. B., Pfefferbaum, B., Sherman, M. D., Melson, A. G., & Brand, M. W. (2013). Family communication across the military deployment experience: Child and spouse report of communication frequency and quality and associated emotions, behaviors, and reactions. *Journal of Loss and Trauma, 18*(2), 103–119. https://doi.org/10.1080/15325024.2012.684576

Hu, T., Guo, H., Sun, H., Nguyen, T.-v. T., & Luo, J. (2017, May). *Spice up your chat: The intentions and sentiment effects of using emojis* [Paper presentation]. 11th International AAAI Conference on Web and Social Media (ICWSM 2017), Montreal, CA.

Hughes, A. (2019). *A small group of prolific users account for a majority of political tweets sent by U.S. adults.* Pew Research Center. https://www.pewresearch.org/fact-tank/2019/10/23/a-small-group-of-prolific-users-account-for-a-majority-of-political-tweets-sent-by-u-s-adults/

Hum, N. J., Chamberlin, P. E., Hambright, B. L., Portwood, A. C., Schat, A. C., & Bevan, J. L. (2011). A picture is worth a thousand words: A content analysis of Facebook profile photographs. *Computers in Human Behavior, 27*(5), 1828–1833. https://doi.org/10.1016/j.chb.2011.04.003

Hunter, L., Muilenburg, L., & Burnside, R. (2012, June 26). *Second language practice using blogs and pen pals with native speakers* [Paper presentation]. EdMedia 2012–World Conference on Educational Media and Technology, Denver, CO.

Hutchens, M. J., Cicchirillo, V. J., & Hmielowski, J. D. (2015). How could you think that?!?!: Understanding intentions to engage in political flaming. *New Media & Society, 17*(8), 1201–1219. https://doi.org/10.1177/1461444814522947

Hutchinson, A. (2019). *Facebook reaches 2.38 billion users, beats revenue estimates in latest update.* Social Media Today. https://www.socialmediatoday.com/news/facebook-reaches-238-billion-users-beats-revenue-estimates-in-latest-upda/553403/

Jackson, N. (2005). Vote winner or a nuisance: Email and elected politicians' relationship with their constituents. *Journal of Nonprofit & Public Sector Marketing, 14*(1–2), 91–108. https://doi.org/10.1300/J054v14n01_06

Jackson, T., Dawson, R., & Wilson, D. (2001). The cost of email interruption. *Journal of Systems and Information Technology, 5*(1), 81–92. https://doi.org/10.1108/13287260180000760

Jin, B. (2013). How lonely people use and perceive Facebook. *Computers in Human Behavior,*

29(6), 2463–2470. https://doi.org/10.1016/j.chb.2013.05.034

Jin, W., Sun, Y., Wang, N., & Zhang, X. (2017). Why do users purchase virtual products in MMORPG? An integrative perspective of social presence and user engagement. *Internet Research*, 27(2), 406–427. https://doi.org/10.1108/IntR-04-2016-0091

Johnson, A. J., Lane, B., Tornes, M., King, S., Wright, K. B., Carr, C. T., Piercy, C., & Rozzell, B. (2013, June). *The social support process and Facebook: Soliciting support from strong and weak ties* [Paper presentation]. Annual meeting of the International Communication Association, London, England.

Johnson, B. K., & Van der Heide, B. (2015). Can sharing affect liking? Online taste performances, feedback, and subsequent media preferences. *Computers in Human Behavior*, 46, 181–190. https://doi.org/10.1016/j.chb.2015.01.018

Jones, S. E., & Yarbrough, A. E. (1985). A naturalistic study of the meanings of touch. *Communications Monographs*, 52(1), 19–56. https://doi.org/10.1080/03637758509376094

Kalman, Y., Aguilar, A., & Ballard, D. (2018, January 3). *The role of chronemic agency in the processing of a multitude of mediated conversation threads* [Paper Presentation]. 51st Hawaii International Conference on System Sciences (HICSS), Hilton Waikoloa Vilage, HI.

Kalman, Y. M., & Rafaeli, S. (2005, January 6). *Email chronemics: Unobtrusive profiling of response times* [Paper presentation]. 38th Annual Hawaii International Conference on System Sciences (HICSS), Big Island, HI.

Kalsnes, B., Larsson, A. O., & Enli, G. (2017). The social media logic of political interaction: Exploring citizens' and politicians' relationship on Facebook and Twitter. *First Monday*, 22(2). https://doi.org/10.5210/fm.v22i2.6348

Kammrath, L. K., Armstrong III, B. F., Lane, S. P., Francis, M. K., Clifton, M., McNab, K. M., & Baumgarten, O. M. (2020). What predicts who we approach for social support? Tests of the attachment figure and strong ties hypotheses. *Journal of Personality and Social Psychology*, 118(3), 481–500. https://doi.org/10.1037/pspi0000189

Kaplan, A. M., & Haenlein, M. (2010). Users of the world, unite! The challenges and opportunities of social media. *Business Horizons*, 53(1), 59–68. https://doi.org/10.1016/j.bushor.2009.09.003

Kaye, J. J., Levitt, M. K., Nevins, J., Golden, J., & Schmidt, V. (2005, April). *Communicating intimacy one bit at a time* [Paper presentation]. Human Factors in Computing Systems (CHI'05), Portland, OR.

Kaye, L. K., Malone, S. A., & Wall, H. J. (2017). Emojis: Insights, affordances, and possibilities for psychological science. *Trends in Cognitive Sciences*, 21(2), 66–68. https://doi.org/10.1016/j.tics.2016.10.007

Kennedy, M. (2020). "If the rise of the TikTok dance and e-girl aesthetic has taught us anything, it's that teenage girls rule the Internet right now": TikTok celebrity, girls and the Coronavirus crisis. *European Journal of Cultural Studies*, 23(6), 1069-1076. https://doi.org/10.1177/1367549420945341

Kent, M. L., & Taylor, M. (2002). Toward a dialogic theory of public relations. *Public Relations Review*, 28(1), 21–37. https://doi.org/10.1016/S0363-8111(02)00108-X

Ke$ha. (2010). We r who we r [Song]. On *Cannibal*. RCA Records.

Keller, F. B., Schoch, D., Stier, S., & Yang, J. (2020). Political astroturfing on Twitter: How to coordinate a disinformation campaign. *Political Communication*, 37(2), 256–280. https://doi.org/10.1080/10584609.2019.1661888

Khalil, H., & Ebner, M. (2015). "How Satisfied Are You With Your MOOC?": A research study about interaction in huge online courses. *Journalism and Mass Communication*, 5(12), 629–639. https://doi.org/10.17265/2160-6579/2015.12.003

Khan, M. L. (2017). Social media engagement: What motivates user participation and consumption on YouTube? *Computers in Human Behavior*, 66, 236–247. https://doi.org/10.1016/j.chb.2016.09.024

Kim, J. (2014). Interactivity, user-generated content and video game: An ethnographic study of *Animal Crossing: Wild World*. *Continuum*, 28(3), 357–370. https://doi.org/10.1080/10304312.2014.893984

Kim, K. J. (2016). Interacting socially with the Internet of Things (IoT): Effects of source attribution and specialization in human-IoT interaction. *Journal of Computer-Mediated Communication*, 21(6), 420–435. https://doi.org/10.1111/jcc4.12177

Kim, S. J. (2008). A framework for advertising in the digital age. *Journal of Advertising Research*, 48(3), 310–312. https://doi.org/10.2501/S0021849908080367

Kissel, P., & Büttgen, M. (2015). Using social media to communicate employer brand identity: The impact on corporate image and employer attractiveness. *Journal of Brand Management*, 22(9), 755–777. https://doi.org/10.1057/bm.2015.42

Kittur, A., & Kraut, R. E. (2008, November). *Harnessing the wisdom of crowds in Wikipedia: Quality through coordination* [Paper presentation]. Computer-Supported Cooperative Work, San Diego, CA.

Klinger, U. (2013). Mastering the art of social media: Swiss parties, the 2011 national election and digital challenges. *Information, Communication & Society*, 16(5), 717–736. https://doi.org/10.1080/1369118X.2013.782329

Klofstad, C. A. (2009). Civic talk and civic participation: The moderating effect of individual predispositions. *American Politics Research*, 37(5), 856–878. https://doi.org/10.1177/1532673X09333960

Knapp, M. L. (1978). *Social intercourse: From greeting to goodbye*. Allyn and Bacon.

Knapp, M. L., & Daly, J. A. (2011). Background and current trends in the study of interpersonal communication. In M. L. Knapp & J. A. Daly (Eds.), *The Sage handbook of interpersonal communication* (pp. 3–22). Sage.

Knapp, M. L., & Vangelisti, A. L. (2005). Relationship stages: A communication perspective. In M. L. Knapp & A. L. Vangelisti (Eds.), *Interpersonal communication and human relationships* (pp. 151–158). Allyn & Bacon.

Kooti, F., Aiello, L. M., Grbovic, M., Lerman, K., & Mantrach, A. (2015, May 18–22). *Evolution of conversations in the age of email overload* [Paper presentation]. 24th International Conference on World Wide Web (WWW'15), Florence, Italy.

Korzenny, F. (1978). A theory of electronic propinquity: Mediated communication in organizations. *Communication Research*, 5, 3–24. https://doi.org/10.1177/009365027800500101

Korzenny, F., & Bauer, C. (1981). Testing the theory of electronic propinquity: Organizational teleconferencing. *Communication Research*, 8(4), 479–498. https://doi.org/10.1177/009365028100800405

Kovic, M., Rauchfleisch, A., Sele, M., & Caspar, C. (2018). Digital astroturfing in politics: Definition, typology, and countermeasures. *Studies in Communication Sciences*, 18(1), 69–85. https://doi.org/10.24434/j.scoms.2018.01.005

Kraemer, R., Whiteman, G., & Banerjee, B. (2013). Conflict and astroturfing in Niyamgiri: The importance of national advocacy networks in anti-corporate social movements. *Organization Studies*, 34(5–6), 823–852. https://doi.org/10.1177/0170840613479240

Krämer, N. C., Neubaum, G., Hirt, M., Knitter, C., Ostendorf, S., & Zeru, S. (2017). "I see you, I know you, it feels good": Qualitative and quantitative analyses of ambient awareness as a potential mediator of social networking sites usage and well-being. *Computers in Human Behavior*, 77–85. https://doi.org/10.1016/j.chb.2017.08.024

Krämer, N. C., Rösner, L., Eimler, S. C., Winter, S., & Neubaum, G. (2014). Let the weakest link go! Empirical explorations on the relative importance of weak and strong ties on social networking sites. *Societies*, 4(4), 785–809. https://doi.org/10.3390/soc4040785

Krämer, N. C., von der Pütten, A. M., & Eimler, S. C. (2012). Human-agent and human-robot interaction theory: Similarities to and differences from human-human interaction. In M. Zacarias & J. de Oliveira (Eds.), *Human-computer interaction: The agency perspective* (pp. 215–240). Springer.

Krebs, S. A., Hobman, E. V., & Bordia, P. (2006). Virtual teams and group member dissimilarity: Consequences for the development of trust. *Small Group Research*, 37(6), 721–741. https://doi.org/10.1177/1046496406294886

Kroher, M., & Wolbring, T. (2015). Social control, social learning, and cheating: Evidence from lab and online experiments on dishonesty. *Social Science Research*, 53, 311–324. https://doi.org/10.1016/j.ssresearch.2015.06.003

Kruikemeier, S., Aparaschivei, A. P., Boomgaarden, H. G., Van Noort, G., & Vliegenthart, R. (2015). Party and candidate websites: A comparative explanatory analysis. *Mass Communication and Society*, 18(6), 821–850. https://doi.org/10.1080/15205436.2015.1051233

Kruikemeier, S., Van Noort, G., Vliegenthart, R., & De Vreese, C. H. (2013). Getting closer: The effects of personalized and interactive online political communication. *European Journal of Communication*, 28(1), 53–66. https://doi.org/10.1177/0267323112464837

Krulwich, R. (2011, February 17). *A (shockingly) short history of "Hello."* NPR - Krulwich Wonders. https://www.npr.org/sections/krulwich/2011/02/17/133785829/a-shockingly-short-history-of-hello

Kruse, L. M., Norris, D. R., & Flinchum, J. R. (2018). Social media as a public sphere? Politics on social media. *Sociological Quarterly*, 59(1), 62–84. https://doi.org/10.1080/00380253.2017.1383143

Kujath, C. L. (2011). Facebook and MySpace: Complement or substitute for face-to-face interaction? *Cyberpsychology, Behavior, and Social Networking*, 14(1–2), 75–78. https://doi.org/10.1089/cyber.2009.0311

Kumar, P. C., Vitak, J., Chetty, M., & Clegg, T. L. (2019). The platformization of the classroom: Teachers as surveillant consumers. *Surveillance & Society*, 17(1/2), 145–152. https://doi.org/10.24908/ss.v17i1/2.12926

Kumar, S. (2015). Contagious memes, viral videos and subversive parody: The grammar of contention on the Indian Web. *International Communication Gazette, 77*(3), 232–247. https://doi.org/10.1177/1748048514568758

Kwon, K. H. (2019). Public referral, viral campaign, and celebrity participation: A social network analysis of the Ice Bucket Challenge on YouTube. *Journal of Interactive Advertising, 19*(2), 87–99. https://doi.org/10.1080/15252019.2018.1561342

Labrecque, L. I. (2014). Fostering consumer-brand relationships in social media environments: The role of parasocial interaction. *Journal of Interactive Marketing, 28*(2), 134–148. https://doi.org/10.1016/j.intmar.2013.12.003

Lachlan, K., Spence, P., & Lin, X. (2017). Natural disasters Twitter and stakeholder communication. In L. Austin & Y. Jin (Eds.), *Social media and crisis communication* (pp. 296–305). Routledge.

Laërtius, D. (1925). *Βίοι καὶ γνῶμαι τῶν ἐν φιλοσοφίᾳ εὐδοκιμησάντων* [*Lives of eminent philosophers*] (R. D. Hicks, Trans.; *Vol. 2*). Harvard University Press.

Lampe, C. A. C., Ellison, N., & Steinfield, C. (2007, April). *A familiar face(book): Profile elements as signals in an online social network* [Paper presentation]. SIGCHI Conference on Human Factors in Computing Systems (CHI'07), San Jose, CA.

Lane, B. L. (2018). Still too much of a good thing? The replication of Tong, Van der Heide, Langwell, and Walther (2008). *Communication Studies, 69*(3), 294–303. https://doi.org/10.1080/10510974.2018.1463273

Lane, B. L., Piercy, C. W., & Carr, C. T. (2016). Making it Facebook official: The warranting value of online relationship status disclosures on relational characteristics. *Computers in Human Behavior, 56*, 1–8. https://doi.org/10.1016/j.chb.2015.11.016

Lane, D. S., & Dal Cin, S. (2018). Sharing beyond slacktivism: The effect of socially observable prosocial media sharing on subsequent offline helping behavior. *Information, Communication & Society, 21*(11), 1523–1540. https://doi.org/10.1080/1369118X.2017.1340496

Langer, E. J., Blank, A., & Chanowitz, B. (1978). The mindlessness of ostensibly thoughtful action: The role of "placebic" information in interpersonal interaction. *Journal of Personality and Social Psychology, 36*(6), 635–642. https://doi.org/10.1037/0022-3514.36.6.635

Laroche, M., Habibi, M. R., Richard, M.-O., & Sankaranarayanan, R. (2012). The effects of social media based brand communities on brand community markers, value creation practices, brand trust and brand loyalty. *Computers in Human Behavior, 28*(5), 1755–1767. https://doi.org/10.1016/j.chb.2012.04.016

Larsson, A. O., & Kalsnes, B. (2014). "Of course we are on Facebook": Use and non-use of social media among Swedish and Norwegian politicians. *European Journal of Communication, 29*(6), 653–667. https://doi.org/10.1177/0267323114531383

Lawrence, E., Sides, J., & Farrell, H. (2010). Self-segregation or deliberation? Blog readership, participation, and polarization in American politics. *Perspectives on Politics, 8*(1), 141–157. https://doi.org/10.1017/S1537592709992714

Leavitt, A. (2015, March 14). *This is a throwaway account: Temporary technical identities and perceptions of anonymity in a massive online community* [Paper presentation]. 18th ACM Conference on Computer Supported Cooperative Work & Social Computing (CSCW'15), Vancouver, BC.

LeBon, G. (1895). *The crowd: A study of the popular mind* (1947, Trans.). Ernest Benn.

Ledbetter, A. M. (2010). Content-and medium-specific decomposition of friendship relational maintenance: Integrating equity and media multiplexity approaches. *Journal of Social and Personal Relationships, 27*(7), 938–955. https://doi.org/10.1177/0265407510376254

Lee, D. Y., & Lehto, M. R. (2013). User acceptance of YouTube for procedural learning: An extension of the technology acceptance model. *Computers & Education, 61*, 193–208. https://doi.org/10.1016/j.compedu.2012.10.001

Lee, E.-J. (2004). Effects of visual representation on social influence in computer-mediated communication: Experimental tests of the social identity model of deindividuation effects. *Human Communication Research, 30*(2), 234–259. https://doi.org/10.1111/j.1468-2958.2004.tb00732.x

Lee, E.-J. (2010). The more humanlike, the better? How speech type and users' cognitive style affect social responses to computers. *Computers in Human Behavior, 26*(4), 665–672. https://doi.org/10.1016/j.chb.2010.01.003

Lee, E.-J., & Jang, J.-w. (2013). Not so imaginary interpersonal contact with public figures on social network sites: How affiliative tendency moderates its effects. *Communication Research, 40*(1), 27–51. https://doi.org/10.1177/0093650211431579

Lee, E.-J., & Oh, S. Y. (2012). To personalize or depersonalize? When and how politicians' personalized tweets affect the public's reactions. *Journal of Communication, 62*(6), 932–949. https://doi.org/10.1111/j.1460-2466.2012.01681.x

Lee, E.-J., & Shin, S. Y. (2012). Are they talking to me? Cognitive and affective effects of interactivity in politicians' Twitter communication. *Cyberpsychology, Behavior, and Social Networking, 15*(10), 515–520. https://doi.org/10.1089/cyber.2012.0228

Lee, K. M. (2006). Effects of Internet use on college students' political efficacy. *CyberPsychology & Behavior, 9*(4), 415–422. https://doi.org/10.1089/cpb.2006.9.415

Lee, K., Webb, S., & Ge, H. (2014). *The dark side of micro-task marketplaces: Characterizing fiverr and automatically detecting crowdturfing* [Paper presentation]. 8th International AAAI Conference on Weblogs and Social Media (ICWSM), Ann Arbor, MI.

Lee, S., & Xenos, M. (in press). Incidental news exposure via social media and political participation: Evidence of reciprocal effects. *New Media & Society.* https://doi.org/10.1177/1461444820962121

LeFebvre, L. E. (2017). Phantom lovers: Ghosting as a relationship dissolution strategy in the technological age. In N. M. Punyanunt-Carter & J. S. Wrench (Eds.), *The impact of social media in modern romantic relationships* (pp. 219–235). Lexington Books.

LeFebvre, L. E., Allen, M., Rasner, R. D., Garstad, S., Wilms, A., & Parrish, C. (2019). Ghosting in emerging adults' romantic relationships: The digital dissolution disappearance strategy. *Imagination, Cognition and Personality, 39*(2), 125–150. https://doi.org/10.1177/0276236618820519

Leonardi, P. M. (2015). Ambient awareness and knowledge acquisition: Using social media to learn "who knows what" and "who knows whom." *MIS Quarterly, 39*(4), 747–762. https://doi.org/10.25300/MISQ/2015/39.4.1

Leonardi, P. M., Huysman, M., & Steinfield, C. (2013). Enterprise social media: Definition, history, and prospects for the study of social technologies in organizations. *Journal of Computer-Mediated Communication, 19*(1), 1–19. https://doi.org/10.1111/jcc4.12029

Levendusky, M. S., Druckman, J. N., & McLain, A. (2016). How group discussions create strong attitudes and strong partisans. *Research & Politics, 3*(2). https://doi.org/10.1177/2053168016645137

Levitan, L., & Wronski, J. (2014). Social context and information seeking: Examining the effects of network attitudinal composition on engagement with political information. *Political Behavior, 36*(4), 793–816. https://doi.org/10.1007/s11109-013-9247-z

Levordashka, A., & Utz, S. (2016). Ambient awareness: From random noise to digital closeness in online social networks. *Computers in Human Behavior, 60,* 147–154. https://doi.org/10.1016/j.chb.2016.02.037

Lewin, K. (1945). The research center for group dynamics at Massachusetts Institute of Technology. *Sociometry, 8*(2), 126–136. https://doi.org/10.2307/2785233

Lima, V. M., Irigaray, H. A. R., & Lourenco, C. (2019). Consumer engagement on social media: Insights from a virtual brand community. *Qualitative Market Research: An International Journal, 2*(1), 14–32. https://doi.org/10.1108/QMR-02-2017-0059

Limperos, A. M., Tamul, D. J., Woolley, J. K., Spinda, J. S. W., & Sundar, S. S. (2014). "It's Not Who You Know, but Who You Add": An investigation into the differential impact of friend adding and self-disclosure on interpersonal perceptions on Facebook. *Computers in Human Behavior, 35,* 496–505. https://doi.org/10.1016/j.chb.2014.02.037

Lin, R., & Utz, S. (2017). Self-disclosure on SNS: Do disclosure intimacy and narrativity influence interpersonal closeness and social attraction? *Computers in Human Behavior, 70,* 426–436. https://doi.org/10.1016/j.chb.2017.01.012

Ling, G. S. F., Rhee, P., & Ecklund, J. M. (2010). Surgical innovations arising from the Iraq and Afghanistan wars. *Annual Review of Medicine, 61,* 457–468. https://doi.org/10.1146/annurev.med.60.071207.140903

Liu, B. F., Fraustino, J. D., & Jin, Y. (2016). Social media use during disasters: How information form and source influence intended behavioral responses. *Communication Research, 43*(5), 626–646. https://doi.org/10.1177/0093650214565917

Liu-Thompkins, Y. (2012). Seeding viral content: The role of message and network factors. *Journal of Advertising Research, 52*(4), 465–478. https://doi.org/10.2501/JAR-52-4-465-478

Livingstone, S., & Helsper, E. (2007). Gradations in digital inclusion: Children, young people and the digital divide. *New Media & Society, 9*(4), 671–696. https://doi.org/10.1177/1461444807080335

Lorenz, T. (2020, June 21). TikTok teens and K-Pop stans say they sank Trump rally. *New York Times.* https://www.nytimes.com/2020/06/21/style/tiktok-trump-rally-tulsa.html

Lowry, P. B., Roberts, T. L., Romano Jr., N. C. Cheney, P. D., & Hightower, R. T. (2006). The impact of group size and social presence on small-group communication: Does computer-mediated communication make a difference? *Small Group Research, 37*(6), 631–661. https://doi.org/10.1177/1046496406294322

Luca, M. (2016). *Reviews, reputation, and revenue: The case of Yelp.Com* [Working paper]. Harvard University. https://papers.ssrn.com/sol3/papers.cfm?abstract_id=1928601

Luque, A. E., Corales, R., Fowler, R. J., DiMarco, J., Van Keken, A., Winters, P., Keefer, M. C., & Fiscella, K. (2013). Bridging the digital divide in HIV care: A pilot study of an iPod personal health record. *Journal of the International Association of Providers of AIDS Care (JIAPAC)*, 12(2), 117–121. https://doi.org/10.1177/1545109712457712

Lyu, S. O. (2016). Travel selfies on social media as objectified self-presentation. *Tourism Management*, 54, 185–195. https://doi.org/10.1016/j.tourman.2015.11.001

Ma, A. (2018). China has started ranking citizens with a creepy "social credit" system: Here's what you can do wrong, and the embarrassing, demeaning ways they can punish you. *Business Insider*. https://www.businessinsider.com/china-social-credit-system-punishments-and-rewards-explained-2018-4

MacGeorge, E. L., Feng, B., & Burleson, B. R. (2011). Supportive communication. In M. K. Knapp & J. A. Daly (Eds.), *Handbook of interpersonal communication* (pp. 317–354). Sage.

Majchrzak, A., Cherbakov, L., & Ives, B. (2009). Harnessing the power of the crowds with corporate social networking tools: How IBM does it. *MIS Quarterly Executive*, 8(2), 103–108.

Makki, T. W., DeCook, J. R., Kadylak, T., & Lee, O. J. (2018). The social value of Snapchat: An exploration of affiliation motivation, the technology acceptance model, and relational maintenance in Snapchat use. *International Journal of Human-Computer Interaction*, 34(5), 410–420. https://doi.org/10.1080/10447318.2017.1357903

Malinowski, B. (1972). Phatic communion. In J. Laver & S. Hutchinson (Eds.), *Communication in face-to-face interaction* (pp. 146–152). Penguin.

Malinowski, E. (2010). Ole Miss' Admiral Ackbar campaign fizzles. *Wired*. https://www.wired.com/2010/09/ole-miss-admiral-ackbar/

Marciano, L., Schulz, P. J., & Camerini, A.-L. (2020). Cyberbullying perpetration and victimization in youth: A meta-analysis of longitudinal studies. *Journal of Computer-Mediated Communication*, 25(2), 163–181. https://doi.org/10.1093/jcmc/zmz031

Maréchal, N. (2018, August 14). *From Russia with crypto: A political history of Telegram* [Paper presentation]. 8th {USENIX} Workshop on Free and Open Communications on the Internet (FOCI'18), Baltimore, MD.

Margolis, M., Resnick, D., & Tu, C.-c. (1997). Campaigning on the Internet: Parties and candidates on the World Wide Web in the 1996 primary season. *Harvard International Journal of Press/Politics*, 2(1), 59–78. https://doi.org/10.1177/1081180X97002001006

Marimow, A. E. (2019). President Trump cannot block his critics on Twitter, federal appeals court rules. *Washington Post*. https://www.washingtonpost.com/local/legal-issues/president-trump-cannot-block-his-critics-on-twitter-federal-appeals-court-rules/2019/07/09/d07a5558-8230-11e9-95a9-e2c830afe24f_story.html

Marwick, A. E., & boyd, d. (2011). I tweet honestly, I tweet passionately: Twitter users, context collapse, and the imagined audience. *New Media & Society*, 13(1), 114–133. https://doi.org/10.1177/1461444810365313

Maslow, A. (1965, May 22–28). *Self-actualization and beyond* [Paper presentation]. Conference on the Training of Counselors of Adults, Catham, MA.

Matei, S. A., & Britt, B. C. (2017). *Structural differentiation in social media: Adhocracy, entropy, and the "1 % effect."* Springer.

McCafferty, D. (2011). Activism vs. slacktivism. *Communications of the ACM*, 54(12), 17–19. https://doi.org/10.1145/2043174.2043182

McCain, J. L., & Campbell, W. K. (2018). Narcissism and social media use: A meta-analytic review. *Psychology of Popular Media Culture*, 7(3), 308. https://doi.org/10.1037/ppm0000137

McEwan, B. (2020). Social media and relationships. In J. van den Bulck (Ed.), *The international encyclopedia of media psychology*. Wiley.

McKew, M. (2019). Did Russia affect the 2016 election? It's now undeniable. *Wired*. https://www.wired.com/story/did-russia-affect-the-2016-election-its-now-undeniable/

McLaren, J., & Zappalà, G. (2002). The "Digital Divide" among financially disadvantaged families in Australia. *First Monday*, 7(11). https://firstmonday.org/ojs/index.php/fm/article/view/1003/924

McLean, S. (2005). *The basics of interpersonal communication*. Pearson.

McLuhan, M. (1994). *Understanding media: The extensions of man* (2nd ed.). MIT Press.

McMurtry, K. (2014). Managing email overload in the workplace. *Performance Improvement*, 53(7), 31–37. https://doi.org/10.1002/pfi.21424

McQuail, D. (2010). *McQuail's mass communication theory* (6th ed.). Sage.

Mehdizadeh, S. (2010). Self-presentation 2.0: Narcissism and self-esteem on Facebook. *Cyberpsychology, Behavior, and Social Networking*,

13(4), 357–364. https://doi.org/10.1089/cyber.2009.0257

Mellner, C. (2016). After-hours availability expectations, work-related smartphone use during leisure, and psychological detachment: The moderating role of boundary control. *International Journal of Workplace Health Management, 9*(2), 146–164. https://doi.org/10.1108/IJWHM-07-2015-0050

Merolla, A. J. (2010). Relational maintenance during military deployment: Perspectives of wives of deployed US soldiers. *Journal of Applied Communication Research, 38*(1), 4–26. https://doi.org/10.1080/00909880903483557

Messaging Anti-abuse Working Group. (2010). *2010 MAAWG email security awareness and usage report.* https://www.m3aawg.org/system/files/2010_maawg-consumer_survey.pdf

Metzger, M. J., Flanagin, A. J., & Zwarun, L. (2003). College student Web use, perceptions of information credibility, and verification behavior. *Computers & Education, 41*(3), 271–290. https://doi.org/10.1016/S0360-1315(03)00049-6

Miczo, N., Mariani, T., & Donahue, C. (2011). The strength of strong ties: Media multiplexity, communication motives, and the maintenance of geographically close friendships. *Communication Reports, 24*(1), 12–24. https://doi.org/10.1080/08934215.2011.555322

Mikal, J. P., Rice, R. E., Abeyta, A., & DeVilbiss, J. (2013). Transition, stress and computer-mediated social support. *Computers in Human Behavior, 29*(5), A40–A53. https://doi.org/10.1016/j.chb.2012.12.012

Mikal, J. P., Rice, R. E., Kent, R. G., & Uchino, B. N. (2016). 100 million strong: A case study of group identification and deindividuation on Imgur.com. *New Media & Society, 18*(11), 2485–2506. https://doi.org/10.1177/1461444815588766

Miller, K. (2009). *Organizational communication: Approaches and processes* (5th ed.). Wadsworth.

Miller-Ott, A. E., Kelly, L., & Duran, R. L. (2014). Cell phone usage expectations, closeness, and relationship satisfaction between parents and their emerging adults in college. *Emerging Adulthood, 2*(4), 313–323. https://doi.org/10.1177/2167696814550195

Misoch, S. (2015). Stranger on the Internet: Online self-disclosure and the role of visual anonymity. *Computers in Human Behavior, 48*, 535–541. https://doi.org/10.1016/j.chb.2015.02.027

Mocanu, D., Rossi, L., Zhang, Q., Karsai, M., & Quattrociocchi, W. (2015). Collective attention in the age of (mis) information. *Computers in Human Behavior, 51*, 1198–1204. https://doi.org/10.1016/j.chb.2015.01.024

Moldovan, S., Steinhart, Y., & Lehmann, D. R. (2019). Propagators, creativity, and informativeness: What helps ads go viral. *Journal of Interactive Marketing, 47*, 102–114. https://doi.org/10.1016/j.intmar.2019.02.004

Molesworth, M. (2006). Real brands in imaginary worlds: Investigating players' experiences of brand placement in digital games. *Journal of Consumer Behaviour, 5*(4), 355–366. https://doi.org/10.1002/cb.186

Montgomery, K. C., & Chester, J. (2009). Interactive food and beverage marketing: Targeting adolescents in the digital age. *Journal of Adolescent Health, 45*(3), S18–S29. https://doi.org/10.1016/j.jadohealth.2009.04.006

Morozov, E. (2009). The brave new world of slacktivism. *Foreign Policy.* from http://neteffect.foreignpolicy.com/posts/2009/05/19/the_brave_new_world_of_slacktivism

Mueller, F. F., Vetere, F., Gibbs, M. R., Kjeldskov, J., Pedell, S., & Howard, S. (2005, April 2–7). *Hug over a distance* [Paper presentation]. SIGCHI Conference on Human Factors in Computing Systems (CHI'05), Portland, OR.

Mullen, B., Dovidio, J. F., Johnson, C., & Copper, C. (1992). In-group-out-group differences in social projection. *Journal of Experimental Social Psychology, 28*(5), 422–440. https://doi.org/10.1016/0022-1031(92)90040-Q

Muñiz, A. M., Jr., & O'Guinn, T. C. (2001). Brand community. *Journal of Consumer Research, 27*(4), 412–431. https://doi.org/10.1086/319618

Munson, S. A. (2008, September 8–10). *Motivating and enabling organizational memory with a workgroup wiki* [Paper presentation]. 4th International Symposium on Wikis (WikiSym'08), Porto, Portugal.

Munson, S. A., & Resnick, P. (2010, April 10–15). *Presenting diverse political opinions: How and how much* [Paper presentation]. SIGCHI Conference on Human Factors in Computing Systems (CHI'10), Atlanta, GA.

Mutz, D. C. (2002). The consequences of cross-cutting networks for political participation. *American Journal of Political Science, 46*(4), 838–855. https://doi.org/10.2307/3088437

Nace, T. (2019, September 2). Alabama's NWS had a perfect response to Trump's Dorian tweet. *Forbes.* https://www.forbes.com/sites/trevornace/2019/09/02/alabamas-nws-had-a-perfect-response-to-trumps-dorian-tweet/#6bb2f04d26a7

Napoli, P. M. (1998). The Internet and the forces of "massification." *Electronic Journal of Communication, 8*(2). http://www.cios.org/EJCPUBLIC/008/2/00828.HTML

Nardi, B., & Harris, J. (2006, November 4–8). *Strangers and friends: Collaborative play in*

World of Warcraft [Paper presentation]. Conference on Computer Supported Cooperative Work (CSCW06), Banff, AB.

Nass, C., Steuer, J., & Tauber, E. R. (1994, April). *Computers are social actors* [Paper presentation]. SIGCHI Conference on Human Factors in Computing Systems (CHI'94), Boston, MA.

National Communication Association. (2020). *NCA interest groups*. National Communication Association. https://www.natcom.org/about-nca/membership-and-interest-groups/nca-interest-groups

Neilsen, J. (2006). *The 90-9-1 rule for participation inequality in social media and online communities*. Neilsen Norman Group. https://www.nngroup.com/articles/participation-inequality/

Neubaum, G., Sobieraj, S., Raasch, J., & Riese, J. (2020). Digital destigmatization: How exposure to networking profiles can reduce social stereotypes. *Computers in Human Behavior*, *112*, 106461. https://doi.org/10.1016/j.chb.2020.106461

Newman, L. H. (2019). The FCC's push to purge Huawei from US networks. *Wired*. https://www.wired.com/story/fcc-rip-replace-huawei-zte/

Nicholls, S. B., & Rice, R. E. (2017). A dual-identity model of responses to deviance in online groups: Integrating social identity theory and expectancy violations theory. *Communication Theory*, *27*, 243–268. https://doi.org/10.1111/comt.12113

Nicholson, C. (2011, December 15). *How to steal an identity in seven easy steps*. SmartPlanet. http://www.smartplanet.com/blog/thinking-tech/how-to-steal-an-identity-in-seven-easy-steps/9487

Nickerson, D. W. (2006). Volunteer phone calls can increase turnout: Evidence from eight field experiments. *American Politics Research*, *34*(3), 271–292. https://doi.org/10.1177/1532673X05275923

Nisar, T. M., & Whitehead, C. (2016). Brand interactions and social media: Enhancing user loyalty through social networking sites. *Computers in Human Behavior*, *62*, 743–753. https://doi.org/10.1016/j.chb.2016.04.042

Nitschke, P., Donges, P., & Schade, H. (2016). Political organizations' use of websites and Facebook. *New Media & Society*, *18*(5), 744–764. https://doi.org/10.1177/1461444814546451

Nowak, G. J., Evans, N. J., Wojdynski, B. W., Ahn, S. J. G., Len-Rios, M. E., Carera, K., Hale, S., & McFalls, D. (2020). Using immersive virtual reality to improve the beliefs and intentions of influenza vaccine avoidant 18-to-49-year-olds: Considerations, effects, and lessons learned. *Vaccine*, *38*(5), 1225–1233. https://doi.org/10.1016/j.vaccine.2019.11.009

Nowak, K. L., & Fox, J. (2018). Avatars and computer-mediated communication: A review of the definitions, uses, and effects of digital representations. *Review of Communication Research*, *6*, 30–53. https://doi.org/10.12840/issn.2255-4165.2018.06.01.015

Nowak, K. L., & Rauh, C. (2005). The influence of the avatar on online perceptions of anthropomorphism, androgyny, credibility, homophily, and attraction. *Journal of Computer-Mediated Communication*, *11*(1), 153–178. https://doi.org/10.1111/j.1083-6101.2006.tb00308.x

Nuzzi, O. (2016, April 13, 2017). *How Pepe the Frog became a Nazi Trump supporter and Alt-Right symbol*. Daily Beast. https://www.thedailybeast.com/how-pepe-the-frog-became-a-nazi-trump-supporter-and-alt-right-symbol

Oberst, U., Wegmann, E., Stodt, B., Brand, M., & Chamarro, A. (2017). Negative consequences from heavy social networking in adolescents: The mediating role of fear of missing out. *Journal of Adolescence*, *55*, 51–60. https://doi.org/10.1016/j.adolescence.2016.12.008

O'Keefe, D. J. (2002). *Persuasion: Theory & research*. Sage.

Olafsson, S., O'Leary, T., Lee, C., & Bickmore, T. (2018, November). Virtual counselor for patients in medication-assisted treatment for opioid use. Graphical and Robotic Embodied Agents for Therapeutic Systems (GREATS18), Sydney, NSW, Australia.

O'Leary, D. E. (2011). The use of social media in the supply chain: Survey and extensions. *Intelligent Systems in Accounting, Finance and Management*, *18*(2–3), 121–144. https://doi.org/10.1002/isaf.327

Ong, E. Y. L., Ang, R. P., Ho, J., Lim, J. C. Y., Goh, D. H., Lee, C. S., & Chua, A. Y. K. (2011). Narcissism, extraversion and adolescents' self-presentation on Facebook. *Personality and Individual Differences*, *50*(2), 180–185. https://doi.org/10.1016/j.paid.2010.09.022

Orosz, G., Szekeres, Á., Kiss, Z. G., Farkas, P., & Roland-Lévy, C. (2015). Elevated romantic love and jealousy if relationship status is declared on Facebook. *Frontiers in Psychology*, *6*, 214. https://doi.org/10.3389/fpsyg.2015.00214

O'Sullivan, P. B., & Carr, C. T. (2018). Mass-personal communication: A model bridging the mass-interpersonal divide. *New Media & Society*, *20*(3), 1161–1180. https://doi.org/10.1177/1461444816686104

Paczkowski, W. F., & Kuruzovich, J. (2016). Checking email in the bathroom: Monitoring email responsiveness behavior in the workplace. *American Journal of Management*, *16*(2). https://doi.org/10.33423/ajm.v16i2.1878

Paisley, B. (2007). Online [Song]. On *5th Gear*. Artista Nashville.

Parigi, P., & Gong, R. (2014). From grassroots to digital ties: A case study of a political consumerism movement. *Journal of Consumer Culture, 14*(2), 236–253. https://doi.org/10.1177/1469540514526280

Parks, M. R. (2011). Boundary conditions for the application of three theories of computer-mediated communication to MySpace. *Journal of Communication, 61*(4), 557–574. https://doi.org/10.1111/j.1460-2466.2011.01569.x

Parks, M. R., & Floyd, K. (1996). Making friends in cyberspace. *Journal of Computer-Mediated Communication, 1*(4). https://doi.org/10.1111/j.1083-6101.1996.tb00176.x

Parks, M. R., & Roberts, L. D. (1998). "Making MOOsic": The development of personal relationships online and a comparison to their offline counterparts. *Journal of Social and Personal Relationships, 15*(4), 517–537. https://doi.org/10.1177/0265407598154005

Parks, P., Cruz, R., & Ahn, S. J. G. (2014). Don't hurt my Avatar: The use and potential of digital self-representation in risk communication. *International Journal of Robots, Education and Art, 4*(2), 10–20.

Papacharissi, Z. (2002). The self online: The utility of personal home pages. *Journal of Broadcasting & Electronic Media, 46*(3), 346–368. https://doi.org/10.1207/s15506878jobem4603_3

Papacharissi, Z. (2012). Audiences as media producers: Content analysis of 260 blogs. In M. Tremayne (Ed.), *Blogging, citizenship, and the future of media* (pp. 37–54). Routledge.

Papacharissi, Z., & Rubin, A. M. (2000). Predictors of Internet use. *Journal of Broadcasting & Electronic Media, 44*(2), 175–196. https://doi.org/10.1207/s15506878jobem4402_2

Patil, S., & Davies, P. (2014). Use of Google Translate in medical communication: Evaluation of accuracy. *BMJ, 349*, g7392. https://doi.org/10.1136/bmj.g7392

Pavalanathan, U., & Eisenstein, J. (2016). *More emojis, less :): The competition for paralinguistic function in microblog writing. First Monday, 21*(11). https://doi.org/10.5210/fm.v21i11.6879

Peña, J., & Chen, M. (2017a). Playing with power: Power poses affect enjoyment, presence, controller responsiveness, and arousal when playing natural motion-controlled video games. *Computers in Human Behavior, 71*, 428–435. https://doi.org/10.1016/j.chb.2017.02.019

Peña, J., & Chen, M. (2017b). With great power comes great responsibility: Superhero primes and expansive poses influence prosocial behavior after a motion-controlled game task. *Computers in Human Behavior, 76*, 378–385. https://doi.org/10.1016/j.chb.2017.07.039

Pearce, K. E., Freelon, D., & Kendzior, S. (2014). The effect of the Internet on civic engagement under authoritarianism: The case of Azerbaijan. *First Monday, 19*(6). https://doi.org/10.5210/fm.v19i6.5000

Pearce, K. E., & Rice, R. E. (2013). Digital divides from access to activities: Comparing mobile and personal computer Internet users. *Journal of Communication, 63*(4), 721–744. https://doi.org/10.1111/jcom.12045

Pearl, D. (2019, July 31). *L'Oréal is bringing beauty online with the help of augmented reality and AI*. Ad Week. https://www.adweek.com/brand-marketing/loreal-is-bringing-beauty-online-with-the-help-of-augmented-reality-and-ai/

Pennington, N. (2020). An examination of relational maintenance and dissolution through social networking sites. *Computers in Human Behavior, 105*, 106–196. https://doi.org/10.1016/j.chb.2019.106196

Pew Research Center. (2019, June 12). *Mobile fact sheet*. Pew Research Center. http://www.pewinternet.org/fact-sheet/mobile/

Piercy, C. W., & Lee, S. K. (2019). A typology of job search sources: Exploring the changing nature of job search networks. *New Media & Society, 21*(6), 1173–1191. https://doi.org/10.1177/1461444818808071

Planalp, S., Rutherford, D. K., & Honeycutt, J. M. (1988). Events that increase uncertainty in personal relationships II: Replication and extension. *Human Communication Research, 14*(4), 516–547. https://doi.org/10.1111/j.1468-2958.1988.tb00166.x

Platt, C. (1999, November 1). You've got smell. *Wired*. https://www.wired.com/1999/11/digiscent/

Polzer, J. T., Crisp, C. B., Jarvenpaa, S. L., & Kim, J. W. (2006). Extending the faultline model to geographically dispersed teams: How collocated subgroups can impair group functioning. *Academy of Management Journal, 46*, 679–692. https://doi.org/10.5465/amj.2006.22083024

Poole, M. S., & Hirokawa, R. Y. (1996). Introduction: Communication and group decision making. In R. Y. Hirokawa & M. S. Poole (Eds.), *Communication and group decision making* (2nd ed., pp. 3–18). Sage.

Postmes, T., Spears, R., & Lea, M. (1998). Breaching or building social boundaries? SIDE-effects of computer-mediated communication. *Communication Research, 25*(6), 689–715. https://doi.org/10.1177/009365098025006006

Powell, W. W., Horvath, A., & Brandtner, C. (2016). Click and mortar: Organizations on the Web.

Research in Organizational Behavior, 36, 101–120. https://doi.org/10.1016/j.riob.2016.07.001

Prensky, M. (2001). Digital natives, digital immigrants Part 2: Do they really think differently? *On the Horizon, 9*(5), 1–6. http://www.alberto-mattiacci.it/docs/did/Digital_Natives_Digital_Immigrants.pdf

Prensky, M. (2009). H. sapiens digital: From digital immigrants and digital natives to digital wisdom. *Innovate: Journal of Online Education, 5*(3). https://www.learntechlib.org/p/104264/

Prentice, D. A., Miller, D. T., & Lightdale, J. R. (1994). Asymmetries in attachments to groups and to their members: Distinguishing between common-identity and common-bond groups. *Personality and Social Psychology Bulletin, 20*(5), 484–493. https://doi.org/10.1177/0146167294205005

Public Relations Society of America. (2019). *About public relations*. Public Relations Society of America. https://www.prsa.org/about/all-about-pr

Purbrick, M. (2019). A report of the 2019 Hong Kong protests. *Asian Affairs, 50*(4), 465–487. https://doi.org/10.1080/03068374.2019.1672397

Pursel, B. K., Zhang, L., Jablokow, K. W., Choi, G. W., & Velegol, D. (2016). Understanding MOOC students: Motivations and behaviours indicative of MOOC completion. *Journal of Computer Assisted Learning, 32*(3), 202–217. https://doi.org/10.1111/jcal.12131

Putnam, R. D. (2001). *Bowling alone: The collapse and revival of American community*. Simon and Schuster.

Qiu, J., Li, Y., Tang, J., Lu, Z., Ye, H., Chen, B., Yang, Q., & Hopcroft, J. E. (2016, April 11–15). *The lifecycle and cascade of WeChat social messaging groups* [Paper presentation]. 25th International Conference on World Wide Web (WWW'16), Montreal, QC.

Quattrociocchi, W., Scala, A., & Sunstein, C. R. (2016). *Echo chambers on Facebook*. SSRN. https://www.researchgate.net/profile/Walter_Quattrociocchi2/publication/331936299_Echo_Chambers_on_Facebook/links/5c93b14b299bf111693e20f4/Echo-Chambers-on-Facebook.pdf

Quinlan, S., Shephard, M., & Paterson, L. (2015). Online discussion and the 2014 Scottish independence referendum: Flaming keyboards or forums for deliberation? *Electoral Studies, 38*, 192–205. https://doi.org/10.1016/j.electstud.2015.02.009

Rafaeli, S., & Sudweeks, F. (1997). Networked interactivity. *Journal of Computer-Mediated Communication, 2*(4). https://doi.org/10.1111/j.1083-6101.1997.tb00201.x

Rains, S. A., Akers, C., Pavlich, C. A., Tsetsi, E., & Appelbaum, M. (2019). Examining the quality of social support messages produced face-to-face and in computer-mediated communication: The effects of hyperpersonal communication. *Communication Monographs, 86*(3), 271–291. https://doi.org/10.1080/03637751.2019.1595076

Ramirez, A., Jr., Sumner, E. M., & Spinda, J. (2017). The relational reconnection function of social network sites. *New Media & Society, 19*(6), 807–825. https://doi.org/10.1177/1461444815614199

Ramirez, A., Jr., Walther, J. B., Burgoon, J. K., & Sunnafrank, M. (2002). Information-seeking strategies, uncertainty, and computer-mediated communication: Toward a conceptual model. *Human Communication Research, 28*(2), 213–228. https://doi.org/10.1093/hcr/28.2.213

Ramirez, A., Jr., & Zhang, S. (2007). When online meets offline: The effect of modality switching on relational communication. *Communication Monographs, 74*(3), 287–310. https://doi.org/10.1080/03637750701543493

Rea, J., Behnke, A., Huff, N., & Allen, K. (2015). The role of online communication in the lives of military spouses. *Contemporary Family Therapy, 37*(3), 329–339. https://doi.org/10.1007/s10591-015-9346-6

Reeves, B., & Nass, C. (1996). *The media equation: How people treat computers, television, and new media like real people and places*. Cambridge University Press.

Regan, P. C. (2017). *The mating game: A primer on love, sex, and marriage* (3rd ed.). Sage.

Register, L. M., & Henley, T. B. (1992). The phenomenology of intimacy. *Journal of Social and Personal Relationships, 9*(4), 467–481. https://doi.org/10.1177/0265407592094001

Reich, S. M., Subrahmanyam, K., & Espinoza, G. (2012). Friending, IMing, and hanging out face-to-face: Overlap in adolescents' online and offline social networks. *Developmental Psychology, 48*(2), 356–368. https://doi.org/10.1037/a0026980

Reicher, S. D., Spears, R., & Postmes, T. (1995). A social identity model of deindividuation phenomena. *European Review of Social Psychology, 6*(1), 161–198. https://doi.org/10.1080/14792779443000049

Reilly, A. H., & Hynan, K. A. (2014). Corporate communication, sustainability, and social media: It's not easy (really) being green. *Business Horizons, 57*(6), 747–758. https://doi.org/10.1016/j.bushor.2014.07.008

Ren, Y., Kraut, R., & Kiesler, S. (2007). Applying common identity and bond theory to design of online communities. *Organization Studies*, 28(3), 377–408. https://doi.org/10.1177/0170840607076007

Rezabek, L., & Cochenour, J. (1998). Visual cues in computer-mediated communication: Supplementing text with emoticons. *Journal of Visual Literacy*, 18(2), 201–215. https://doi.org/10.1080/23796529.1998.11674539

Rheingold, H. (1993). *The virtual community: Finding connection in a computerized world.* Addison-Wesley Longman.

Rheingold, H. (2003). *Smart mobs: The next social revolution.* Perseus.

Rice, R. E. (1980). Computer conferencing. In B. Dervin & M. J. Voigt (Eds.), *Progress in communication sciences* (Vol. 7, pp. 215–240). Ablex.

Rickwood, D. J. (2010). Promoting youth mental health through computer-mediated communication. *International Journal of Mental Health Promotion*, 12(3), 32–44. https://doi.org/10.1080/14623730.2010.9721817

Ridgely, C. (2019, August 19). *Disney flexes all of its brands on Twitter with hilarious Disney+ thread.* ComicBook.com. https://comicbook.com/movies/2019/08/19/disney-plus-flex-twitter-thread-marvel-pixar-star-wars/#5

Rochadiat, A. M. P., Tong, S. T., & Novak, J. M. (2018). Online dating and courtship among Muslim American women: Negotiating technology, religious identity, and culture. *New Media & Society*, 20(4), 1618–1639. https://doi.org/10.1177/1461444817702396

Rogers, E. M. (1999). Anatomy of the two subdisciplines of communication study. *Human Communication Research*, 25(4), 618–631. https://doi.org/10.1111/j.1468-2958.1999.tb00465.x

Rogers, K. (2016, March 21). Boaty McBoatface: What you get when you let the Internet decide. *New York Times.* https://www.nytimes.com/2016/03/22/world/europe/boaty-mcboatface-what-you-get-when-you-let-the-internet-decide.html

Rojas, H., & Puig-I-Abril, E. (2009). Mobilizers mobilized: Information, expression, mobilization and participation in the digital age. *Journal of Computer-Mediated Communication*, 14(4), 902–927. https://doi.org/10.1111/j.1083-6101.2009.01475.x

Rösner, L., & Krämer, N. C. (2016). Verbal venting in the social web: Effects of anonymity and group norms on aggressive language use in online comments. *Social Media + Society*, 2(3). https://doi.org/10.1177/2056305116664220

Rovers, A. F., & Van Essen, H. A. (2005, March). *FootIO-Design and evaluation of a device to enable foot interaction over a computer network* [Paper presentation]. First Joint Euro-haptics Conference and Symposium on Haptic Interfaces for Virtual Environment and Teleoperator Systems, Pisa, Italy.

Rovers, L., & Van Essen, H. A. (2004, June). *Design and evaluation of hapticons for enriched instant messaging* [Paper presentation]. EuroHaptics 2004, Munich, Germany.

Rowe, I. (2015). Civility 2.0: A comparative analysis of incivility in online political discussion. *Information, Communication & Society*, 18(2), 121–138. https://doi.org/10.1080/1369118X.2014.940365

Rozzell, B., Piercy, C., Carr, C. T., King, S., Lane, B., Tornes, M., Johnson, A. J., & Wright, K. B. (2014, June). Notification pending: Online social support from close and nonclose relational ties via Facebook. *Computers in Human Behavior*, 38, 272–280. https://doi.org/10.1016/j.chb.2014.06.006

Ruesch, J., & Bateson, G. (1968). *Communication: The social matrix of psychiatry.* W. W. Norton.

Rui, J. R. (2020). How a social network profile affects employers' impressions of the candidate: An application of norm evaluation. *Management Communication Quarterly*, 34(3), 328–349. https://doi.org/10.1177/0893318920916723

Ruppel, E. K. (2015). The affordance utilization model: Communication technology use as relationships develop. *Marriage & Family Review*, 51(8), 669–686. https://doi.org/10.1080/01494929.2015.1061628

Ryan, T., & Xenos, S. (2011). Who uses Facebook? An investigation into the relationship between the Big Five, shyness, narcissism, loneliness, and Facebook usage. *Computers in Human Behavior*, 27(5), 1658–1664. https://doi.org/10.1016/j.chb.2011.02.004

Rybalko, S., & Seltzer, T. (2010). Dialogic communication in 140 characters or less: How Fortune 500 companies engage stakeholders using Twitter. *Public Relations Review*, 36(4), 336–341. https://doi.org/10.1016/j.pubrev.2010.08.004

Sam, H. K., Othman, A. E. A., & Nordin, Z. S. (2005). Computer self-efficacy, computer anxiety, and attitudes toward the Internet: A study among undergraduates in Unimas. *Journal of Educational Technology & Society*, 8(4), 205–219.

Sandle, T. (2016, July 22). UN thinks Internet access is a human right. *Business Insider.* https://www.businessinsider.com/un-says-internet-access-is-a-human-right-2016-7

Santana, A. D. (2014). Virtuous or vitriolic: The effect of anonymity on civility in online newspaper reader comment boards. *Journalism Practice*, 8(1), 18–33. https://doi.org/10.1080/17512786.2013.813194

Sarbaugh-Thompson, M., & Feldman, M. S. (1998). Electronic mail and organizational communication: Does saying "hi" really matter? *Organization Science, 9*(6), 685–698. https://doi.org/10.1287/orsc.9.6.685

Scheerder, A., Van Deursen, A., & Van Dijk, J. (2017). Determinants of Internet skills, uses and outcomes. A systematic review of the second-and third-level digital divide. *Telematics and Informatics, 34*(8), 1607–1624. https://doi.org/10.1016/j.tele.2017.07.007

Schlenker, B. R., & Trudeau, J. V. (1990). Impact of self-presentations on private self-beliefs: Effects of prior self-beliefs and misattribution. *Journal of Personality and Social Psychology, 58*(1), 22–32. https://doi.org/10.1037/0022-3514.58.1.22

Schouten, A. P., Heerkens, M., Veringa, I., & Antheunis, M. L. (2014, May). *Strike a pose: How pose and expression in online profile pictures affect impressions of interpersonal attraction and intelligence* [Paper presentation]. 64th annual meeting of the International Communication Association, Seattle, WA.

Schuller, B., Steidl, S., Batliner, A., Burkhardt, F., Devillers, L., Müller, C., & Narayanan, S. (2013). Paralinguistics in speech and language—State-of-the-art and the challenge. *Computer Speech & Language, 27*(1), 4–39. https://doi.org/10.1016/j.csl.2012.02.005

Schulman, N. (2014). *In real life: Love, lies & identity in the digital age.* Grand Central Publishing.

Scott, G. G., Boyle, E. A., Czerniawska, K., & Courtney, A. (2018). Posting photos on Facebook: The impact of narcissism, social anxiety, loneliness, and shyness. *Personality and Individual Differences, 133*, 67–72. https://doi.org/10.1016/j.paid.2016.12.039

Scott, G. G., & Fullwood, C. (2020). Does recent research evidence support the hyperpersonal model of online impression management? *Current Opinion in Psychology, 36*, 106–111. https://doi.org/10.1016/j.copsyc.2020.05.005

Schwartz, M., & Hayes, J. (2008). A history of transatlantic cables. *IEEE Communications Magazine, 46*(9), 42–48. https://doi.org/10.1109/MCOM.2008.4623705

Seltzer, T., & Mitrook, M. A. (2007). The dialogic potential of weblogs in relationship building. *Public Relations Review, 33*(2), 227–229. https://doi.org/10.1016/j.pubrev.2007.02.011

Selwyn, N. (2004). The information aged: A qualitative study of older adults' use of information and communications technology. *Journal of Aging Studies, 18*, 369–384. https://doi.org/10.1016/j.jaging.2004.06.008

Selwyn, N., Gorard, S., Furlong, J., & Madden, L. (2003). Older adults' use of information and communications technology in everyday life. *Ageing and Society, 23*(5), 561–582. https://doi.org/10.1017/S0144686X03001302

SensoryCo. (n.d.). Taking amusement to the next level with sensory effects at theme parks. https://sensoryco4d.com/amusement-sensory-effects-at-theme-parks/

Shannon, C. E., & Weaver, W. (1949). *The mathematical theory of communication.* University of Illinois Press.

Sharabi, L. L., & Caughlin, J. P. (2017a). What predicts first date success? A longitudinal study of modality switching in online dating. *Personal Relationships, 24*(2), 370–391. https://doi.org/10.1111/pere.12188

Sharabi, L., & Caughlin, J. P. (2017b). Usage patterns of social media across stages of romantic relationships. In N. Punyanunt-Carter & J. S. Wrench (Eds.), *The impact of social media in modern relationships* (pp. 15–30). Lexington Books.

Shepherd, M. (2019, August 26). *Crowdfunding statistics (2019): Market size and growth. Fundera.* https://www.fundera.com/resources/crowdfunding-statistics

Sherif, M., Harvey, O. J., White, B. J., Hood, W. R., & Sherif, C. W. (1961). *Intergroup conflict and cooperation: The Robber's Cave Experiment.* University Book Exchange.

Shifman, L. (2012). An anatomy of a YouTube meme. *New Media & Society, 14*(2), 187–203. https://doi.org/10.1177/1461444811412160

Short, J., Williams, E., & Christie, B. (1976). *The social psychology of telecommunications.* Wiley.

Shulman, S., Seiffge-Krenke, I., & Dimitrovsky, L. (1994). The functions of pen pals for adolescents. *Journal of Psychology, 128*(1), 89–100. https://doi.org/10.1080/00223980.1994.9712714

Siegel, J., Dubrovsky, V., Kiesler, S., & McGuire, T. W. (1986). Group processes in computer-mediated communication. *Organizational Behavior and Human Decision Process, 37*(2), 157–187. https://doi.org/10.1016/0749-5978(86)90050-6

Silver, C. (2010). The Facebook privacy war: What is personal data? *Wired.* http://www.wired.com/geekdad/2010/05/the-facebook-privacy-war-what-is-personal-data/

Silver, W. L., & Finger, T. E. (1987). *Neurobiology of taste and smell.* Wiley.

Singhal, S., Neustaedter, C., Ooi, Y. L., Antle, A. N., & Matkin, B. (2017, February). *Flex-N-Feel: The design and evaluation of emotive gloves for couples to support touch over distance* [Paper presentation]. ACM Conference on Computer Supported Cooperative Work and Social Computing (CSCW'17), Portland, OR.

Sivertzen, A.-M., Nilsen, E. R., & Olafsen, A. H. (2013). Employer branding: Employer attractiveness and the use of social media. *Journal of Product & Brand Management, 22*(7), 473–483. https://doi.org/10.1108/JPBM-09-2013-0393

Slegg, J. (2013). *Fortune 500 social media: 77% active on Twitter; 70% on Facebook.* Search engine watch. https://www.searchenginewatch.com/2013/07/25/fortune-500-social-media-77-active-on-twitter-70-on-facebook/

Sly, L. (2018, January 29). U.S. soldiers are revealing sensitive and dangerous information by jogging. *Washington Post.* https://www.washingtonpost.com/world/a-map-showing-the-users-of-fitness-devices-lets-the-world-see-where-us-soldiers-are-and-what-they-are-doing/2018/01/28/86915662-0441-11e8-aa61-f3391373867e_story.html

Smeltzer, L. R. (1986). An analysis of receivers' reactions to electronically mediated communication. *Journal of Business Communication, 23*(4), 37–54. https://doi.org/10.1177/002194368602300405

Smith, A. (2015, April 1). U.S. smartphone use in 2015. *Internet & Tech.* https://www.pewinternet.org/2015/04/01/us-smartphone-use-in-2015/

Smith, A. C., Bensink, M., Armfield, N., Stillman, J., & Caffery, L. (2005). Telemedicine and rural health care applications. *Journal of Postgraduate Medicine, 51*(4), 286–293.

Smith, S. W., Hitt, R., Park, H. S., Walther, J., Liang, Y. J., & Hsieh, G. (2016). An effort to increase organ donation registration through intergroup competition and electronic word of mouth. *Journal of Health Communication, 21*(3), 376–386. https://doi.org/10.1080/10810730.2015.1095815

Smith, W. R., Stephens, K. K., Robertson, B. R., Li, J., & Murthy, D. (2018, May 20–23). *Social media in citizen-led disaster response: Rescuer roles, coordination challenges, and untapped potential* [Paper presentation]. 15th International Conference on Information Systems for Crisis Response and Management (ISCRAM) Conference, Rochester, NY.

Smock, A. (2012). *Leveraging social media for learning: Communities of practice on Flickr* [Unpublished doctoral dissertation]. Michigan State University.

Snyder, M., Tanke, E. D., & Berscheid, E. (1977). Social perceptions and interpersonal behavior: On the self-fulfilling nature of social stereotypes. *Journal of Personality and Social Psychology, 35*(9), 656–666. https://doi.org/10.1037/0022-3514.35.9.656

Sobieraj, S., & Berry, J. M. (2011). From incivility to outrage: Political discourse in blogs, talk radio, and cable news. *Political Communication, 28*(1), 19–41. https://doi.org/10.1080/10584609.2010.542360

Solove, D. J. (2011). *Nothing to hide: The false tradeoff between privacy and security.* Yale University Press.

Song, Q., Wang, Y., Chen, Y., Benitez, J., & Hu, J. (2019). Impact of the usage of social media in the workplace on team and employee performance. *Information & Management, 56*(8), 103–160. https://doi.org/10.1016/j.im.2019.04.003

Sood, S., Mbarika, V., Jugoo, S., Dookhy, R., Doarn, C. R., Prakash, N., & Merrell, R. C. (2007). What is telemedicine? A collection of 104 peer-reviewed perspectives and theoretical underpinnings. *Telemedicine and e-Health, 13*(5), 573–590. https://doi.org/10.1089/tmj.2006.0073

Spence, P. R., Westerman, D., Edwards, C., & Edwards, A. (2014). Welcoming our robot overlords: Initial expectations about interaction with a robot. *Communication Research Reports, 31*(3), 272–280. https://doi.org/10.1080/08824096.2014.924337

Spitzberg, B. H. (2014). Toward a model of meme diffusion (M^{3}D). *Communication Theory, 24*(3), 311–339. https://doi.org/10.1111/comt.12042

Sprecher, S. (2014). Initial interactions online-text, online-audio, online-video, or face-to-face: Effects of modality on liking, closeness, and other interpersonal outcomes. *Computers in Human Behavior, 31*(1), 190–197. https://doi.org/10.1016/j.chb.2013.10.029

Sprecher, S., Zimmerman, C., & Fehr, B. (2014). The influence of compassionate love on strategies used to end a relationship. *Journal of Social and Personal Relationships, 31*(5), 697–705. https://doi.org/10.1177/0265407513517958

Sproull, L., & Kiesler, S. (1986). Reducing social context cues: Electronic mail in organizational communication. *Management Science, 32*(11), 1492–1512. https://doi.org/10.1287/mnsc.32.11.1492

Steeves, V. (2019). *Summary of research on youth online privacy.* Ottawa, ON: Office of the Privacy Commissioner of Canada. https://www.priv.gc.ca/media/1731/yp_201003_e.pdf

Stephens, K. K., Cho, J. K., & Ballard, D. I. (2012). Simultaneity, sequentiality, and speed: Organizational messages about multiple-task completion. *Human Communication Research, 38*(1), 23–47. https://doi.org/10.1111/j.1468-2958.2011.01420.x

Stephens, K. K., & Davis, J. (2009). The social influences on electronic multitasking in organizational meetings. *Management*

Communication Quarterly, 23(1), 63–83. https://doi.org/10.1177/0893318909335417

Stieglitz, S., & Dang-Xuan, L. (2012, January). *Political communication and influence through microblogging: An empirical analysis of sentiment in Twitter messages and retweet behavior* [Paper presentation]. 45th Hawaii International Conference on System Sciences (HICSS), Kauai, HI.

Stromer-Galley, J., Bryant, L., & Bimber, B. (2015). Context and medium matter: Expressing disagreements online and face-to-face in political deliberations. *Journal of Public Deliberation, 11*(1), article 1. https://doi.org/10.16997/jdd.218

Su, J., Shukla, A., Goel, S., & Narayanan, A. (2017, April 3–7). *De-anonymizing web browsing data with social networks* [Paper presentation]. 26th International Conference on World Wide Web (WWW'07), Perth, Australia.

Subrahmanyam, K., Reich, S. M., Waechter, N., & Espinoza, G. (2008). Online and offline social networks: Use of social networking sites by emerging adults. *Journal of Applied Developmental Psychology, 29*(6), 420–433. https://doi.org/10.1016/j.appdev.2008.07.003

Suler, J. (2004). The online disinhibition effect. *CyberPsychology & Behavior, 7*(3), 321–326. https://doi.org/10.1089/1094931041291295

Sumner, E. M., Hayes, R. A., Carr, C. T., & Wohn, D. Y. (2020). Assessing the cognitive and communicative properties of Facebook Reactions and Likes as lightweight feedback cues. *First Monday, 25*(2). https://doi.org/10.5210/fm.v25i2.9621

Sumner, W. G. (1906). *Folkways*. Ginn.

Sussman, N. M., & Tyson, D. H. (2000). Sex and power: Gender differences in computer-mediated interactions. *Computers in Human Behavior, 16*(4), 381–394. https://doi.org/10.1016/S0747-5632(00)00020-0

Suzuki, S. (1998). In-group and out-group communication patterns in international organizations: Implications for social identity theory. *Communication Research, 25*(2), 154–182. https://doi.org/10.1177/009365098025002002

Tajfel, H., Billig, M. G., Bundy, R. P., & Flament, C. (1971). Social categorization and intergroup behaviour. *European Journal of Social Psychology, 1*(2), 149–178. https://doi.org/10.1002/ejsp.2420010202

Tajfel, H., & Turner, J. C. (1986). The social identity theory of intergroup behavior. In S. Worchel & W. G. Austin (Eds.), *The psychology of intergroup relations* (pp. 7–24). Nelson-Hall.

Tandoc, E. C., Jr., & Jenkins, J. (2018). Out of bounds? How Gawker's outing a married man fits into the boundaries of journalism. *New Media & Society, 20*(2), 581–598. https://doi.org/10.1177/1461444816665381

Tanis, M., & Postmes, T. (2003). Social cues and impression formation in CMC. *Journal of Communication, 53*(4), 676–693. https://doi.org/10.1111/j.1460-2466.2003.tb02917.x

Tapscott, D., & Williams, A. D. (2008). *Wikinomics: How mass collaboration changes everything*. Portfolio.

Tassabehji, R., & Vakola, M. (2005). Business email: The killer impact. *Communications of the ACM, 48*(11), 64–70. https://doi.org/10.1145/1096000.1096006

Tates, K., Antheunis, M. L., Kanters, S., Nieboer, T. E., & Gerritse, M. B. E. (2017). The effect of screen-to-screen versus face-to-face consultation on doctor-patient communication: An experimental study with simulated patients. *Journal of Medical Internet Research, 19*(12), e421. https://doi.org/10.2196/jmir.8033

Taylor, A. S., & Vincent, J. (2005). An SMS history. In L. Hamill, A. Lasen & D. Diaper (Eds.), *Mobile world* (pp. 75–91). Springer.

Taylor, D. C., & Aday, J. B. (2016). Consumer generated restaurant ratings: A preliminary look at OpenTable.com. *Journal of New Business Ideas & Trends, 14*(1), 14–22.

Taylor, S. H., & Ledbetter, A. M. (2017). Extending media multiplexity theory to the extended family: Communication satisfaction and tie strength as moderators of violations of media use expectations. *New Media & Society, 19*(9), 1369–1387. https://doi.org/10.1177/1461444816638458

Tech for Campaigns. (2019). *2018 elections spending: Federal and state*. https://www.techforcampaigns.org/2018-political-digital-advertising-report

Tellis, G. J., MacInnis, D. J., Tirunillai, S., & Zhang, Y. (2019). What drives virality (sharing) of online digital content? The critical role of information, emotion, and brand prominence. *Journal of Marketing, 83*(4), 1–20. https://doi.org/10.1177/0022242919841034

Terblanche, N. S. (2011). You cannot run or hide from social media—ask a politician. *Journal of Public Affairs, 11*(3), 156–167. https://doi.org/10.1002/pa.404

Thorpe, H., & Ahmad, N. (2015). *Youth, action sports and political agency in the Middle East: Lessons from a grassroots parkour group in Gaza. International Review for the Sociology of Sport, 50*(6), 678–704. https://doi.org/10.1177/1012690213490521

Tidwell, N. D., Eastwick, P. W., & Finkel, E. J. (2013). Perceived, not actual, similarity predicts initial attraction in a live romantic context: Evidence from the speed-dating paradigm. *Personal*

Relationships, 20(2), 199–215. https://doi.org/10.1111/j.1475-6811.2012.01405.x

Tidwell, L. C., & Walther, J. B. (2006). Computer-mediated communication effects on disclosure, impressions, and interpersonal evaluations: Getting to know one another a bit at a time. *Human Communication Research, 28*(3), 317–348. https://doi.org/10.1111/j.1468-2958.2002.tb00811.x

Thomas, E. J., & Fink, C. F. (1963). Effects of group size. *Psychological Bulletin, 60*(4), 371–384. https://doi.org/10.1037/h0047169

Tokunaga, R. S. (2011). Social networking site or social surveillance site? Understanding the use of interpersonal electronic surveillance in romantic relationships. *Computers in Human Behavior, 27*(2), 705–713. https://doi.org/10.1016/j.chb.2010.08.014

Tong, S. T. (2013). Facebook use during relationship termination: Uncertainty reduction and surveillance. *Cyberpsychology, Behavior, and Social Networking, 16*(11), 788–793. https://doi.org/10.1089/cyber.2012.0549

Tong, S. T., Heinemann-Lafave, D., Jeon, J., Kolodziej-Smith, R., & Warshay, N. (2013). The use of pro-ana blogs for online social support. *Eating Disorders, 21*(5), 408–422. https://doi.org/10.1080/10640266.2013.827538

Tong, S. T., Van der Heide, B., Langwell, L., & Walther, J. B. (2008). Too much of a good thing? The relationship between number of friends and interpersonal impressions on Facebook. *Journal of Computer-Mediated Communication, 13*(3), 531–549. https://doi.org/10.1111/j.1083-6101.2008.00409.x

Tong, S. T., & Walther, J. B. (2011). Relational maintenance and CMC. In K. B. Wright & L. M. Webb (Eds.), *Computer-mediated communication in personal relationships* (pp. 98–118). Peter Lang.

Tong, S. T., & Westerman, D. K. (2016, January). *Relational and masspersonal maintenance: Romantic partners' use of social network websites* [Paper presentation]. 49th Hawaii International Conference on System Sciences (HICSS'16), Koloa, HI.

Trager, G. L. (1958). Paralanguage: A first approximation. *Studies in Linguistics, 13*, 1–11.

Tronstad, R. (2008). Character identification in *World of Warcraft*: The relationship between capacity and appearance. In H. G. Corneliussen & J. W. Rettberg (Eds.), *Digital culture, play, and identity: A* World of Warcraft *reader* (pp. 249–264). MIT Press.

Treré, E. (2020). The banality of WhatsApp: On the everyday politics of backstage activism in Mexico and Spain. *First Monday, 25*(1). https://doi.org/10.5210/fm.v25i1.10404

Trikha, R. (2015). *The history of "Hello, World."* HackerRank. https://blog.hackerrank.com/the-history-of-hello-world/

Tromble, R. (2018). The great leveler? Comparing citizen–politician Twitter engagement across three Western democracies. *European Political Science, 17*, 223–239. https://doi.org/10.1057/s41304-016-0022-6

Tsimonis, G., & Dimitriadis, S. (2014). Brand strategies in social media. *Marketing Intelligence & Planning, 32*(3), 328–344. https://doi.org/10.1108/MIP-04-2013-0056

Tufekci, Z. (2017). *Twitter and tear gas: The power and fragility of networked protests*. Yale University Press.

Tufekci, Z., & Wilson, C. (2012). Social media and the decision to participate in political protest: Observations from Tahrir Square. *Journal of Communication, 62*(2), 363–379. https://doi.org/10.1111/j.1460-2466.2012.01629.x

Turkle, S. (1995). *Life on the screen: Identity in the age of the Internet*. Simon & Schuster.

Turkle, S. (2005). *The second self: Computers and the human spirit*. MIT Press.

Turner, J. W., Grube, J. A., & Meyers, J. (2001). Developing an optimal match within online communities: An exploration of CMC support communities and traditional support. *Journal of Communication, 51*(2), 231–251. https://doi.org/10.1111/j.1460-2466.2001.tb02879.x

Twitter. (2020). *FAQ*. Investor relations. https://investor.twitterinc.com/contact/faq/default.aspx

US Department of State. (2019). *Reports and Statistics: U.S. Passports*. Bureau of Consular Affairs. https://travel.state.gov/content/travel/en/about-us/reports-and-statistics.html

Utz, S. (2010). Show me your friends and I will tell you what type of person you are: How one's profile, number of friends, and type of friends influence impression formation on social network sites. *Journal of Computer-Mediated Communication, 15*(2), 314–335. https://doi.org/10.1111/j.1083-6101.2010.01522.x

Utz, S., & Beukeboom, C. J. (2011). The role of social network sites in romantic relationships: Effects on jealousy and relationship happiness. *Journal of Computer-Mediated Communication, 16*(4), 511–527. https://doi.org/10.1111/j.1083-6101.2011.01552.x

Utz, S., & Breuer, J. (2019). The relationship between networking, LinkedIn use, and retrieving informational benefits. *Cyberpsychology, Behavior, and Social Networking, 22*(3), 180–185. https://doi.org/10.1089/cyber.2018.0294

Utz, S., Muscanell, N., & Khalid, C. (2015). Snapchat elicits more jealousy than Facebook: A comparison of Snapchat and Facebook use. *Cyberpsychology, Behavior, and Social Networking, 18*(3), 141–146. https://doi.org/10.1089/cyber.2014.0479

Valenzuela, S., Halpern, D., Katz, J. E., & Miranda, J. P. (2019). The paradox of participation versus misinformation: Social media, political engagement, and the spread of misinformation. *Digital Journalism, 7*(6), 802–823. https://doi.org/0.1080/21670811.2019.1623701

Van Buskirk, E. (2010). Report: Facebook CEO Mark Zuckerberg doesn't believe in privacy. *Wired.* https://www.wired.com/2010/04/report-facebook-ceo-mark-zuckerberg-doesnt-believe-in-privacy/

Van Deursen, A. J. A. M., & Helsper, E. J. (2015). The third-level digital divide: Who benefits most from being online? In L. Robinson, S. R. Cotten, J. Schultz, T. M. Hale, & A. Williams (Eds.), *Communication and information technologies annual* (Vol. 10, pp. 29–52). Emerald Group.

Van Dijk, J. (2005). *The deepening divide: Inequality in the information society.* Sage.

Van Dijk, J. (2020). *The digital divide.* Polity Press.

Van Gelder, L. (1996). The strange case of the electronic lover. In R. King (Ed.), *Computerization and controversy: Value conflict and social choices* (2nd ed., pp. 533–546). Morgan Kaufmann.

Van Osch, Y., Blanken, I., Meijs, M. H. J., & Van Wolferen, J. (2015). A group's physical attractiveness is greater than the average attractiveness of its members: The group attractiveness effect. *Personality and Social Psychology Bulletin, 41*(4), 559–574. https://doi.org/10.1177/0146167215572799

Vardi, M. Y. (2012). Will MOOCs destroy academia? *Communications of the ACM, 55*(11), 5. https://doi.org/10.1145/2366316.2366317

Vick, K. (2017). The digital divide: A quarter of the nation is without broadband. *Time.* https://time.com/4718032/the-digital-divide/

Viswanath, B., Mislove, A., Cha, M., & Gummadi, K. P. (2009, August). *On the evolution of user interaction in Facebook* [Paper presentation]. 2nd ACM workshop on Online social networks (WOSN'09), Barcelona, Spain.

Vitak, J., & Ellison, N. B. (2013). "There's a network out there you might as well tap": Exploring the benefits of and barriers to exchanging informational and support-based resources on Facebook. *New Media & Society, 15*, 243–259. https://doi.org/10.1177/1461444812451566

Vitak, J., Zube, P., Smock, A., Carr, C. T., Ellison, N., & Lampe, C. (2011). It's complicated: Facebook users' political participation in the 2008 election. *Cyberpsychology, Behavior, and Social Networking, 14*(3), 107–114. https://doi.org/10.1089/cyber.2009.0226

Vivek, S. D., Beatty, S. E., & Morgan, R. M. (2012). Customer engagement: Exploring customer relationships beyond purchase. *Journal of Marketing Theory and Practice, 20*(2), 122–146. https://doi.org/10.2753/MTP1069-6679200201

Vogels, E. A. (2020). *About half of never-married Americans have used an online dating site or app.* Pew Research Center. https://www.pewresearch.org/fact-tank/2020/03/24/the-never-been-married-are-biggest-users-of-online-dating/

Von der Pütten, A. M., Krämer, N. C., & Eimler, S. C. (2011, November 14–18). *Living with a robot companion: Empirical study on the interaction with an artificial health advisor* [Paper presentation]. 13th international conference on multimodal interfaces, Alicante, Spain.

Waldman, D. A., Atwater, L. E., & Antonioni, D. (1998). Has 360 degree feedback gone amok? *Academy of Management Perspectives, 12*(2), 86–94. https://doi.org/10.5465/ame.1998.650519

Walker, A. G., & Smither, J. W. (1999). A five-year study of upward feedback: What managers do with their results matters. *Personnel Psychology, 52*(2), 393–423. https://doi.org/10.1111/j.1744-6570.1999.tb00166.x

Wallsten, K. (2014). Microblogging and the news: Political elites and the ultimate retweet. In J. Bishop (Ed.), *Political campaigning in the information age* (pp. 128–147). IGI Global.

Walrave, M., Poels, K., Antheunis, M. L., Van den Broeck, E., & Van Noort, G. (2018). Like or dislike? Adolescents' responses to personalized social network site advertising. *Journal of Marketing Communications, 24*(6), 599–616. https://doi.org/10.1080/13527266.2016.1182938

Walther, J. B. (1992). Interpersonal effects in computer-mediated interaction: A relational perspective. *Communication Research, 19*(1), 52–90. https://doi.org/10.1177/009365092019001003

Walther, J. B. (1996). Computer-mediated communication: Impersonal, interpersonal, and hyperpersonal interaction. *Communication Research, 23*(1), 3–43. https://doi.org/10.1177/009365096023001001

Walther, J. B. (1997). Group and interpersonal effects in international computer-mediated collaboration. *Human Communication Research, 23*(3), 342–369. https://doi.org/10.1111/j.1468-2958.1997.tb00400.x

Walther, J. B. (2007). Selective self-presentation in computer-mediated communication: Hyperpersonal dimensions of technology, language, and cognition. *Computers in Human Behavior*, 23(5), 2538–2557. https://doi.org/10.1016/j.chb.2006.05.002

Walther, J. B. (2011). Theories of computer-mediated communication and interpersonal relations. In M. L. Knapp & J. A. Daly (Eds.), *The Sage handbook of interpersonal communication* (pp. 443–480). Sage.

Walther, J. B., Anderson, J. F., & Park, D. W. (1994). Interpersonal effects in computer-mediated communication. *Communication Research*, 21(4), 460–487. https://doi.org/10.1177/009365094021004002

Walther, J. B., & Bazarova, N. N. (2008). Validation and application of electronic propinquity theory to computer-mediated communication in groups. *Communication Research*, 35(5), 622–645. https://doi.org/10.1177/0093650208321783

Walther, J. B., & Burgoon, J. K. (1992). Relational communication in computer-mediated interaction. *Human Communication Research*, 19(1), 50–88. https://doi.org/10.1111/j.1468-2958.1992.tb00295.x

Walther, J. B., & Carr, C. T. (2010). Internet interaction and intergroup dynamics: Problems and solutions in computer-mediated communication. In H. Giles, S. Reid, & J. Harwood (Eds.), *Dynamics of intergroup communication* (pp. 208–220). Peter Lange.

Walther, J. B., Carr, C. T., Choi, S., DeAndrea, D., Kim, J., Tong, S., & Van Der Heide, B. (2010). Interaction of interpersonal, peer, and media influence sources online: A research agenda for technology convergence. In Z. Papacharissi (Ed.), *The networked self* (pp. 17–38). Routledge.

Walther, J. B., & D'Addario, K. P. (2001). The impacts of emoticons on message interpretation in computer-mediated communication. *Social Science Computer Review*, 19(3), 324–347. https://doi.org/10.1177/089443930101900307

Walther, J. B., DeAndrea, D., Kim, J., & Anthony, J. C. (2010). The influence of online comments on perceptions of antimarijuana public service announcements on YouTube. *Human Communication Research*, 36(4), 469–492. https://doi.org/10.1111/j.1468-2958.2010.01384.x

Walther, J. B., Hoter, E., Ganayem, A., & Shonfeld, M. (2015). Computer-mediated communication and the reduction of prejudice: A controlled longitudinal field experiment among Jews and Arabs in Israel. *Computers in Human Behavior*, 52, 550–558. https://doi.org/10.1016/j.chb.2014.08.004

Walther, J. B., Liang, Y. J., DeAndrea, D. C., Tong, S. T., Carr, C. T., Spottswood, E. L., & Amichai-Hamburger, Y. (2011). The effect of feedback on identity shift in computer-mediated communication. *Media Psychology*, 14(1), 1–26. https://doi.org/10.1080/15213269.2010.547832

Walther, J. B., Liang, Y. J., Ganster, T., Wohn, D. Y., & Emington, J. (2012). Online reviews, helpfulness ratings, and consumer attitudes: An extension of congruity theory to multiple sources in Web 2.0. *Journal of Computer-Mediated Communication*, 18(1), 97–112. https://doi.org/10.1111/j.1083-6101.2012.01595.x

Walther, J. B., Loh, T., & Granka, L. (2005). Let me count the ways: The interchange of verbal and nonverbal cues in computer-mediated and face-to-face affinity. *Journal of Language and Social Psychology*, 24, 36–65. https://doi.org/10.1177/0261927X04273036

Walther, J. B., & Parks, M. R. (2002). Cues filtered out, cues filtered in: Computer-mediated communication and relationships. In M. L. Knapp & J. A. Daly (Eds.), *Handbook of interpersonal communication* (3rd ed., pp. 529–563). Sage.

Walther, J. B., Van der Heide, B., Hamel, L. M., & Shulman, H. C. (2009). Self-generated versus other-generated statements and impressions in computer-mediated communication: A test of warranting theory using Facebook. *Communication Research*, 36(2), 229–253. https://doi.org/10.1177/0093650208330251

Walther, J. B., Van der Heide, B., Kim, S.-Y., Westerman, D., & Tong, S. T. (2008). The role of friends' appearance and behavior on evaluations of individuals on Facebook: Are we known by the company we keep? *Human Communication Research*, 34(1), 28–49. https://doi.org/10.1111/j.1468-2958.2007.00312.x

Walther, J. B., Van der Heide, B., Tong, S. T., Carr, C. T., & Atkin, C. K. (2010). Effects of interpersonal goals on inadvertent intrapersonal influence in computer-mediated communication. *Human Communication Research*, 36(3), 323–347. https://doi.org/10.1111/j.1468-2958.2010.01378.x

Walther, J. B., & Whitty, M. T. (2021). Language, psychology, and new new media: The hyperpersonal model of mediated communication at twenty-five years. *Journal of Language and Social Psychology*, 40(1), 120–135. https://doi.org/10.1177/0261927X20967703

Wang, Y., & Yang, Y. (2020). Dialogic communication on social media: How organizations use Twitter to build dialogic relationships with their publics. *Computers in Human Behavior*, 104, 106183. https://doi.org/10.1016/j.chb.2019.106183

Warschauer, M. (2002, July 2). Reconceptualizing the digital divide. *First Monday, 7*(7). https://firstmonday.org/ojs/index.php/fm/article/view/967/888/

Wattles, J. (2015). *Peeple co-founder pushes back against backlash over app.* CNN Business. https://money.cnn.com/2015/10/04/technology/peeple-app-social-suspended/

Weisband, S. P., Schneider, S. K., & Connolly, T. (1995). Computer-mediated communication and social information: Status salience and status differences. *Academy of Management Journal, 38*(4), 1124–1151. https://doi.org/10.5465/256623

Weisskirch, R. S., & Delevi, R. (2012). Its ovr b/n u n me: Technology use, attachment styles, and gender roles in relationship dissolution. *Cyberpsychology, Behavior, and Social Networking, 15*(9), 486–490. https://doi.org/10.1089/cyber.2012.0169

Wellman, B. (2012). Is Dunbar's number up? *British Journal of Psychology, 103*(2), 174–176. https://doi.org/10.1111/j.2044-8295.2011.02075.x

Wenberg, J., & Wilmot, W. (1973). *The personal communication process.* Wiley.

Wesson, D. A. (1992). The handshake as non-verbal communication in business. *Marketing Intelligence & Planning, 10*(9), 41–46. https://doi.org/10.1108/02634509210020132

Whittaker, E., & Kowalski, R. M. (2015). Cyberbullying via social media. *Journal of School Violence, 14*(1), 11–29. https://doi.org/10.1080/15388220.2014.949377

WhoIsHostingThis. (2019, September 29). *Where in the world does the Internet live?* https://www.whoishostingthis.com/blog/2013/12/06/internet-infographic/

Wijenayake, S., Van Berkel, N., Kostakos, V., & Goncalves, J. (2019). Measuring the effects of gender on online social conformity. *Proceedings of the ACM on Human-Computer Interaction: Vol. 3* (article 145). https://doi.org/10.1145/3359247

Williams, D. (2006). Why game studies now? Gamers don't bowl alone. *Games and Culture, 1*(1), 13–16. https://doi.org/10.1177/1555412005281774

Williams, D., Ducheneaut, N., Xiong, L., Zhang, Y., Yee, N., & Nickell, E. (2006). From tree house to barracks the social life of guilds in *World of Warcraft. Games and Culture, 1*(4), 338–361. https://doi.org/10.1177/1555412006292616

Wingate, V. S., Minney, J. A., & Guadagno, R. E. (2013). Sticks and stones may break your bones, but words will always hurt you: A review of cyberbullying. *Social Influence, 8*(2–3), 87–106. https://doi.org/10.1080/15534510.2012.730491

Wintour, P. (2018, January 10). Russian bid to influence Brexit vote detailed in new US Senate report. *The Guardian.* https://www.theguardian.com/world/2018/jan/10/russian-influence-brexit-vote-detailed-us-senate-report

Wise, K., Alhabash, S., & Park, H. (2010). Emotional responses during social information seeking on Facebook. *Cyberpsychology, Behavior, and Social Networking, 13*(5), 555–562. https://doi.org/10.1089/cyber.2009.0365

Wise, K., Hamman, B., & Thorson, K. (2006). Moderation, response rate, and message interactivity: Features of online communities and their effects on intent to participate. *Journal of Computer-Mediated Communication, 12*(1), 24–41. https://doi.org/10.1111/j.1083-6101.2006.00313.x

Wright, K. B. (2002). Social support within an on-line cancer community: An assessment of emotional support, perceptions of advantages and disadvantages, and motives for using the community from a communication perspective. *Journal of Applied Communication Research, 30*(3), 195–209. https://doi.org/10.1080/00909880216586

Wright, K. B. (2004). On-line relational maintenance strategies and perceptions of partners within exclusively internet-based and primarily internet-based relationships. *Communication Studies, 55*(2), 239–253. https://doi.org/10.1080/10510970409388617

Wright, K. B., & Bell, S. B. (2003). Health-related support groups on the Internet: Linking empirical findings to social support and computer-mediated communication theory. *Journal of Health Psychology, 8*(1), 39–54. https://doi.org/10.1177/1359105303008001429

Wohn, D., Velasquez, A., Bjornrud, T., & Lampe, C. (2012, May 5–10). *Habit as an explanation of participation in an online peer-production community* [Paper presentation]. SIGCHI Conference on Human Factors in Computing Systems (CHI'12), Austin, TX.

Wood, D., McMinn, S., & Feng, E. (2019). *China used Twitter to disrupt Hong Kong protests, but efforts began years earlier.* NPR: World. https://www.npr.org/2019/09/17/758146019/china-used-twitter-to-disrupt-hong-kong-protests-but-efforts-began-years-earlier

Wynne, R. (2014). The real difference between PR and advertising. *Forbes.* https://www.forbes.com/sites/robertwynne/2014/07/08/the-real-difference-between-pr-and-advertising-credibility/#283682452bb9

Xia, F., Yang, L. T., Wang, L., & Vinel, A. (2012). *Internet of Things. International Journal of Communication Systems, 25*(9), 1101–1102. https://doi.org/10.1002/dac.2417

Xie, B., & Jaeger, P. T. (2008). Older adults and political participation on the Internet: A cross-cultural comparison of the USA and China. *Journal of Cross-Cultural Gerontology, 23*(1), 1–15. https://doi.org/10.1007/s10823-007-9050-6

Yan, J., Liu, N., Wang, G., Zhang, W., Jiang, Y., & Chen, Z. (2009, April). *How much can behavioral targeting help online advertising?* [Paper presentation]. 18th international conference on World Wide Web (WWW'09), Madrid, Spain.

Yao, Y., Viswanath, B., Cryan, J., Zheng, H., & Zhao, B. Y. (2017, October 30–November 3). *Automated crowdturfing attacks and defenses in online review systems* [Paper presentation]. ACM SIGSAC Conference on Computer and Communications Security (CSS'17), Dallas, TX.

Yee, N., & Bailenson, J. (2007). The Proteus effect: The effect of transformed self-representation on behavior. *Human Communication Research, 33*(3), 271–290. https://doi.org/10.1111/j.1468-2958.2007.00299.x

Yee, N., Bailenson, J. N., & Ducheneaut, N. (2009). The Proteus effect: Implications of transformed digital self-representation on online and offline behavior. *Communication Research, 36,* 285–312. https://doi.org/10.1177/0093650208330254

Yee, N., Bailenson, J. N., Urbanek, M., Chang, F., & Merget, D. (2007). The unbearable likeness of being digital: The persistence of nonverbal social norms in online virtual environments. *CyberPsychology & Behavior, 10,* 115–121. https://doi.org/10.1089/cpb.2006.9984

Yu, R. P. (2020). Use of messaging apps and social network sites among older adults: A mixed-method study. *International Journal of Communication, 14,* 4453–4473. https://ijoc.org/index.php/ijoc/article/viewFile/14435/3194

Yu, R. P., Ellison, N. B., McCammon, R. J., & Langa, K. M. (2016). Mapping the two levels of digital divide: Internet access and social network site adoption among older adults in the USA. *Information, Communication & Society, 19*(10), 1445–1464. https://doi.org/10.1080/1369118X.2015.1109695

Zafarani, R., Tang, L., & Liu, H. (2015). User identification across social media. *ACM Transactions on Knowledge Discovery from Data (TKDD), 10*(2), article 16. https://doi.org/10.1145/2747880

Zappavigna, M. (2014). Coffeetweets: Bonding around the bean on Twitter. In P. Seargeant & C. Tagg (Eds.), *The language of social media: Identity and community on the Internet* (pp. 139–160). Palgrave Macmillan.

Zdechlik, M. (2018, December 27). *Patients are turning to GoFundMe to fill health insurance gaps.* Shots: Health News from NPR. https://www.npr.org/sections/health-shots/2018/12/27/633979867/patients-are-turning-to-gofundme-to-fill-health-insurance-gaps

Zhang, D., Zhou, L., Kehoe, J. L., & Kilic, I. Y. (2016). What online reviewer behaviors really matter? Effects of verbal and nonverbal behaviors on detection of fake online reviews. *Journal of Management Information Systems, 33*(2), 456–481. https://doi.org/10.1080/07421222.2016.1205907

Zhang, M., & Wolff, R. S. (2004). Crossing the digital divide: Cost-effective broadband wireless access for rural and remote areas. *IEEE Communications Magazine, 42*(2), 99–105. https://doi.org/10.1109/MCOM.2003.1267107

Zhang, Q., Marksbury, N., & Heim, S. (2010, January). *A case study of communication and social interactions in learning in second life* [Paper presentation]. 43rd Hawaii International Conference on System Sciences (HICSS), Honolulu, HI.

Zheng, S., Rosson, M. B., Shih, P. C., & Carroll, J. M. (2015, March 14–18). *Understanding student motivation, behaviors and perceptions in MOOCs* [Paper presentation]. 18th ACM conference on computer supported cooperative work & social computing (CSCW'15), Vancouver, BC.

Zhou, Z., Zhang, Q., Su, C., & Zhou, N. (2012). How do brand communities generate brand relationships? Intermediate mechanisms. *Journal of Business Research, 65*(7), 890–895. https://doi.org/10.1016/j.jbusres.2011.06.034

Zimbardo, P. G. (1969). The human choice: Individuation, reason, and order versus deindividuation, impulse and chaos. In W. J. Arnold & D. Levine (Eds.), *Nebraska symposium on motivation.* University of Nebraska Press.

Zimmer, M., Kumar, P., Vitak, J., Liao, Y., & Chamberlain Kritikos, K. (2020). "There's nothing really they can do with this information": Unpacking how users manage privacy boundaries for personal fitness information. *Information, Communication & Society, 23*(7), 1020–1037. https://doi.org/10.1080/1369118X.2018.1543442

Zompetti, J. P. (2017). *Divisive discourse: The extreme rhetoric of contemporary American politics* (2nd ed.). Cognella Academic.

Index

CPSIA information can be obtained
at www.ICGtesting.com
Printed in the USA
BVHW050509240421
605575BV00001B/1

9 781538 131718